T0139117

Human–Machine Interface Technology Advancements and Applications

Human–Machine Interface Technology Advancements and Applications focuses on analysis, design, and evaluation perspectives in HMI technological breakthroughs and applications. It covers a wide range of ideas, methodologies, approaches, and instruments to give the reader a thorough understanding of the field's current academic and industry practice and debate. Physical, cognitive, social, and emotional factors are all considered in the work, which is exemplified by key application fields such as aerospace, automobile, medicine, and defense. This book covers AI and machine learning methodologies as well as biological signals and HMI applications. Nanotechnology, user interface design, and interactive systems are also featured. The MATLAB approach to signal processing applications is also included.

This book discusses advances in the field of human–machine interfaces and provides practical knowledge in biomedical signal processing, AI, and machine learning. It discusses augmented reality/virtual reality-based HMI applications. It examines advances in nanotechnology, user interface design, and interactive systems.

This book is intended to serve as a research guide that will both inform readers about the fundamentals of HMI from academic and industrial perspectives and provide a glimpse into how human-centered designers, such as engineers and human factors specialists, will attempt to design and develop human–machine systems in the future.

Materials, Devices, and Circuits: Design and Reliability

Series Editors
Shubham Tayal, K. K. Paliwal, and Amit Kumar Jainy

For more information about this series, please visit: https://www.routledge.com/Materials-Devices-and-Circuits/book-series/MDCDR

Contents

Designed cover image: © Shutterstock

First edition published 2024
by CRC Press
6000 Broken Sound Parkway NW, Suite 300, Boca Raton, FL 33487-2742

and by CRC Press
4 Park Square, Milton Park, Abingdon, Oxon, OX14 4RN

CRC Press is an imprint of Taylor & Francis Group, LLC

© 2024 selection and editorial matter, Ravichander Janapati, Usha Desai, Shrirang A. Kulkarni, Shubham Tayal; individual chapters, the contributors

Library of Congress Cataloging-in-Publication Data

Names: Janapati, Ravichander, editor.
Title: Human-machine interface technology advancements and applications /
edited by Ravichander Janapati, Usha Desai, Shrirang A Kulkarni, Shubham Tayal.
Description: First edition. | Boca Raton: CRC Press, [2024] | Series:
Materials, devices, and circuits: design and reliability | Includes
bibliographical references and index.
Identifiers: LCCN 2023000405 (print) | LCCN 2023000406 (ebook) | ISBN
9781032351520 (hardback) | ISBN 9781032354231 (paperback) | ISBN
9781003326830 (ebook)
Subjects: LCSH: Human-machine systems.
Classification: LCC TA167 .H8667 2023 (print) | LCC TA167 (ebook) | DDC
620.8/2--dc23/eng/20230130
LC record available at https://lccn.loc.gov/2023000405
LC ebook record available at https://lccn.loc.gov/2023000406

ISBN: 978-1-032-35152-0 (hbk)
ISBN: 978-1-032-35423-1 (pbk)
ISBN: 978-1-003-32683-0 (ebk)

DOI: 10.1201/9781003326830

Typeset in Times
by Deanta Global Publishing Services, Chennai, India

Human–Machine Interface Technology Advancements and Applications

Edited by
Ravichander Janapati,
Usha Desai,
Shrirang Ambaji Kulkarni, and
Shubham Tayal

CRC Press
Taylor & Francis Group
Boca Raton London New York

CRC Press is an imprint of the
Taylor & Francis Group, an **informa** business

Preface

Anything that could give rise to smarter-than-human intelligence — in the form of Artificial Intelligence, brain-computer interfaces, or neuroscience-based human intelligence enhancement — wins hands down beyond contest as doing the most to change the world. Nothing else is even in the same league.

— Eliezer Yudkowsky

The book entitled "Human–Machine Interface (HMI) Technology Advancements and Applications" delves into the latest developments in HMI technology, algorithms, and applications in health care and technology. The topics of cognition, artificial intelligence (AI), and applications of machine learning are covered. This book explains how to create machine learning applications in processing the brain–computer information, physiological, pathological, nanotechnology, natural language processing (NLP), smart manufacturing, and hybrid electrical vehicle (HEV) domains. It also covers nanotechnology advancements, user interface design, and interactive system user experience, as well as MATLAB-based methodologies for biomedical signal processing applications.

This book provides an exposure to biomedical signal processing, AI, brain–computer Interfacing, and applications. The expected outcome from this book is to provide knowledge to researchers, with the fundamentals as well as latest advancements in the field of AI and HMI research. The content of this book will assist the readers to continue their research with better concepts in the relevant fields. Further, AI and HMI are very potential areas of study where students can be guided to take up projects at undergraduate, postgraduate, and PhD levels. Hence exposure of AI and HMI to the Faculty of Engineering, Technology, Pharmacy, and Management will make them competent to guide students on good projects and dissertations.

In this book, the original chapter contributions focus on HMI from analysis, design, and evaluation perspectives in HMI technological breakthroughs and applications. It covers a wide range of ideas, methodologies, approaches, and instruments to give the reader a thorough understanding of the field's current academic and industry practice and debate. Physical, cognitive, social, and emotional factors are all considered in the work, which is exemplified by key application fields such as aerospace, automobile, medicine, and defense. This book covers AI and machine learning methodologies as well as biological signals and HMI applications. Nanotechnology, user interface design, and interactive systems are also featured. The MATLAB approach to signal processing applications is also included. Above all, this book is intended to serve as a research guide that will both inform readers about the fundamentals of HMI from academic and industrial perspectives and provide a glimpse into how human-centered designers, such as engineers and human factors specialists, will attempt to design and develop human–machine systems in the future.

HMI Technology Advancements and Applications has 15 original chapters that examine the HMI from the perspectives of research, implementation, and analysis.

This book discusses advances in the field of HMI and provides practical knowledge in biomedical signal processing, AI, and machine learning, as well as observes advances in nanotechnology, user interface design, and interactive systems.

The proposed book covers a wide range of ideas, methodologies, approaches, and instruments to provide the reader with a comprehensive understanding of current academic and industry practice and debate in the field. As evidenced by key application fields such as aerospace, automobile, medicine, and defense, the work considers physical, cognitive, social, and emotional factors.

This book is intended to serve as a research adviser that will both inform readers about the fundamentals of HMI from academic and industrial perspectives and provide a glimpse into how human-centered designers, such as engineers and human factors specialists, will attempt to design and develop human–machine systems in the future.

Editors' Biographies

Dr. Ravichander Janapati is Associate Professor at Department of ECE, SR University, Warangal. He is a senior member of IEEE. He graduated in Electronics and Communication Engineering from JNTU Hyderabad and received his PhD in the Adhoc Wireless Sensor Networks field and M. Tech in Digital Electronics and Communication Systems from the Jawaharlal Nehru Technical University, Anantapur. His areas of interest are wireless sensor networks and brain–computer interface.

Dr. Usha Desai is Professor and Head at Department of ECE, SR University, Warangal. She received her PhD in the area of Biomedical Signal Processing from REVA University, Bengaluru in 2017 and M. Tech in Digital Electronics from Visvesvaraya Technological University, Belagavi, Karnataka.

Dr. Shrirang A. Kulkarni is Post-Doctoral Fellow with the School of Global Health Management and Informatics, University of Central Florida. Before joining UCF, he worked as an Associate Professor with the Department of Computer Science and Engineering at National Institute of Engineering, Mysore, India. He earned his PhD in Computer and Information Science from Visvesvaraya Technological University in the year 2012. His current research work includes developing new algorithms for analysis of healthcare data using machine learning. He is also a senior member of Association for Computing Machinery (ACM).

Dr. Shubham Tayal is Assistant Professor in the Department of Electronics and Communication Engineering at SR University, Warangal, India. He has more than 6 years of academic/research experience teaching at UG and PG levels. He received his PhD in Microelectronics and VLSI Design from National Institute of Technology, Kurukshetra, M. Tech (VLSI Design) from YMCA University of Science and Technology, Faridabad, and B. Tech (Electronics and Communication Engineering) from MDU, Rohtak. He is the editor/coeditor of eight books in total from CRC Press (Taylor & Francis Group, USA) and Springer Nature. He is a recipient of Green ThinkerZ International Distinguished Young Researcher Award 2020. His research interests include simulation and modeling of multigate semiconductor devices, device-circuit co-design in digital/analog domain, machine learning, and IoT.

List of Contributors

Avula, Jahnavi
Department of Computer Science Engineering, SR Engineering College, Warangal, Telangana, India.

B. N., Aravind
Department of Electronics and Communication Engineering, Rajeev Institute of Technology, Hassan, Visvesvaraya Technological University, Karnataka, India.

B., Shadaksharappa
Department of Computer Science and Engineering, Sri Sairam College of Engineering, Anekal, Bangaluru, Karnataka, India.

Balar, Pankti
Pharmacy Section, L. M. College of Pharmacy, Ahmedabad, Gujarat, India.

Bhalerao, Shailesh Vitthalrao
Department of Biosciences and Biomedical Engineering, Indian Institute of Technology Indore, Madhya Pradesh, India.

Bhattacharya, Sandip
Department of Electronics and Communication Engineering, SR University, Warangal, Telangana, India.

Billa, Chitharanjan
Software Architect, Oracle & Big Data Consultant (Federal Contractor), 22nd Century Technologies Inc., Virginia/Washington, DC, District of Columbia, USA.

Biswas, Souvik
Department of Electrical Engineering, Techno International New Town, Kolkata, West Bengal, India

Bokka, Prudvi Charan
Department of Computer Science Engineering, SR Engineering College, Warangal, Telangana, India.

Chavali, Murthy
Office of the Dean (Research) & Department of Chemistry, School of Science & Environmental Studies, Faculty of Science & Health Science, MIT World Peace University (MIT-WPU), Survey No: #124, Paud Road, Kothrud, Pune 411 038 Maharashtra, INDIA. NTRC-MCETRC, Tenali, Guntur, Andhra Pradesh, India.

Chavda, Vivek P.
Department of Pharmaceutics and Pharmaceutical Technology, L. M. College of Pharmacy, Ahmedabad, Gujarat, India.

D., Krishna
Department of Electronics and Communication Engineering, PSCMR College of Engineering and Technology, Vijayawada, Andhra Pradesh, India.

Das, Sohan
Department of Electrical Engineering, Techno International New Town, Kolkata, West Bengal, India.

Dayanand, Deepthi
Department of Computer Science, PES University, Banashankari Stage III, Dwaraka Nagar, Banashankari, Bengaluru, Karnataka, India.

Desai, Usha
Department of Electronics and Communication Engineering, SR University, Warangal, Telangana, India.

G., Sridevi
Department of Electronics and Communication Engineering, Raghu Engineering College, Visakhapatnam, India.

Hegde, Nayana
School of Electronics and Communication Engineering, REVA University, Bangalore, India.

Hegde, Rajesh B.
Department of Computer Science and Engineering, Sri Jayachamarajendra College of Engineering, Mysuru, Karnataka, India.

Janapati, Ravichander
Department of Electronics and Communication Engineering, SR University, Warangal, Telangana, India.

K., Sutejas
Department of Computer Science, PES University, Banashankari Stage III, Dwaraka Nagar, Banashankari, Bengaluru, Karnataka, India.

K. V., Suresh
Department of Electronics and Communication Engineering, Siddaganga Institute of Technology, Tumkur, Karnataka, India.

Kandala, Rajesh N. V. P. S.
School of Electronics Engineering, VIT-AP University, Vijayawada, Andhra Pradesh, India.

Krishna, Savitha
Department of Biology, Wilmington College, Wilmington, Ohio, USA.

Kollem, Sreedhar
Department of Electronics and Communication Engineering, SR University, Warangal, Telangana, India.

Kumar, Sunil
School of Computer Science Engineering, Vellore Institute of Technology, Chennai. Tamil Nadu.

M. S., Sudeep
Department of Computer Science and Engineering, Sri Jayachamarajendra College of Engineering, Mysuru, Karnataka, India.

Maheswari, Shishir
Department of Electronics and Communication Engineering, Thapar Institute of Engineering and Technology, Patiala, India.

Maji, Tamal
Department of Electrical Engineering, Techno International New Town, Kolkata, West Bengal, India.

Manvi, Sunilkumar S.
School of Computer Science Engineering, REVA University, Bangalore, India.

Misra, Iti Saha
Department of Electronics & Telecommunication Engineering, Jadavpur University, Kolkata, West Bengal, India.

N., Yashwanth
Department of Electronics and Communication Engineering, Manipal Institute of Technology, Manipal Academy of Higher Education, Manipal, Karnataka, India.

Nagavi, Trisiladevi C.
Department of Computer Science and Engineering, Sri Jayachamarajendra Collegeof Engineering, JSS Science and Technology University Mysuru, Karnataka, India.

P., Pramod Kumar
Department of Computer Science and Artificial Intelligence, SR University, Warangal, Telangana, India.

P., Rajesh Kumar
Department of Electronics and Communication Engineering, Andhra University College of Engineering (Autonomous), Andhra University, Visakhapatnam, India.

P., Ramkumar
Department of Computer Science and Engineering, Sri Sairam College of Engineering, Anekal, Bangaluru, Karnataka, India.

Pachori, Ram Bilas
Department of Electrical Engineering, Indian Institute of Technology Indore, Madhya Pradesh, India.
Center for Advanced Electronics, Indian Institute of Technology Indore, Madhya Pradesh, India.

Padmanabhan, Shobana
School of Computer Science and Engineering, RV University, RV Vidyanikethan Post 8th Mile, Mysuru Road, Bengaluru 560 059 Karnataka, INDIA.

Parupally, Sindh
Department of Computer Science Engineering, SR Engineering College, Warangal, Telangana, India.

Patel Disha B.
Pharmacy Section, L. M. College of Pharmacy, Ahmedabad, Gujarat, India.

Patel, Mital
Department of Pharmaceutics and Pharmaceutical Technology, L. M. College of Pharmacy, Ahmedabad, Gujarat, India.

Patel, Ritu J.
Pharmacy Section, L. M. College of Pharmacy, Ahmedabad, Gujarat, India.

Patel, Piyush
Electronics and Communication Engineering, L. D. college of Engineering, Ahmedabad, Gujarat, India.

Prasad, Ch Rajendra
Department of Electronics and Communication Engineering, SR University, Warangal, Telangana, India.

Roy, Asitava Deb
Department of Pathology and Laboratory Medicine, All India Institute of Medical Sciences, Deoghar, Jharkhand.

S., Ravi
Department of Electronics and Communication Engineering, Seshadri Rao Gudlavalleru Engineering College, Vijayawada, Andhra Pradesh, India.

Samala, Srinivas
Department of Electronics and Communication Engineering, SR University, Warangal, Telangana, India.

Syamasree Biswas Raha
Department of Electrical Engineering, Techno International New Town, Kolkata, West Bengal, India.

Sanchith
Department of Computer Science and Engineering, Sri Jayachamarajendra College of Engineering, Mysuru, Karnataka, India.

Sarkar, Sukrit
Department of Electrical Engineering, Techno International New Town, Kolkata, West Bengal, India.

Shishir, Vrishank
Department of Computer Science, PES University, Banashankari Stage III, Dwaraka Nagar, Banashankari, Bengaluru, Karnataka, India.

Sukka, Prashanth
Department of Computer Science Engineering, SR Engineering College, Warangal, Telangana, India.

Urs, Nataraj H. D.
School of Electronics and Communication Engineering, Reva University, Bangalore, India.

V., Prajith
Department of Computer Science and Engineering, Sri Jayachamarajendra College of Engineering, Mysuru, Karnataka, India.

T., Ananda Babu
Department of Electronics and Communication Engineering, Seshadri Rao Gudlavalleru Engineering College, Vijayawada, Andhra Pradesh, India.

Thakkar, Kunjal H.
Pharmacy Section, L. M. College of Pharmacy, Ahmedabad, Gujarat, India.

Y., Mamillu
Department of Electronics and Communication Engineering, Seshadri Rao Gudlavalleru Engineering College, Vijayawada, Andhra Pradesh, India.

1 Neural Networks for Human–Machine Interface

Deepthi Dayanand, K. Sutejas, Vrishank Shishir,
Shobana Padmanabhan, and Murthy Chavali

CONTENTS

1.1 DEEP LEARNING AND NEURAL NETWORKS

The neural network in the human brain is the most complex one in existence, evolving over millions of years. The fundamental unit of the network is the "neuron." There are billions of neurons in the network, and they are highly interconnected. The neural network works as follows to process very fast complex sensory input. A neuron on receiving an input signal, in the form of an electrical impulse, processes the signal and fires signals to the next neuron. This is the output of the conducting stage. A spike of the electrical signal is generated when the input signal is sufficiently strong. The site of transmission of signals from one neuron to another is the "synapse," which converts the activity from the axon into electrical signals. These properties of neurons have been studied to understand and create an artificial neural network (ANN) in computer systems [1–8].

Now consider the human visual system and its function in the human brain. Most people effortlessly recognize handwritten digits. This is achieved through five visual

DOI: 10.1201/9781003326830-1

1

cortices among which V1 is the primary cortex, containing over 140 million neurons with billions of interconnections. The visual cortices V1–V5 progressively perform more complex image processing. This is the equivalent of a supercomputer.

When we are born, we have a huge mass of untrained neural network, which gets trained and perfected over the years [8–12]. Visual patterns are hard to recognize, and the way of recognizing patterns varies from person to person but the basic intuition goes something like "an eight consists of two loops but nine has only one." That is, with exposure, neural networks get "trained" to recognize visual patterns. The more input over time, the more accurate the training is.

A subset of machine learning called "deep learning" distinguishes itself by how it approaches challenges. Machine learning requires skilled personnel, someone who is an expert in that particular domain, as opposed to deep learning, which comprehends features on its own accord, negating the necessity for domain knowledge. But as the machine learns on its own, deep learning algorithms require more time when compared to traditional machine learning

algorithms. On the brighter side, since the machine's domain knowledge does indeed run "deep," deep learning algorithms run significantly faster on testing datasets.

1.1.1 How Deep Learning Works

Deep learning models learn in a plethora of ways, ranging from supervised and unsupervised learning to reinforcement learning. Iterating through the examples, supervised learning categorizes and makes predictions using labeled datasets. Since the datasets need to be labeled and monitored, this requires some human involvement [13–26]. On the contrary, unsupervised learning uses clustering and association to group data according to patterns, through analysis. With reinforcement learning, as the name suggests, the model uses "reinforcement" in the form of rewards and punishment.

ANNs, also called deep learning neural networks, are mere imitations of the human brain. They strive to solve the key activities performed by the human brain, namely identification, categorization, and characterization. Deep learning neural networks use their layers of artificial neurons to optimize their predictions, using forward propagation. Backpropagation, as seen before, uses gradient descent to calculate errors and change the function's weights and biases to train the model. Accuracy is also time-dependent and improves with time and the number of iterations. The input and output layers of the model are also called the visible layers. The input layer receives the data for classification, while the output layer contains the final prediction by the model.

To solve certain issues or datasets, there are various forms of neural networks, often, highly complex. Consider two examples:

- **CNN—convolutional neural networks**

Most CNNs are used for image classification and computer vision applications.

- **RNN—recurrent neural networks**

Most RNNs are used in natural language processing applications. They are known for using sequential data input.

1.1.2 Deep Learning Methods

Deep learning methods involve a myriad of techniques including learning rate decay, transfer learning, dropout, and starting from scratch [27–29].

Learning Rate Decay

The learning rate is a hyperparameter that is responsible for the regulation of change the model undergoes depending on the predicted errors. A hyperparameter is a component that defines the system or establishes conditions for its operations before the learning process. Excessive learning rates may result in unstable training processes and inadequate weights, whereas slow learning rates may result in the possibility of getting stuck.

The learning rate decay method seeks to adjust the learning rate for the most optimal solution. This works on improving the performance of the model and also reducing the time required for training the model. It is also known as learning rate annealing. The most frequently used adaptations of learning rate comprise methods to slow down the rate of learning over time.

Transfer Learning

Transfer learning is a method used on pre-trained models to refine them even further. New data consisting of previously unidentified classifications is pushed onto the model to test the model on accuracy with respect to categorization. This method drastically reduces computation time.

Training from Scratch

Training from scratch is a method that uses an extensive labeled dataset, and as the name suggests, the model is trained from scratch using a whole new network architecture. This method is used for new applications that require niche functionalities. Unfortunately, this method uses a lot of data and requires weeks to be trained.

Dropout

Dropout is a method that is used in case of overfitting issues present in the neural network that contains a great number of parameters. It works by arbitrarily removing units and connections during the training phase of the model. This maximizes the performance of the neural network, especially under supervised learning.

1.1.3 Applications of Deep Learning

Some examples of applications of deep learning include the following [30–32]:

1. Law enforcement: Deep learning algorithms evaluate transactional data and learn some common underlying patterns within it. In case of a spike of transactions that do not coincide with the usual behavior of the model, it can be used to alert for signs of fraud or other illegal misconduct. Deep learning

also plays a role in investigative analysis by extracting patterns from multiple sources of audio, video, and text. Moreover, speech recognition and computer vision add to the efficiency of the analysis.

2. Financial services: Financial services frequently make use of predictive analysis to evaluate business risks, uncover frauds, monitor trading options, and help clients with managing their portfolios.

3. Health care: Deep learning and image recognition are playing pivotal roles in health care to evaluate medical reports and images. It furthers the leaps of advancement in medical imaging and radiology.

4. Customer service: Deep learning technology is used widely in businesses' customer care procedures. Using chatbots requires a simple type of artificial intelligence, which is utilized in many different applications, businesses, and customer support websites. Traditional chatbots, which are frequently found in menus resembling call centers, use natural language and even facial recognition. However, more advanced chatbot solutions make an effort to ascertain whether there are many answers to ambiguous questions through machine learning. The chatbot then attempts to immediately respond to these inquiries or direct the interaction to a human user depending on the responses it has received. By providing speech recognition capabilities, virtual assistants expand the concept of a chatbot. This invents a fresh technique to interact with users in a tailored way.

1.1.4   Limitations and Drawbacks

Deep learning does not depend on human intervention as much as traditional computing; it relies on observations made from the training data. Hence, the information contained in the training data becomes their perception of the world. In the case of small datasets, it would be difficult for deep learning models to learn from a source that is not representative of general cases. Another major concern with respect to deep learning algorithms is the presence of human societal biases. When a model is trained on data that is influenced by social factors and prejudiced biases, it will replicate similar biases in its predictions. Maintaining the pace of the learning rate factors into another limitation. If the learning rate is too high, the model tends to converge, and if the learning rate is too slow, then the chances of the model getting stuck in itself increase. Limitations may also result from deep learning models' hardware specifications. To ensure increased effectiveness and lower time consumption, multicore high-performing graphics processing units (GPUs) are needed. Moreover, each set end contributes toward expenses and consume a lot of energy. Furthermore, these require additional hardware specifications for memory requirements.

The powerful models of learning can be achieved only through extensive training on large volumes of data. These also call for more parameters, which in turn call for more data requirements. Deep learning models are rigid and incapable of multitasking after they have been trained. Only one specific problem can they effectively and precisely solve. Even resolving a similar issue would necessitate retraining the whole system from scratch. Limitations of deep learning models cover data manipulation algorithms and any form of reasoning. In tasks like these, there will always be a demand for deep learning capabilities from us humans ourselves.

1.2 RECREATION OF HUMAN NEURAL NETWORKS IN MACHINES

The recreation of human neural networks in computer systems is the ANN. ANNs also consist of a large number of neurons, with a large number of interconnections among them, to achieve the powerful functionality of human neural networks. Each neuron in this has a modifiable weight, which is the equivalent of the synapse of the neurons in the human brain. The artificial neuron fires an electrical output in the form of a single number. Some of the earliest learning algorithms were derived from the concept of teaching a child—training a biological brain through examples and mistakes, to make better decisions in the future. In this section, we will learn exactly how an ANN gets designed.

1.2.1 ARTIFICIAL NEURONS

As the name suggests, artificial neurons are related to biological neurons. The idea for an artificial neuron, which is the building block for deep learning, comes from the study of the human brain. It can be said without a doubt that the human brain is the most complex network with over 86 billion neurons. And to further the point that it is neurons that provide the cognitive abilities, we can compare the human brain with that of a rodent's brain. A rodent's brain weighs 1,500 g, and a human brain weighs 1,508 g. But a rodent's brain has a mere 12 billion neurons whereas a human brain has 86 billion neurons. More the density and number of neurons, the higher the cognitive abilities.

The biological neuron is comprised of the following major parts: dendrite—the part responsible for receiving incoming signals, soma—the part responsible for processing input signals and deciding if the neuron needs to fire and output signal, axon—the part responsible for getting processed signals from one neuron to another neuron or the required locations, and synapse—the connection between the current neuron's axon and another neuron's dendrite. The functioning of the neuron is a culmination of the activities of all these parts. The dendrite receives signals which can be either excitatory—making the neuron fire or inhibitory—keeping the neuron from firing. The soma processes the inputs and decides whether or not to fire. If it is to fire, the nerve impulse or action potential is conducted down the axon and the information is passed.

Artificial neurons, known as perceptrons, are the building blocks of neural networks. They have been created to mimic the functionality of the neurons in the human brain. The major functions of a perceptron are the input layer—taking input from the user or a previous layer, the activation function—processing input using a function called an activation function to decide whether or not the neuron should fire, and the output layer–sending 0 for not firing and 1 for firing. The output is propagated to the next layer or returned as an output. Hence, we can draw a comparison as follows: the input layer is equivalent to the dendrite, the activation function—the node of the neuron—is equivalent to the soma, and the output layer is equivalent to the axon and the interconnections among the neurons forming the network to the synapse.

Now the question arises as to how the node or the activation function knows when to fire. To answer that, we need to understand the functioning of the artificial neuron at a deeper level, starting with "threshold" functions.

1.2.2 PERCEPTRON AND SIGMOID NEURONS

As discussed in the previous section, a perceptron is the most basic building block of an ANN. It has an input layer, a node, and an output layer. The node contains a function known as the activation function, which decides whether or not the neuron should fire.

Consider a simple problem as follows: You want to watch a movie. You decide whether or not you want to watch the movie based on the reviews by three of your friends, A, B, and C. Among them, you value person A's opinion 0.1, B's opinion 0.6, and C's opinion 0.3—the numbers, called weights, being arbitrary to represent the fact that each person's opinion has a different bearing on you watching the movie. You would watch the movie only if the ratings given by your friend matter— opinion*rating > 1.

Consider the following model of a perceptron (See figure 1.1 and figure 1.2):

The working of the above perceptron is as follows: It takes input from three people and weights represented by the variables w_1, w_2, and w_3. The variable 'y' represents the decision taken regarding whether to watch the movie or not. The perceptron takes the weighted sum of the inputs and then compares it against a *threshold* value. In the above diagram, $w_0 *x_0$ represents this threshold, represented by θ. Hence, for our simple example, if the perceptron has learned the problem correctly, it should look something like this.

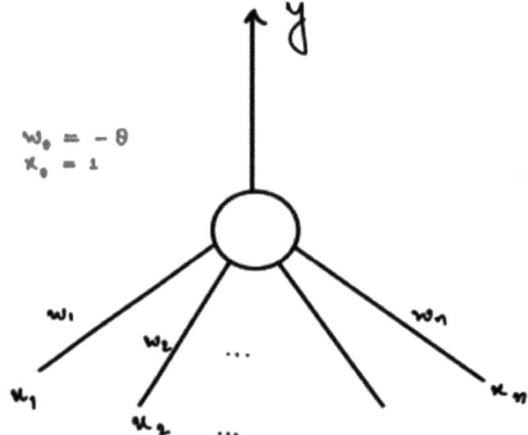

FIGURE 1.1   A perceptron.

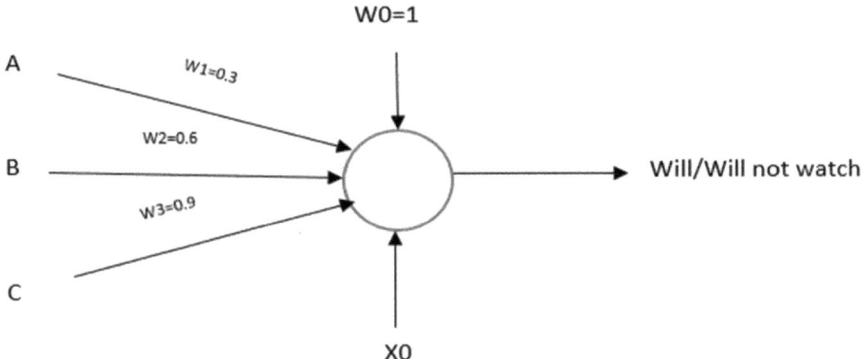

FIGURE 1.2 A perceptron that predicts moviegoing decision.

This perception, when it calculates the weighted sum, predicts correctly whether or not you would go to the movie. Mathematically,

$$\sum_{i=0}^{n} \omega_i \cdot x_i - \theta < 0 \qquad (1.1)$$

$$\sum_{i=0}^{n} \omega_i \cdot x_i - \theta \geq 0 \qquad (1.2)$$

Algorithm: Perceptron learning algorithm

P<- inputs with label = 1
N<- inputs with label = 0
Initialize weights 'w' randomly
while not convergence do
Pick random x∈pUN
if x∈p and w.x < 0 then
w = w + x
end
if x∈N and w.x ≥ 0 then
w = w − x
end
end

The above algorithm converges when all the inputs have been classified correctly. The algorithm uses concepts of linear algebra to learn the weights. Let $\mathbf{w}$ be the vector of all the weights and $\mathbf{x}$ be the vector of all the inputs. We can say that the perceptron has learned when $\mathbf{w}^T\mathbf{x} = 0$. In other words, $\mathbf{w}$ and $\mathbf{x}$ are perpendicular and $\mathbf{w}^T\mathbf{x}$ divides the positive and negative points clearly. The $\mathbf{w}$ vector is initially set to random values. Through the algorithm, the angle between $\mathbf{w}$ and $\mathbf{w}^T\mathbf{x}$ is made closer to 90° by incrementing or decrementing with the value of **x**. In terms of linear algebra, cos(p) = 1, where cos(p) represents the angle between **w** and **x**. When we add **x** to **w**, we increase cos(p) and vice versa. Over a few passes, the perceptron arrives at the correct set of weights.

The issue with this is that it is a very harsh classifier. Even if the weighted sum comes up to somewhere around 0.51, with 0.5 being the threshold value, the neuron will fire. And if it is 0.49, it will not fire. Hence, the new proposed neuron, the *sigmoid* neuron, uses the sigmoid logistic function to help smooth out the curve.

$$y = \frac{1}{1 + e^{-\left(w_0 + \sum_{i=1}^{n} w_i x_i\right)}} \tag{1.3}$$

Other benefits of using the sigmoid function include the following:

- The sigmoid activation function is differentiable and continuous, and hence we can use calculus to help with future operations such as *backpropagation*.
- The predicted (i.e., output) values are now between 0 and 1, which can be used as a probability indicator.

The sigmoid neuron is represented as follows (See figure 1.3):

The above discussion has shown us the characteristics of the most basic building block of a neural network. Now, we will look at certain properties of neural networks and then look at more complex connected neural networks.

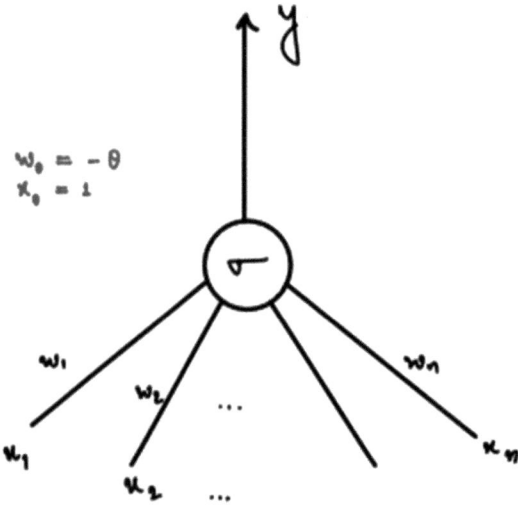

FIGURE 1.3   A sigmoid neuron.

1.2.3 Universal Approximation Theorem

One of the major specialties of neural networks has been demonstrated by the universal approximation theorem. It says that any function can be approximated by a neural network to any arbitrary accuracy. Mathematically,

$$|f(x) - f'(x)| = E$$

where 'f' is the actual function, f' is the function represented by the neural network, and E is an arbitrary positive constant.

We can look at a very simple understanding of this by using any basic Boolean operation and a single/network of perceptrons. Let us take the NAND function as an example. With logic gates, the NAND function is represented by Figure 1.4. This can also be represented by a connection of perceptions as shown below:

With the weights = −2 and bias = 3. The bias term is an adjustable numerical value added to a perceptron's weighted sum to increase the accuracy of the model's decision (prediction). NAND gates are universal gates for computing functions as we can recreate most functions using a combination of NAND gates. Therefore, we can infer that perceptrons are universal for calculating functions as well. Now that we know the building blocks of an ANN and its capabilities, let us look at the architecture of a simple neural network (See figure 1.5).

1.2.4 The Architecture of a Neural Network

Consider the simple neural network depicted in Figure 1.6. The neural network, just like a perceptron, has major components in it. Each set of neurons in the same line forms what is known as a 'layer' of the neural network. From this, we can conclude that the above neural network has three layers. The first layer is known as the 'input layer'. This is where the initial parameters and the random weights are passed into the model. The second layer is known as the hidden layer. It is called the hidden layer

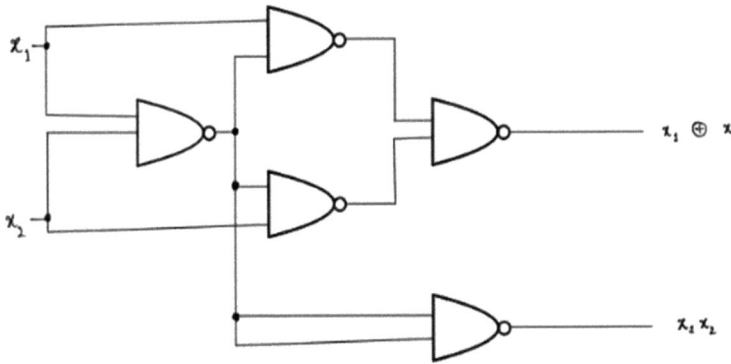

FIGURE 1.4 Logic gates implementing NAND functionality.

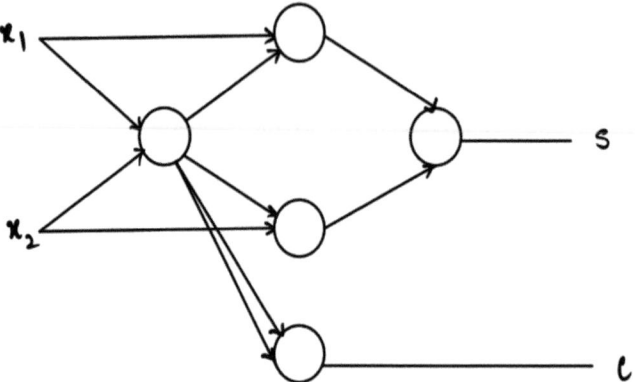

FIGURE 1.5   A network of perceptrons implementing NAND functionality.

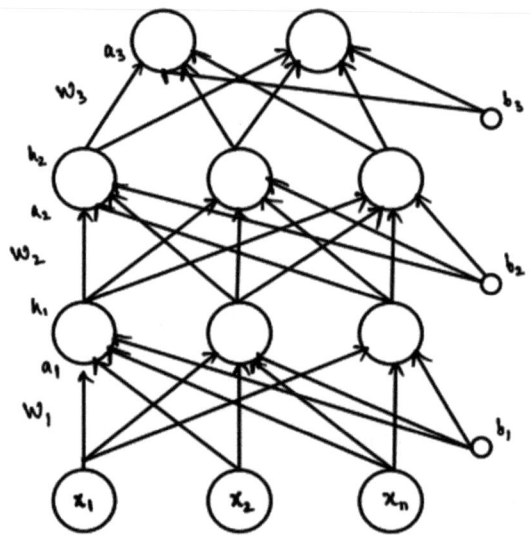

FIGURE 1.6   A neural network.

as the hidden layer estimates the output as some function of the input, but the function or the distribution of the data is unknown or hidden. Exactly how the individual neurons and their activations estimate the function is also hidden in a multitude of real values, which is difficult to correlate with real-world meaning. The third and final layer is known as the output layer, which, as the name conveys, provides the output. The layers are shown in Figure 1.7.

Now the question arises as to how the number of layers or the number of neurons per layer is determined. It is however not simple. We break it into three parts mainly. Let us consider the problem of classifying an image into any one of seven categories.

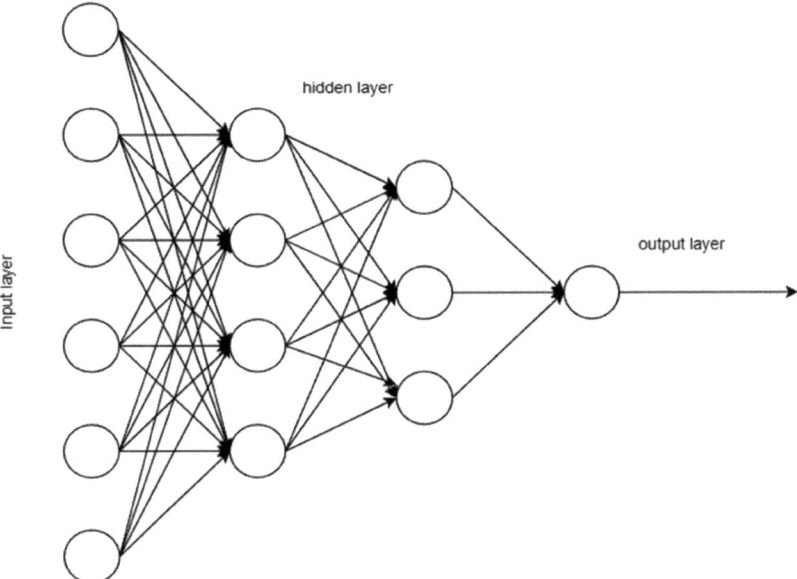

FIGURE 1.7   Layers in a neural network.

Let us assume that the provided image is free of any aberrations. Further, the images used for training the model are all of the same sizes of 64 × 64 pixels.

Now, the first layer or input layer needs to take all the input parameters as its input. Since we are dealing with images, the most normal way to input them would be as a matrix of pixels. A 64 × 64 image would have 4,096 pixels. This means that for a network to classify such an image, the input layer would need to have 4,096 neurons. The output layer would need to classify the image into one of seven categories. As one would expect, we would need seven neurons in this layer, each representing one of the output classes.

As for the hidden layers, the task of determining the number of neurons or the number of layers is not an easy one. This is done by a lot of trial and error at times, though there are some rules of thumb for this. The issue with having too few hidden layers is that it will not be able to predict accurately. The issue with having too many layers is overfitting, i.e., it will learn the training data very well but will perform poorly on the testing data. One such rule of thumb is the following equation:

$$N_h = \frac{N_s}{\left(* \left(N_i + N_o\right)\right)} \tag{1.4}$$

N_i = number of input neurons
N_o = number of output neurons
N_s = number of samples in training dataset
α = an arbitrary scaling factor usually 2–10

Along the same line, determining the number of hidden layers can somewhat follow this guideline:

0: Capable of representing only linearly separable functions or decisions. Very similar to a single sigmoid neuron.

1: This can approximate any function that contains a continuous mapping from one finite set to another finite set. The mapping of functions from two finite domains.

2: Two layers have the ability to represent an "arbitrary decision boundary" to "arbitrary accuracy" with "rational activation functions" and can approximate any "smooth mapping" to any accuracy.

These above rules can help, though not always. The best way to determine what is the best hidden layer design for your model is by trial and error since it vastly depends on the input data and the degrees of freedom, along with the task at hand. The working of the above model can be approximated to an aggregation of the processes done in a sigmoid neuron.

Consider the following neural network shown in Figure 1.8. Let the inputs, as shown, be x_1, x_2, x_3, and so on, and let there be two hidden layers, with three neurons each, and let the output layer have two neurons. Mathematically, the pre-activation

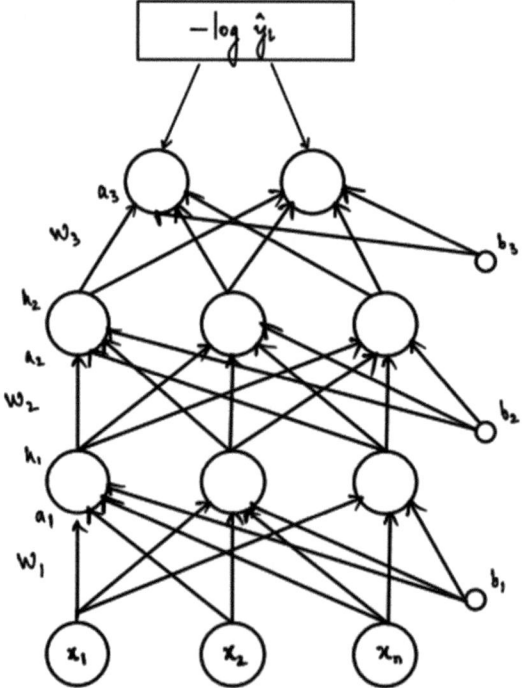

FIGURE 1.8 A neural network with two hidden layers.

functions (a_i) are represented by the equation below where h_i is the post-activation part of the neuron and W_i represents the weights.

$$a_i = b_i + W_i h_{I-1}(x) \qquad (1.5)$$

The activation layer h_i is calculated, where g represents the activation function.

$$h_i = g(\, a_i(x)) \qquad (1.6)$$

Let the activation function for the final layer be represented by the function O. Then, the output values will be:

$$y = O(a_L(x)) \qquad (1.7)$$

where a_L represents the pre-activation of the L'th layer of the neural network. The above process helps us to see how each layer can be related to the other. If we instead wanted to find the operations done by the neural network as a single function, i.e., the function that relates x to y:

$$\{f \rightarrow f(x) = y\} \qquad (1.8)$$

This can be obtained by a series of the above operations done in repetition, and in doing so, we arrive at the following final form. As mentioned above, the above equation relates x with y directly, which shows in one equation the entire effect of such a neural network.

$$f(x) = O(W^3 g(W^2 g(W^1 x + b_1) + b_2) + b_3) \qquad (1.9)$$

However, an important point to be noted is that the above equation does not represent the function that the neural network is approximating. It is merely a set of numerical values—weights and biases—which when combined in the above form approximate the function to an arbitrary constant, as per the universal approximation theorem. What this means is, if we develop a neural network to approximate $f(x) = x^2$, then the above function would not expand to x^2. It would be a set of matrix and vector operations with values of weights and biases such that the numerical value of the answer would be an approximation of x^2. Now that we have understood the architecture of a neural network, let us look at one of the most important algorithms in the learning of weights, the gradient descent algorithm.

1.2.5 GRADIENT DESCENT AND COST FUNCTIONS

If we recall the perceptron learning algorithm, the weights, which were initially set to random values, are adjusted in each iteration such that the model's prediction comes closer and closer to the required answer. In linear algebra and geometry terms, try to make $\mathbf{w.x} = \mathbf{0}$, i.e., make the vector $\mathbf{w}$ perpendicular to $\mathbf{x}$. The gradient descent algorithm does the same but in n dimensions. In the above example of a feedforward neural network, upon further analysis, we realize that "x lies in $\mathbf{R^n}$," "b_1, b_2 lie in n," "b_3 lies in k," and the pre-activation of the final layer lies in $\mathbf{R^k}$. Now, to find the value of $\mathbf{w}$ for $\mathbf{w.x} = \mathbf{0}$, we need to use a technique that applies the perceptron learning algorithm in not one but n dimensions.

Thus, we can define gradient descent as:

> Gradient descent is an iterative first-order optimization algorithm used to find a local minimum/maximum of a given function.

Gradient descent uses calculus to reduce the error to as low as possible. Hence, we need to use a cost function that is continuous and differentiable to be our loss function. Another requirement of the cost function is that it has to be convex in nature. Now we come to what exactly gradient means.

The gradient is simply the slope of a curve or function at any given point in a particular direction. Mathematically, representing the gradient for a vector by:

$$\nabla f\left(p\right) = \begin{pmatrix} \dfrac{\partial f\left(p\right)}{\partial x_1} \\ \vdots \\ \dfrac{\partial f\left(p\right)}{\partial x_n} \end{pmatrix} \tag{1.10}$$

The cost function or the loss function used in this context is a function that determines how far away the predicted value is from the actual value, i.e., it quantifies the error. Some standard loss functions are discussed below.

Quadratic loss function and mean square error function

$$C_{MST}\left(W, B, S^r, E^r\right) = \frac{1}{2}\left(\sum_j (a_j^L - E_j^r)^2\right) \tag{1.11}$$

The above equation is a general form of a quadratic cost function.

The arguments in the LHS are:

W – Weights of neural network
B – Bias of neural network
S^r – Input of neural network
E^r – Expected output of the neural network

The quadratic loss function basically takes the difference between the true value and the predicted value but relates the error to the square of this value. A well-known example of this is the mean square error (MSE) function.

The MSE function, shown below, calculates the difference between every data point and its estimation (prediction) by the model, squares the difference, and sums the squares of the differences of all the data points and halves the sum.

$$MSE = \frac{1}{n}\sum_{i=1}^{n}\left(Y_i - \hat{Y}_i\right)^2 \tag{1.12}$$

Cross-entropy loss function

Cross-entropy builds on the idea of entropy from information theory. Entropy can be thought of as the number of bits required to transmit a randomly selected event from a probability distribution. It can be calculated for a random variable X with a set of discrete states x, each having a probability P(x), as the negative of the sum of the product of the probability of each x and the log of that probability. Cross-entropy can then be thought of as the number of bits required to represent or transmit an average event from one distribution compared to another distribution. Let P denote a target or underlying probability distribution, and let Q denote an approximation of the target distribution. Subsequently, the cross-entropy of Q from P is the number of additional bits to represent an event using Q instead of P. It can be calculated using the probabilities of the events from P and Q as shown below:

$$H(p,q) = -\sum_{x \in X} p(x)\log\big(q(x)\big) \tag{1.13}$$

Of the loss functions discussed here, the most commonly used one, is the cross-entropy function, since its values always lie in the [0,1] range, which helps when we are dealing with probabilistic interpretations.

Now that we have understood what the gradient descent algorithm uses, let us look at the algorithm itself. The equation for gradient descent can be represented as shown below where p_n represents the n'th point, $f(p_n)$ represents the direction of descent, $\acute{\eta}$ represents the extent of descent (aka learning rate), and p_{n+1} represents the n+1th point.

$$p_{n+1} = p_n - \eta \nabla f\left(p_n\right) \tag{1.14}$$

Intuitively, gradient descent can be understood as follows. At every point, we check the direction of the maximum gradient. The direction of the maximum gradient provides the direction in which there is a steeper change in the value of the cost function. Since we subtract this from the current value, we take a small step toward that particular local minimum of the cost function. This will land us at a local minimum of the cost function after multiple iterations. A point to note here is that we cannot land at the global minimum using gradient descent. We can equate this to a ball rolling down a hill. Let us imagine a situation where there is no gravity. Let there exists a structure as shown below. Let a ball be placed at the point shown. Now, our task is to find the lowest point of the valley. Intuitively, if we think about it, we would see the direction in which the steepness (or slope) is higher, and move the ball in the

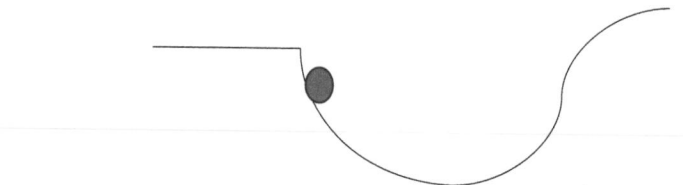

FIGURE 1.9   Illustrating gradient descent.

opposite direction. Further, as we get closer to the target (lowest point), the slope keeps on decreasing and hence we would know that we need to move the ball shorter distances.

The above process in which one iteratively moves the ball around to find the steepest point by moving it away from the direction of increasing slope, in proportion to the rate of increase, is the gradient descent algorithm. There are multiple variations of the gradient descent algorithm. They are discussed below (See figure 1.9).

- Batch gradient descent

  This variation of gradient descent uses all of the training examples in every iteration of the algorithm. As one may infer, it does not work well if the data size is exceedingly large, as it becomes time-consuming.

- Stochastic gradient descent

  This variation of gradient descent uses only one example of training in every iteration. Thus, this updates the parameters after every iteration.

- Mini batch gradient descent

  This variation of gradient descent is a mixture of the previous two. In this, data is divided into small batches and then gradient descent is performed on each of these batches. Thus, it works well with even large training datasets.

  Now that we have understood the gradient descent algorithm, we shall take a look at backpropagation.

Backpropagation

Backpropagation is the algorithm used to effectively train the neural network through computations of intermediate derivatives and calculating their effect on the final loss. Let us take the following example to intuitively understand backpropagation. Consider the following neural network (See figure 1.10):

In the above network, let us say during a certain iteration of the training, the model predicts some output y' for some input x' and let y be the actual output. Now, let us evaluate the loss using any function g. Let us represent the loss by g(y', y). Let

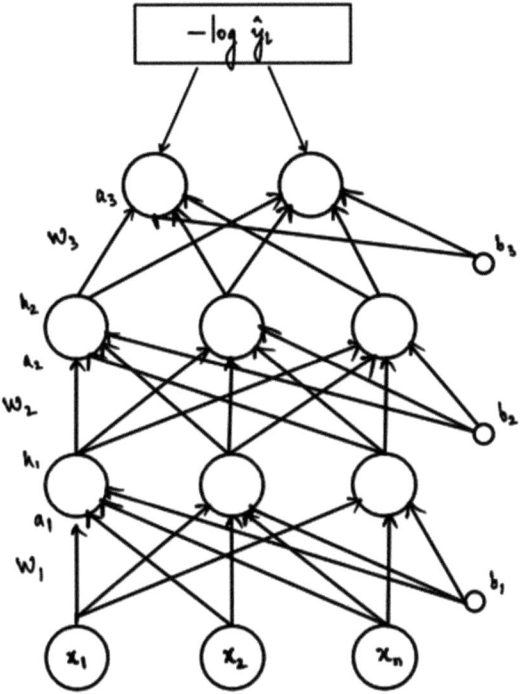

FIGURE 1.10   An example of neural network to understand backpropagation.

$g(y', y) != 0$, i.e., if there is some error in the model's prediction. Now, let us think of an intuitive way to reduce the error. We see from the diagram that the output can be influenced by any of the weights and biases. Therefore, we can choose their values in such a way as to minimize the error in the prediction.

Let us imagine a situation where these different layers and neurons could communicate with each other in a certain language. But the only constraint is that any layer can only communicate with the layer with which it is connected. Let us start with the output layer as it is the one that outputs the predicted value.

The output layer will first ask the last hidden layer why there was an error in the output. This layer will in turn check with its weights and biases. But, as we can infer, these are influenced by the layer below it. Therefore, the hidden layer says that it did not cause the error and, in turn, asks the layer below it. This process continues until the network finds the weights that have the most impact on the output. Now, this weight is changed to better the output to reduce the error in the prediction.

Mathematically, we want to find the derivative of each set of weights and biases with the output loss function and then use the weights that have the maximum effect, i.e., the highest rate of change with respect to the loss. Since at every layer, the output is a function of its pre-activation functions, to find the derivative of every weight with the loss function, we would have to traverse/backpropagate through the entire

$$\underbrace{\frac{\partial \mathcal{L}(\theta)}{\partial W_{111}}}_{\text{Talk to the weight directly}} = \underbrace{\frac{\partial \mathcal{L}(\theta)}{\partial \hat{y}} \frac{\partial \hat{y}}{\partial a_3}}_{\text{Talk to the output layer}} \quad \underbrace{\frac{\partial a_3}{\partial h_2} \frac{\partial h_2}{\partial a_2} \quad \frac{\partial a_2}{\partial h_1} \frac{\partial h_1}{\partial a_2}}_{\text{Talk to the previous hidden layers}} \quad \underbrace{\frac{\partial a_1}{\partial W_{111}}}_{\substack{\text{Now talk to} \\ \text{the weights}}}$$

FIGURE 1.11   Chain rule for gradient calculation.

network, starting from the output loss, all the way to the input layer. Suppose that the error lies in W_{11}. To find this, we would use the chain rule as depicted in the equation below where a_i represents the pre-activation at the ith layer and h_i represents the post-activation at the ith layer (See figure 1.11).

The above computation is for a very simple network. As the equation shows, the gradients for the output, hidden, and input layers are done separately. This process is what increases the efficiency and accuracy of a neural network's training process.

1.3   HANDWRITING RECOGNITION USING DEEP NEURAL NETWORKS

Here is a brief overview of how neural networks are able to recognize human handwriting and the natural language of human beings. Humans can understand, comprehend, and make sense of what they see, with the help of what they have learned and experienced. Most of it is done unconsciously. Humans understand and analyze handwriting through intuition and pattern recognition. The visual system in our brains is responsible for reading and understanding handwritten letters, characters, and digits. We can train a neural network to understand the pattern as well. However, it is a difficult task for computers.

Before the use of deep learning, optical character recognition (OCR) software was being used for human handwriting recognition. The OCR software was first developed by Ray Kurzweil in 1974. One of its drawbacks was that it was not able to recognize unknown characters. The problem occurs in the following two key areas:

- **Character extraction**—the OCR fails to recognize characters with no distinct separation between letters, as in cursive writing.
- **Feature extraction**—properties like aspect ratio, pixel distribution, number of strokes, distance from the image center, etc., are hardcoded to match the input symbols, which means that these properties had to be added manually. This leads to an increased application (software) development time.

Handwritten text analysis is the ability of a machine to analyze and recognize human handwriting and transcribe it into text. Handwriting analysis thus falls into the category of text recognition problems. Early text recognition technologies can be traced back to trying to solve the problem of transferring text from one source to another source. One way of solving this involved scanning the light source coming from a character and classifying it based on the outcome of intensity. Each character scanned would be identified by classifying it into its corresponding alphabet.

With the advent of technology, this process was refined and improved, and the optical era grew into the image processing era. Here is where machine learning and neural networks exponentially improved the accuracy of the classification of text. We require the computer to recognize something scribbled by a person and transcribe it into text. Considering the example of a page of paper with handwritten text. Here, the computer gets one input, the image of the paper. This needs to be converted into multiple outputs, where each output is the corresponding character recognized from the page. CNNs are the neural networks used for image processing.

However, CNNs are not a good fit for the problem of handwriting recognition, as they are "one-to-one" systems, whereas this problem is a "one-to-many" problem. Another issue faced by CNNs is segmenting each character to accurately classify it. To solve this, we move from using CNNs to another type of neural network called RNNs, as they are considered to be "one-to-many" systems. RNNs convert the input into a sequence of vectors, which are fed into the neural network. The neural network takes into account the current state and also the previous state, to predict the new output sequence. But as RNNs are required to maintain the previous state of the system, training this system cannot be parallelized. Thus, it is a relatively slow process.

Building and training a system to analyze handwritten texts consists of a series of steps. First and foremost, a pertinent dataset must be chosen for training the neural network. This dataset should be split by an appropriate ratio into a training set and a testing set. *Encoders* can be used to preprocess the inputs to the network to create tensors of data. The architecture of the neural network must be finalized before constructing the model. Once the network is set up, testing starts on the model to reach the highest possible accuracy. The MNIST database is a database of handwritten digits that has a training set of 60,000 examples and a test set of 10,000 examples and is used in most introductory projects for handwritten image recognition.

Handwritten text analysis has active areas of research that include signature verification, bank check processing, and postal address interpretation

1.4   HUMAN–COMPUTER INTERFACE AND DEEP LEARNING

Interaction is the key component in how users can utilize any machine to get the required output. "Human–Computer Interaction" or "HCI" (sometimes spelt CHI, putting the computer first) is an entire field of study devoted to this. HCI, which often consists of a display of some form and several additional tools for communicating with computers, is where human–computer interaction occurs. Digital signage is often discussed in terms of its interactivity, but all similar systems have always been interactive in some form. In addition to passing by a digital sign and viewing it, people may respond to QR code signs by scanning them.

Currently, there is arising trend toward hands-free interactions, automation, and personalization as we analyze how people engage with computers and screens. This is largely owing to the exponential growth of computing prowess, especially with regard to how computers have gotten better at understanding us. Similar to how electricity defined the previous century, the technologies emerging currently are going to mold our future in ways unimaginable. The evolution of HCI across time may be traced to technological advances. In light of this, we might be able to predict what the coming ten years will bring with some level of certainty.

Computing started as huge machines, taking up entire rooms. These were electronic digital computers that used vacuum tubes and punch cards. In 1946, the Electronic Numerical Integrator and Computer (ENIAC) was brought to life. About 30 tons and more than 18,000 vacuum tubes made up the ENIAC. It lacked an operating system and could only do one task at once. The integrated circuit underwent yet another revolution in 1963, giving rise to what we now consider to be contemporary computers. In 1980, MS-DOS was released, and in 1981, IBM unveiled the personal computer (PC). Three years later, Apple released the Macintosh icon-based interface system. The Windows operating system from Microsoft was then released in the 1990s. Then came 1984 and Apple unveiled the Macintosh interface system. This made use of the icon-based interface. In 1990, along came the Windows Operating System, courtesy of Microsoft.

The 1980s were all about improving the use of computers, now found in people's homes and offices. It completely changed the industry when Apple introduced the Macintosh in 1984. The interaction became much simpler because using computers no longer required expertise. At the time, the most predominant interfaces included the icons, mouse, and keyboard. The advent of the Windows OS in the 1990s solidified the dominance of icon-based user interfaces.

Subsequently, everyone could access the Internet thanks to the World Wide Web. The fact that individuals were connecting with computers in order to engage with one another was a key component of the new connected web. As computers evolved into tools for both communication and task completion, HCI design principles switched from task completion to interaction promotion. This marked the emergence of social computing. The touch pad appeared, enabling users to interact with applications and data on their screens, using a stylus or even just their fingers.

The 2007 release of the iPhone popularized *touch* as a fresh approach to HCI. It is now required that communication has to be quick, natural, and personalized. Touchscreens have started to appear on more consumer electronics, including desktops, laptops, and tablets. Interactive digital signage is becoming a reality, thanks to touch technologies. Early touch screens were pricey, single-touch displays with a professional level of quality. When the multi-touch screen was introduced, formerly static surfaces could suddenly be manipulated like an iPhone. In the last ten years, touchscreens have multiplied exponentially.

The goal of modern HCI techniques is to personalize interaction tools and processes as much as possible while providing viewers with more options and a more humanist perspective. In addition to the exponential growth of HCI possibilities, sign systems are adopting technologies created for personal devices more swiftly. Today, we may interact with a screen without touching it at all thanks to technology like voice-activated interfaces (VUIs). It only requires that you utter orders or queries, and it will then fulfil your requests. It closely resembles home smart speakers like Amazon's Alexa and Microsoft's Cortana, which currently support this kind of interaction. It attempts to please a crowd accustomed to using smartphone apps like Siri and Google Assistant.

With augmented reality (AR), computers are temporarily integrated into the display and used to overlay data on real-world objects or digital signs. In this technique, interactive elements can be included in static presentations. Additionally, it gives digital signage ecosystems access to gameplay approaches.

Facial recognition using artificial intelligence is the most popular rise of interacting with machines. Cameras may be incorporated into or attached to screens, and when a user approaches them, the sign "looks" at them to detect their facial expression. The camera can now observe more information thanks to even more advanced face recognition technology. Based on predetermined sets of criteria, it then determines what information would be pertinent for that viewer.

We will begin to view content that is uniquely suited to us by merely approaching the screen and gazing at it as AI programs and machine learning get more advanced. This kind of HCI is almost entirely undetectable and appears more organic. Additionally, we have prototypes of software that can detect a user's eye location on a screen. The writing can be made larger or clearer by using a specific blink pattern. Interactions can happen considerably more quickly than they do presently by adopting EGT (eye-gaze tracking)-only or even mixed systems.

REFERENCES

1. Human-Machine Interface (HMI) Defined-. https://www.inductiveautomation.com/resources/article/what-is-hmi
2. G. D. Abowd, R. Beale. Users, systems and interfaces: A unifying framework for interaction. In D. Diaper, N. Hammond, editors, *HCI'91: Usability Now*. British Computer Society Special Interest Group on Human-Computer Interaction, Cambridge, 1991.
3. R. Beale, J. Finlay, editors. *Neural Networks and Pattern Recognition in Human-Computer Interaction*. Ellis Horwood, 1992.
4. J Finlay, R. Beale. Neural networks and pattern recognition in human-computer interaction. *ACM SIGCHI Bulletin*, 25(2), 1993, pp 25–35. https://doi.org/10.1145/155804.155813
5. H. L. Dreyfus, S. E. Dreyfus, T. Athanasiou. *Mind Over Machine: The Power of Human Intuition and Expertise in the Era of the Computer*. Basil Blackwell, Oxford, 1986.
6. R. R. Hoffman. How Can Expertise be Defined? Implications of Research from Cognitive Psychology. In: Williams, R., Faulkner, W., Fleck, J. (eds) *Exploring Expertise*. Palgrave Macmillan, London, 1998. https://doi.org/10.1007/978-1-349-13693-3_4
7. L. Jean, W. Etienne. *Situated Learning: Legitimate Peripheral Participation*. Cambridge University Express, Cambridge, 1991. Wenger E. Communities of practice: Learning, meaning, and identity. Cambridge University Press; 1998.
8. J. Akyeampong, S. Udoka, G. Caruso, M. Bordegoni. Evaluation of hydraulic excavator human-machine interface concepts using NASA TLX. *International Journal of Industrial Ergonomics*, 44(3), pp. 374–382, 2014.
9. B. Li, W. Liu, X. Zhang, Z. Pan, F. Zhao. Effectiveness assessment of human-machine interface in driver assistance system. *Engineering*, 197, pp. 567–574, 2013.

10. S. Sreetharan, M. Schutz. Improving human-computer interface design through the application of basic research on audiovisual integration and amplitude envelope. *Multimodal Technologies and Interaction*, 3(1), p. 4, 2019.

11. I. Dolgov, S. B. Hottman. 11 Human factors in unmanned aircraft systems. *Unmanned Aircraft Systems*, 165–181, 2011.

12. L. Giraudet, J. P. Imbert, M. Bérenger, S. Tremblay, M. Causse. The neuroergonomic evaluation of human-machine interface design in air traffic control using behavioural and EEG/ERP measures. *Behavioural Brain Research*, 294, pp. 246–253, 2016.

13. G. Chao. Human-computer interaction: Process and principles of human-computer interface design. In 2009 International Conference on Computer and Automation Engineering, pp. 230–233, Bangkok, Thailand, 2009.

14. J. H. Goldberg, X. P. Kotval. Computer interface evaluation using eye movements: Methods and constructs. *International Journal of Industrial Ergonomics*, 24(6), pp. 631–645, 1999.

15. A. Géron. *Hands-On Machine Learning with Scikit-Learn, Keras, and TensorFlow: Concepts, Tools, and Techniques to Build Intelligent Systems*. O'Reilly Media, California, 2019.

16. J. S. Ha, P. H. Seong. A human-machine interface evaluation method: A difficulty evaluation method in information searching (DEMIS). *Reliability Engineering and System Safety*, 94(10), pp. 1557–1567, 2009.

17. I. Goodfellow, Y. Bengio, A. Courville. *Deep Learning*. MIT Press Book, 2016.

18. C. Aggarwal. *Neural Networks and Deep Learning: A Textbook*. Springer, 2019.

19. A. Zhang, Z. C. Lipton, M. Li, A. J. Smell. *Dive into Deep Learning [Interactive E-book]* Available: https://d2l.ai

20. R. Janapati, et al. Towards a more theory-driven BCI using source reconstructed dynamics of EEG time-series. *Nano LIFE*, 12(02), p. 2250005, 2022.

21. R. Janapati, et al. Web interface applications controllers used by autonomous EEG-BCI technologies. In AIP Conference Proceedings. Vol. 2418. No. 1. AIP Publishing LLC, 2022.

22. R. Janapati, V. Dalal, and R. Sengupta. Advances in experimental paradigms for EEG-BCI. In Proceedings of the 2nd International Conference on Recent Trends in Machine Learning, IoT, Smart Cities and Applications. Springer, Singapore, 2022.

23. S. R. Chandiran. *Hands-on Deep Learning Algorithm with Python*. Packt Publishing Limited, Birmingham.

24. C. Bishop. *Pattern Recognition and Machine Learning. 2nd Printing*. Springer, New York, 2011.

25. M. Nielsen, *Neural Networks and Deep Learning [Ebook]* Available: http://neuraln etworksanddeeplearning.com/index.html

26. R. Janapati, et al. Signal processing algorithms based on evolutionary optimization techniques in the BCI: A review. Springer - Advances in Intelligent Systems and Computing Series. 2020, ISSN: 2194–5357. *Computational Vision and Bio-Inspired Computing*, vol 1, pp. 165–174, 2021.

27. C. Billa, M. Chavali. Artificial intelligence leveraged internet of medical things and continuous health monitoring and combating pandemics within the internet of medical things framework. Chapter 1 In *Book Emerging Technologies for Combatting Pandemics: AI, IoMT, and Analytics*, edited by M. R. Hossain Mondal, U. Kose, V. B. S. Prasath, P. Podder, S. Bharati, J. Kamruzzaman, Auerbach Publications, 2022, https://doi.org/10.1201/9781003324447-1; eBook ISBN9781003324447

28. R. Janapati, et al. Review on EEG-BCI classification techniques advancements. *IOP Conference Series: Materials Science and Engineering*. 981(3). IOP Publishing, 2020.

29. R. Janapati, et al. Various signals used for device navigation in BCI production. *IOP Conference Series: Materials Science and Engineering.* 981 (3). IOP Publishing, 2020.

30. M. Khapra. *CS6910/ CS7015. Class Lecture, Topic: "Deep Learning"*, Indian Institute of Technology - Madras, Chennai, Jan. 2022.

31. I. E. Lagaris, A. Likas, D. I. Fotiadis, *Artificial Neural Networks for Solving Ordinary and Partial Differential Equations*, University of Ioannina, Ioannina, Greece, 1998

32. S. Liang, R. Srikanth, *Why Deep Neural Networks for Function Approximation?*, Coordinated Science Laboratory and Department of Electrical and Computer Engineering University of Illinois at Urbana-Champaign, Urbana, IL, 2022.

2 An Overview of Recent Approaches in Brain–Computer Interface Systems Using Electroencephalography

Shishir Maheswari, Kandala N. V. P. S. Rajesh, Usha Desai, and T. Sunil Kumar

CONTENTS

2.1 INTRODUCTION

The brain is a processing and controlling unit of the human body. It processes the electrical activity carried by neurons throughout the nervous system and controls various actions such as behavior, sensation, thoughts, and emotion [1]. This electrical activity has been of utmost importance for researchers for the last few decades to understand brain functioning. The brain's electrical activity can be recorded as signals and processed via computers to understand brain functioning better. This technology of combining brain signals and computers is referred to as the brain–computer interface (BCI), mind–machine interface (MMI), or brain–machine interface (BMI) [2].

DOI: 10.1201/9781003326830-2

BCI is used to develop an artificial intelligence system that understands brain functioning. It processes the brain signal and produces real-time output through an activity that does not involve muscular stimulation [1, 3]. BCI aims to develop a system that controls the devices in the surrounding environment in real time. In past and present scenarios, it is one of the most challenging and rapidly developing research areas.

BCI is initially developed to help individuals with partial or complete motor impairments to make them interact and communicate with their surroundings. In today's world, BCI is not only limited to patients with impairments. Instead, it is now also developing for healthy people to control external devices without muscular stimulation. BCI research areas now extend to medical and nonmedical domains such as developing prosthetics devices, humanoid robots, entertainment applications, arts, automobiles, and smart homes [4].

The remaining chapter is organized as follows: The primary steps involved in designing BCI systems are given in Section 2.2. The brief discussion on state-of-the-art BCI systems is provided in Section 2.3. The future scope and conclusions are discussed in Section 2.4.

2.2 FUNDAMENTALS OF BCI SYSTEM DESIGN

The BCI system generally consists of stages such as data acquisition, preprocessing, feature extraction and data analysis, classification, and application interface to achieve the BCI goals. The typical BCI system block diagram is shown in Figure 2.1. Data acquisition provides a way to connect the brain and computer to collect data. Feature extraction and data analysis help to understand the brain signals, and classification enables the machine to learn and make decisions. The application interface is responsible for controlling various output devices.

2.2.1 DATA ACQUISITION

This process collects brain data using invasive or noninvasive methods. In invasive methods, the recording device is inserted inside the brain, a surgical method. The

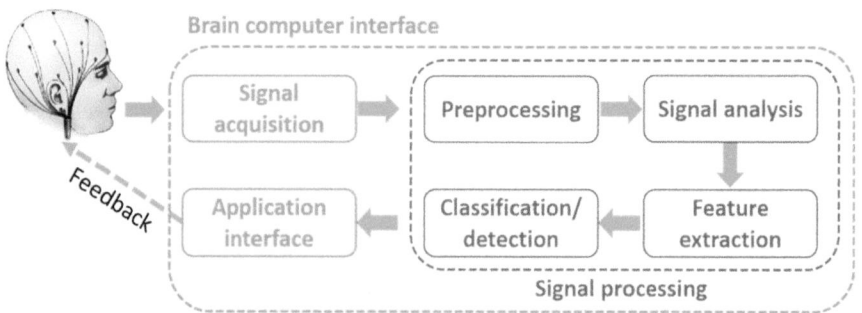

FIGURE 2.1 Block diagram representing general steps in BCI systems.

noninvasive approach is the most widely followed for brain data recording in present scenarios. Electroencephalography (EEG), magnetoencephalography (MEG), magnetic resonance imaging (MRI), functional MRI (fMRI), and positron emission tomography (PET) are various noninvasive methods to record brain data. Among them, MRI, PET, and MEG are famous but expensive. These also have large time constants, which makes them unsuitable for rapid and real-time applications. On the other hand, EEG data recording happens by placing the electrodes on the scalp of a patient [4]. It is a fast, convenient, and economical way of data recording but also contains a slight amount of noise which can be taken care during preprocessing step.

2.2.1.1 Dataset

The literature on BCI systems shows that the EEG is the primary diagnostic tool to analyze brain data. Some of the datasets used for developing BCI systems are mentioned in Table 2.1.

2.2.2 SIGNAL PREPROCESSING

The brain signal recorded from sources may consist of noise. Biological activities such as eye blinking and heart beating may influence the data recording and act as noise. Hence, preprocessing is essential to alleviate noise or artifacts. Many filtering techniques exist for noise reduction, such as time domain, frequency domain, and adaptive filtering. Preprocessing may also involve downsampling, which reduces the length of the original signal without affecting the information contained in it. Downsampling helps in reducing the computations.

2.2.3 SIGNAL ANALYSIS

Signal analysis is the further process after preprocessing. Signal analysis techniques convert or transform the signal into a meaningful representation. The signal can be examined in the time, frequency, or time-frequency domains. Generally, the data is recorded as time samples, so most of the signals are initially available in the time domain. If the signal does not convey much information in the time domain, it can be transformed into

---

TABLE 2.1

Some of the Common EEG-Based Datasets Employed in BCI Systems

Datasets	Description
Imagination of right-hand thumb movement [5]	These are some of the popular motor imagery datasets
Left or right hand MI [6]	and beneficial options for creating BCIs.
DEAP [7, 8]	These are the EEG-based emotions datasets, which depict
SEED [8, 9]	human emotions such as sad, happy, fear, and neutral.
MNIST brain digits [10]	These datasets are the EEG recordings, when a
IMAGENET brain [11]	random image or number is shown to a user.

a frequency or time-frequency domain. The choice of these domains depends upon the researcher's chosen analysis approach. Often, a signal does not provide much information in one domain but can be analyzed conveniently in other domains.

2.2.4   FEATURE EXTRACTION

Features play an essential role in any of the computer-aided approaches. The raw signals have a large number of data samples. If the system or approach processes so many data samples, it may result in more processing time or low system performance. Therefore, the large data samples need to be transformed into new data with few samples. These new data or samples are referred to as features representing meaningful interpretations of original data. The original or transformed signal can extract features such as mean, variance, entropies, energies, skewness, and kurtosis. Additionally, we can also use feature ranking to drop irrelevant features further.

2.2.5   CLASSIFICATION

Classifiers are the supervised learning algorithms of machine learning. Many BCI applications have used various classifiers such as support vector machine (SVM), ensemble, decision tree, random forest (RF), and naïve Bayes. Deep learning algorithms are recently playing an essential role in BCI classification. In conventional methods of machine learning, handcrafted features are supplied. Whereas in deep learning, the features are automatically extracted from the data. In this step, the extracted features are fed to the classifier to further distinguish between the different states of brain signals.

2.2.6   APPLICATION INTERFACE

The application interface receives the response from the classification process. Based on the response, it controls other output devices, such as moving a cursor on the monitor screen or a robotic arm. The application interface also provides feedback to the brain based on the response received from the sensors, which collect data from the outer environment.

2.3   OVERVIEW OF EXISTING APPROACHES

This section provides recent advancements of BCI system developments in the literature briefly.

2.3.1   RECENT LITERATURE

The authors have developed a BCI system to recognize EEG signals' motor and mental imagery tasks using time-frequency domain-based feature extraction [12]. Initially, EEG segments are denoised using a multiscale principal component analysis (MPCA). Later, they prepared time-frequency scalogram images from preprocessed

EEG segments using continuous wavelet transform. These scalograms are then given to various CNN models such as AlexNet, SqueezeNet, ShuffleNet, GoogLeNet, ResNet-18, ResNet-50, ResNet-101, MobileNetV2, InceptionV3, and DenseNet-201 for classification purpose. The study was examined using three public datasets from BCI competition III. The scalogram-based approach has achieved a maximum average classification accuracy of 99.52% with ShuffleNet.

Abenna et al. [13] have developed an automated approach for EEG-based BCI using statistical features extracted from beta waves. In this approach, the authors preprocessed EEG data to reduce the effect of nonstationariness. More specifically, the authors subtracted the mean of the last values from the current sample of EEG signal. Eleven features extracted from beta waves after applying common spatial filtering are fed to a light gradient boosting machine to determine the class of the EEG signal. Authors reported 99% and 95% accuracy on UCI and PhysioNet EEG datasets, respectively.

Kaongoen et al. [14] have investigated speech imagery (SI) for EEG-based BCI for home appliances control. SI is a mental task where the person imagines speaking out without making sounds. To achieve this, the authors have used data acquired from a wearable ear-EEG headphone, and a temporally stacked multi-band covariance matrix (TSMBCM) represents the SI activities. More specifically, features computed from different frequency bands obtained after processing through Butterworth band filters are fed to multilayer extreme learning machine to recognize the EEG segment's class. This approach proved to be promising and achieved the highest true positive rate of 0.85 with a command delivery time of 3.79/s. The authors have also found that adding artifact cancellation algorithms would increase the model's accuracy and be more beneficial for real-life applications.

The authors in [15] have classified motor imagery tasks using temporal and channel attention convolutional network (TCACNet). This TCACNet has an attention mechanism module (AMM) and a convolution neural network (CNN). AMM preprocesses the EEG data, and CNN is used for feature extraction and classification. The proposed scheme is examined on two public datasets: BCI competition dataset IV 2(a) and the High Gamma dataset. This approach achieved average classification accuracies of 86.8% and 96.2% on two datasets, respectively.

Authors in [16] developed a BCI system for detecting imagined word segments from continuous EEG signals. The authors have exploited five feature extraction methods: empirical mode decomposition (EMD), wavelet decomposition, fractal dimension, frequency energies, and chaos theory. The classification is performed using four different classifiers. The proposed method using the RF classifier is examined on three datasets and reported the F1 scores of 0.73, 0.79, and 0.68, respectively.

The approach in [17] aims to optimize the preprocessing stage in motor imagery-based BCI. More specifically, the authors wanted to optimize accuracy and timing costs. For this purpose, the Taguchi method is adopted. This approach employed features extracted using the Hjorth algorithm and SVM to classify the EEG segments. The statistical significance of the extracted features is tested using the analysis of variance (ANOVA) method. The method was verified on a BCI Competition IV-2b dataset and achieved the highest average accuracy of 0.82.

Gaur et al. [18] have proposed a motor imagery classification approach using the Pearson correlation coefficient. The approach finds the correlation between the EEG signals and selects the highly correlated EEG channels for specific tasks, thereby reducing computational complexity. The proposed method used BCI Competition III Dataset IIIa and IVa datasets to analyze the imagined left and right hand and foot tasks. Authors have reported a classification accuracy of 98.34% in motor imagery classification in BCI.

Yu et al. [19] proposed an empirical Fourier decomposition (EFD) and improved EFD (IEFD)-based approach to detect motor imagery and mental imagery EEG tasks. In this approach, the EEG signals are denoised using MPCA, and EFD is applied to decompose the EEG signals. Later IEFD is used to select a single prominent mode. After that, time and frequency domain features are extracted and fed to feedforward neural network. This approach has reported a classification accuracy of 82.70% in the subject-independent scenario.

Tiwari and Chaturvedi [20] have developed a BCI system using dynamic channel relevance (DCR) score-based channel selection. This approach uses multivariate EMD-based features and an SVM classifier to classify four motor imagery tasks: left hand, right hand, tongue, and foot. The proposed method is examined on three open-source datasets: BCI Competition IV-II a, BCI Competition IV dataset I, and BCI Competition III-IVa. Authors have reported a maximum classification accuracy of 85.4% on dataset 1 (BCI Competition IV-II a).

The authors in [21] employed a common spatial pattern-based feature along with linear discriminant analysis (LDA) to classify motor imagery EEG tasks. This approach has achieved an accuracy of approximately 80% when experiments are performed on data (BCI Competition IV-2a) acquired from stroke patients and healthy people.

Liu et al. [22] proposed a recalibration-free cross-device transfer learning framework. Specifically, they developed ALign and Pool for EEG Headset domain Adaptation (ALPHA) to improve the implementation accuracy of steady-state visual evoked potential-based BCI (SSVEP-BCI). ALPHA has provided better performance than the fully calibrated method of training-based task-related component analysis (TRCA).

Authors in [23] have developed a CNN-based prognosis model to predict the outcome of stroke patients' recovery after a two-week rehabilitation training with BCI. Authors have computed EEG power spectrum and functional connectivity and fed them to two-way CNN. This approach predicted stroke patients' recovery with a mean squared error of 0.89.

The approach in [24] has used decision fusion to combine concurrent local field potential (LFP) and scalp EEG signals and developed LFP-EEG-BCI. Their experiments suggested that LFP-EEG-BCI has achieved significantly better performance over LFP-BCI and EEG-BCI when experiments are performed on data acquired from the paraplegic patient.

The authors in [25] have proposed a BCI system based on steady-state visual evoked potential EEG. They have utilized an ensemble method-based deep learning approach for this purpose. Authors have used ear-EEG data from an open dataset and achieved an average classification accuracy of 81.74%.

Authors in [26] have proposed an end-to-end approach using a CNN-based neural network with an attention module for classifying motor imagery tasks. This approach uses time series data for classification purposes. Authors have also created a public dataset that contains both motor imagery and motor execution tasks. In their experiments, they found that it is easier to decode motor execution than motor imagery.

Kant et al. [27] have proposed a transferred learning-based approach for classifying EEG signals into rest, attention, and working classes. Authors have used wavelet packet transform to theta, alpha, and beta waves from EEG signals. Thereafter, scalograms are extracted and used with CNNs to determine the class of EEG signal. They formulated several binary classification problems with the aforementioned three tasks.

Hermosilla et al. [28] have developed a CNN-based approach for classifying motor imagery EEG. Specifically, they employed end-to-end shallow architecture that extracts both temporal and spatial features.

2.4 CONCLUSION AND FUTURE SCOPE

This chapter provided the details of the importance of BCI in real-time applications, such as in medical and nonmedical domains. We have mainly discussed the essential steps in developing BCI machines and the state-of-the-art approaches available in the recent literature that uses various databases for different BCI applications. Though much research has been carried out, there is still scope to improve the research. For example, local binary patterns and their variants have proved to be a significant feature representation in various signal classification applications [29–33], but LBPs are less explored in this area. Also, vision transformer [34, 35]-based deep learning architectures can be an effective alternative to CNNs in developing BCI machines. Also, there is a need to introduce more versatile datasets like patients with neurological disorders to implement BCI systems in a better way.

REFERENCES

1. Houssein, E.H.; Hammad, A.; Ali, A.A. Human emotion recognition from EEG-based brain–computer interface using machine learning: A comprehensive review. *Neural Comput. Appl.* 2022, 34(15), 1–31.
2. Prashant, P.; Joshi, A.; Gandhi, V. Brain computer interface: A review. In Proceedings of the 2015 5th Nirma University International Conference on Engineering (NUiCONE); IEEE, 2015; pp. 1–6.
3. Mak, J.N.; Wolpaw, J.R. Clinical applications of brain-computer interfaces: Current state and future prospects. *IEEE Rev. Biomed. Eng.* 2009, 2, 187–199.
4. Värbu, K.; Muhammad, N.; Muhammad, Y. Past, present, and future of EEG-based BCI applications. *Sensors* 2022, 22, 3331.
5. Neuper, C.; Müller-Putz, G.R.; Scherer, R.; Pfurtscheller, G. Motor imagery and EEG-based control of spelling devices and neuroprostheses. *Prog. Brain Res.* 2006, 159, 393–409.
6. Cho, H.; Ahn, M.; Ahn, S.; Kwon, M.; Jun, S.C. EEG datasets for motor imagery brain–computer interface. *Gigascience* 2017, 6, gix034.

7. Koelstra, S.; Muhl, C.; Soleymani, M.; Lee, J.-S.; Yazdani, A.; Ebrahimi, T.; Pun, T.; Nijholt, A.; Patras, I. Deap: A database for emotion analysis; using physiological signals. *IEEE Trans. Affect. Comput.* 2011, *3*, 18–31.

8. Iyer, A.; Das, S.S.; Teotia, R.; Maheshwari, S.; Sharma, R.R. CNN and LSTM based ensemble learning for human emotion recognition using EEG recordings. *Multimed. Tools Appl.* 2022, 82(4), 1–14.

9. Zheng, W.-L.; Lu, B.-L. Investigating critical frequency bands and channels for EEG-based emotion recognition with deep neural networks. *IEEE Trans. Auton. Ment. Dev.* 2015, *7*, 162–175.

10. Tirupattur, P.; Rawat, Y.S.; Spampinato, C.; Shah, M. Thoughtviz: Visualizing human thoughts using generative adversarial network. In Proceedings of the Proceedings of the 26th ACM International Conference on Multimedia; 2018; pp. 950–958.

11. Spampinato, C.; Palazzo, S.; Kavasidis, I.; Giordano, D.; Souly, N.; Shah, M. Deep learning human mind for automated visual classification. In Proceedings of the IEEE Conference on Computer Vision and Pattern Recognition; 2017; pp. 6809–6817.

12. Sadiq, M.T.; Aziz, M.Z.; Almogren, A.; Yousaf, A.; Siuly, S.; Rehman, A.U. Exploiting pretrained CNN models for the development of an EEG-based robust BCI framework. *Comput. Biol. Med.* 2022, *143*, 105242.

13. Abenna, S.; Nahid, M.; Bouyghf, H.; Ouacha, B. EEG-based BCI: A novel improvement for EEG signals classification based on real-time preprocessing. *Comput. Biol. Med.* 2022, *148*, 105931.

14. Kaongoen, N.; Choi, J.; Jo, S. A novel online BCI system using speech imagery and ear-EEG for home appliances control. *Comput. Methods Programs Biomed.* 2022, *224*, 107022.

15. Liu, X.; Shi, R.; Hui, Q.; Xu, S.; Wang, S.; Na, R.; Sun, Y.; Ding, W.; Zheng, D.; Chen, X. TCACNet: Temporal and channel attention convolutional network for motor imagery classification of EEG-based BCI. *Inf. Process. Manag.* 2022, *59*, 103001.

16. Hernández-Del-Toro, T.; Reyes-García, C.A.; Villaseñor-Pineda, L. Toward asynchronous EEG-based BCI: Detecting imagined words segments in continuous EEG signals. *Biomed. Signal Process. Control* 2021, *65*, 102351.

17. Dagdevir, E.; Tokmakci, M. Optimization of preprocessing stage in EEG based BCI systems in terms of accuracy and timing cost. *Biomed. Signal Process. Control* 2021, *67*, 102548.

18. Gaur, P.; McCreadie, K.; Pachori, R.B.; Wang, H.; Prasad, G. An automatic subject specific channel selection method for enhancing motor imagery classification in EEG-BCI using correlation. *Biomed. Signal Process. Control* 2021, *68*, 102574.

19. Yu, X.; Aziz, M.Z.; Sadiq, M.T.; Fan, Z.; Xiao, G. A new framework for automatic detection of motor and mental imagery EEG signals for robust BCI systems. *IEEE Trans. Instrum. Meas.* 2021, *70*, 1–12.

20. Tiwari, A.; Chaturvedi, A. A novel channel selection method for BCI classification using dynamic channel relevance. *IEEE Access* 2021, *9*, 126698–126716.

21. Gaur, P.; Gupta, H.; Chowdhury, A.; McCreadie, K.; Pachori, R.B.; Wang, H. A sliding window common spatial pattern for enhancing motor imagery classification in EEG-BCI. *IEEE Trans. Instrum. Meas.* 2021, *70*, 1–9.

22. Liu, B.; Chen, X.; Li, X.; Wang, Y.; Gao, X.; Gao, S. Align and pool for EEG headset domain adaptation (ALPHA) to facilitate dry electrode based SSVEP-BCI. *IEEE Trans. Biomed. Eng.* 2021, *69*, 795–806.

23. Lin, P.-J.; Jia, T.; Li, C.; Li, T.; Qian, C.; Li, Z.; Pan, Y.; Ji, L. CNN-based prognosis of BCI rehabilitation using EEG from first session BCI training. *IEEE Trans. Neural Syst. Rehabil. Eng.* 2021, *29*, 1936–1943.

24. Feng, Z.; Sun, Y.; Qian, L.; Qi, Y.; Wang, Y.; Guan, C.; Sun, Y. Design a novel BCI for neurorehabilitation using concurrent LFP and EEG features: A case study. *IEEE Trans. Biomed. Eng.* 2021, *69*, 1554–1563.

25. Zhu, Y.; Li, Y.; Lu, J.; Li, P. EEGNet with ensemble learning to improve the cross-session classification of SSVEP based BCI from Ear-EEG. *IEEE Access* 2021, *9*, 15295–15303.

26. Lashgari, E.; Ott, J.; Connelly, A.; Baldi, P.; Maoz, U. An end-to-end CNN with attentional mechanism applied to raw EEG in a BCI classification task. *J. Neural Eng.* 2021, *18*, 0460e3.

27. Kant, P.; Laskar, S.H.; Hazarika, J. Transfer learning-based EEG analysis of visual attention and working memory on motor cortex for BCI. *Neural Comput. Appl.* 2022, *34*, 20179–20190.

28. Hermosilla, D.M.; Codorniú, R.T.; Baracaldo, R.L.; Zamora, R.S.; Rodriguez, D.D.; Albuerne, Y.L.; Álvarez, J.R.N. Shallow convolutional network excel for classifying motor imagery EEG in BCI applications. *IEEE Access* 2021, *9*, 98275–98286.

29. Kumar, T.S.; Kanhangad, V.; Pachori, R.B. Classification of seizure and seizure-free EEG signals using multi-level local patterns. In Proceedings of the 2014 19th International Conference on Digital Signal Processing; IEEE, 2014; pp. 646–650.

30. Kumar, T.S.; Kanhangad, V. Detection of electrocardiographic changes in partial epileptic patients using local binary pattern based composite feature. *Australas. Phys. Eng. Sci. Med.* 2018, *41*, 209–216.

31. Rajesh, K.N.; Kumar, T.S. Schizophrenia detection in adolescents from EEG signals using symmetrically weighted local binary patterns. In Proceedings of the 2021 43rd Annual International Conference of the IEEE Engineering in Medicine & Biology Society (EMBC); IEEE, 2021; pp. 963–966.

32. Kumar, T.S.; Hussain, M.A.; Kanhangad, V. Classification of voiced and non-voiced speech signals using empirical wavelet transform and multi-level local patterns. In Proceedings of the International Conference on Digital Signal Processing, DSP; Institute of Electrical and Electronics Engineers Inc., 2015; Vol. 2015-September, pp. 163–167.

33. Sairamya, N.J.; Subathra, M.S.P. EEG-based classification of normal and seizure types using relaxed local neighbour difference pattern and artificial neural network. *Knowledge-Based Syst.* 2022, *249*, 108508.

34. Dosovitskiy, A.; Beyer, L.; Kolesnikov, A.; Weissenborn, D.; Zhai, X.; Unterthiner, T.; Dehghani, M.; Minderer, M.; Heigold, G.; Gelly, S. An image is worth 16x16 words: Transformers for image recognition at scale. *arXiv Prepr. arXiv2010.11929* 2020.

35. Wang, W.; Xie, E.; Li, X.; Fan, D.-P.; Song, K.; Liang, D.; Lu, T.; Luo, P.; Shao, L. Pyramid vision transformer: A versatile backbone for dense prediction without convolutions. In Proceedings of the Proceedings of the IEEE/CVF International Conference on Computer Vision; 2021; pp. 568–578.

3 Automatic Detection of Motor Imagery EEG Signals Using Swarm Decomposition for Robust BCI Systems

Shailesh Vitthalrao Bhalerao,
and Ram Bilas Pachori

CONTENTS

3.1 INTRODUCTION

A brain–computer interface (BCI) allows humans to communicate with external assistance tools using brain signals like electroencephalogram (EEG). BCI was initially developed to enable disabled people to operate and control external assistive devices with the use of their brain activity, without the need for actual body motions, even when they are unable to move freely or control some sections of their bodies [1, 2]. The BCI technology has a wide range of uses in neuroprosthetics and biomedical engineering. In motor imagery (MI)-based BCIs, a person is required to carry out the imagination in the brain that corresponds to a particular MI task into control signals for various imaginary tasks, namely the movement of legs, hands, tongue, or feet, etc.

Nowadays, the majority of BCIs operate in a synchronized way, meaning that the user can only transmit commands or messages to computers or other external

DOI: 10.1201/9781003326830-3

devices at particular times. In other words, the information transfer rate of synchronized BCI is determined by the system, not by the users themselves. Users can only choose which command or message to generate; they cannot choose when to control the devices or computer [1, 3–9]. Therefore, this kind of BCI technology does not offer a real-time means of human–machine interaction and introduces several limitations to the use for clinical practices. The main limitation of the BCI-EEG-based system is to retrieve MI task-specific information from the recorded MI-EEG signals, in which signals are always suffered by noise and artifacts. Additionally, it is difficult to categorize MI-EEG signals and is made even more difficult when they are synthesized for use in clinical assistive tools due to the selection of limited BCI channels, a high number of trials, a lack of feature generalization, and the need for a significant amount of classifier training time on extracted features. Therefore, there is a significant issue with the accurate responsiveness of BCI to the MI signals acquisition and controlling of assistive devices for the disabled individual [10–13]. Over the last few years, numerous BCI-related works have been experimented to achieve different MI tasks in which many works are based on novel feature extraction, classification, and decomposition methods [4, 14–19]. Among these techniques, decomposition methods are becoming increasingly popular and play a significant part in the MI-EEG tasks of BCI applications. However, the traditional decomposition methods are not suitable for the MI-EEG application due to a number of shortcomings such as low noise rejection capability, mode mixing issue, and selection of predefined mode number, as highlighted in a lot of literature [4, 16, 17, 19–21].

To overcome these limitations, in this chapter, we have proposed a new classification framework based on the recently proposed technique, swarm decomposition (SWD) [22], for the extraction of decomposed bands from the MI-EEG signals and improved classification performance related to the MI-EEG tasks, namely left-hand (LH) movement, right-hand (RH) movement, both feet (BF) movement, and tongue (T) movement. The SWD approach, which is inspired by swarm intelligence, is constructed using an iterative algorithm and swarm filtering (SwF). SWD has a significant advantage in resolving the issue of mode mixing by utilizing the benefits of noise reduction and intelligent decomposition. The SWD method is data adaptive technique that shows superiority for the analysis of nonlinear and nonstationary signals like EEG [22]. It provides a set of oscillatory components (OCs) that can be considered as narrow-band amplitude and frequency-modulated signals. In this study, we have developed a generic framework (see Section 3.3) for MI-EEG task classification with the use of new hybrid feature (HF). These HFs have been derived using a novel method, combining statistical features, namely rational asymmetry (RASM) and differential asymmetry (DASM), with deep convolutional neural network (CNN) features (DCF) [23, 24]. Motivated by these features [23–26], here we have extracted these features from obtained OCs to get enhanced EEG signals, which correspond to three different bands (6–24 Hz, 8–12 Hz, and 16–24 Hz) [2]. These features are then classified into LH, RH, BF, and T MI-EEG tasks using recently used bidirectional long short-term memory (BiLSTM) [27]. Finally, to investigate the effectiveness of SWD-based classification framework, it has been compared with the recently published state-of-the-art works [2, 3, 17, 18, 28–32].

The following are the significant contributions to the proposed work:

1. This study develops a novel SWD-based classification framework to classify MI-EEG tasks and improves the performance of the deployable BCI system. The proposed model outperforms the other related state-of-the-art work.
2. This study uses the MI-EEG detection model based on newly adopted HFs and BiLSTM and delivers the highest classification accuracy.

Further, this chapter is arranged as follows: Section 3.2 gives short overview of related works. The proposed SWD-based MI-EEG classification framework is detailed in the Section 3.3. Then, the experimental results and discussions are presented in Section 3.4, and finally in Section 3.5, we conclude with its discussion.

3.2   RELATED WORKS

In literature, there have been numerous attempts to identify MI-EEG activity from the recorded EEG signals. The following is a recap of the major published works:

In work, Kevric et al. [1] categorized the controlling commands of BF and LH for wheelchair-based BCI applications with the use of wavelet packet decomposition (WPD). This work addresses issue related to the nonstationary properties of the BCI signals and employs the highest classification accuracy of 92.8%. In Dutta et al. [33], the multivariate extension of empirical mode decomposition (EMD) has been used for the classification of three nonmotor cognitive EEG mental tasks of three subjects. With the extracted multivariate autoregressive (MVAR) features and least squares support vector machine (LS-SVM) classifier, it has delivered a classification accuracy of 94.43% in the 10-fold cross-validation approach. In similar work in [34], the multivariate EMD has been applied to enhance the efficiency of MI task classification.. They employ the time-frequency representation with common spatial pattern (CSP) features, and classification accuracy of 94% is obtained. Further, Siuly et al. [35] proposed an adaptive classification framework, namely a cross-correlation-based logistic regression algorithm in which three different feature parameters are adaptively processed to improve the cross-correlation sequence between MI-EEG task and recorded EEG signals. It employs logistic regression (LR) to classify the HF parameters and provides an average classification accuracy of 93.91%. In [36], the authors have employed a three-stage BCI recognition system in which it is performed on computed unique wavelet scattering coefficients with three distinct recurrent neural network (RNN) architectures: long short-term memory (LSTM), BiLSTM, and gated units. It has achieved classification performance of 90.23% for MI-EEG signals and 84.25% for magnetoencephalogram (MEG) signals and enhanced classification performance with optimizing computational complexity. Further, Meng et al. [37] have adopted a new channel optimization scheme for MI tasks classification and delivered the highest classification accuracy of 87.42% on L_1-norm-based CSP features related to limited EEG channels. This work has proved its potential for BCI application by significantly reducing channel usage and time complexity. In a similar work, CSP features have been utilized, and they enhanced MI-EEG task classification using upper triangle filter bank autoencoder (UTFB-AE) [15]. In contrast to the above techniques, Ma et al. [16] concluded that the extreme learning machine (ELM) with segmentation swap (Seg-Swap) data augmentation technique demonstrates higher MI-EEG classification accuracy of 95.04% than tensor-based schemes (TBS) and CSP with support vector machine (SVM), nonnegative multiway factorization, and power spectral density. Furthermore, a few research [38] exploited spatial filtering with eigencomponent features based on the CSP and achieved average performance above 80% during MI-EEG tasks. Recently authors [39] have introduced a time-frequency approach with the use of variational nonlinear chirp mode decomposition (VN-CMD) and hybrid time-domain features. It achieved an accuracy of 89.6% in the case of multiclass MI-EEG signals. In several studies [40–42], particle swarm optimization (PSO) has been used to address several issues in BCI applications, including adaptive signal filtering and feature selection. Overall, the works mentioned above highlight various experiments on the novel methodology with certain limitations and related research scope in MI-EEG multi-classification problem. In contrast to the work mentioned above, the current study aims to develop a general classification framework for BCI systems based on mental EEG imagery movements.

3.3 PROPOSED METHODOLOGY

Our proposed methodology incorporates a new machine learning approach for MI-EEG task classification utilizing the BCI Competition IV dataset 2A [43]. To clean the EEG data, we have adopted preprocessing into our model. We have employed both stand-alone HF and stand-alone features (RASM, DASM, and DCF) [23, 24], which have been extracted from the dominant OCs. Using the Student's *t*-test feature ranking method, the best combination of features has been carefully selected and then applied to the BiLSTM classifier for multitask MI-EEG classification problem. The automated SWD-based MI-EEG tasks classification model is shown in block diagram in Figure 3.1. It comprises all major implementation steps that show how well the MI-BCI application works, and a description is provided in the following subsections:

3.3.1 DATABASE COLLECTION

We have evaluated our proposed techniques using EEG dataset, which is available from the Berlin BCI group's BCI Competition IV dataset 2A. It is found at https://www.bbci.de/competition/iv/#dataset1 [43–45]. Using BrainAmp MR plus amplifiers these EEG datasets were recorded from nine healthy patients. Without receiving any feedback input, MI was used the entire time. The LH, RH, BF, and T were chosen as the four classes of MI tasks for each subject. Two different types of

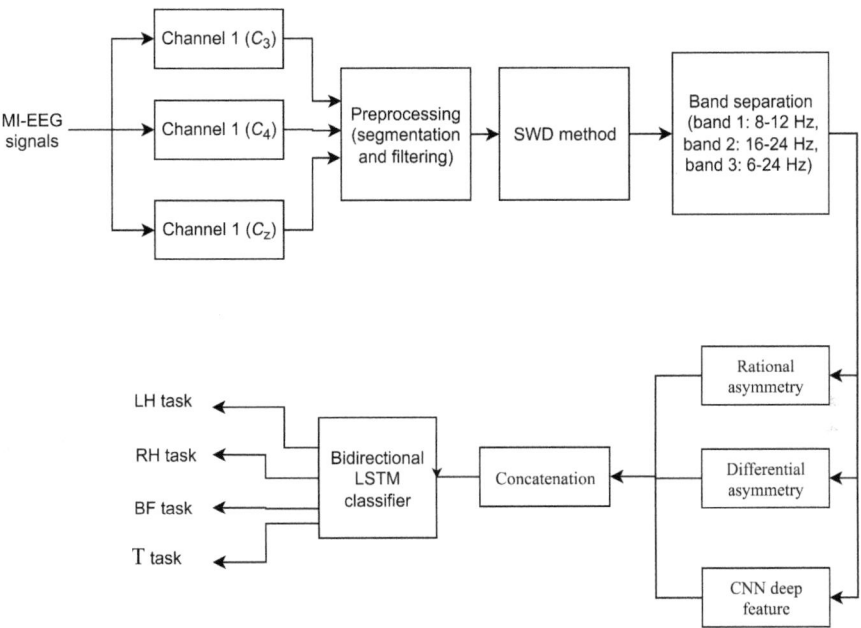

FIGURE 3.1 The overall structure of the SWD-based MI-EEG tasks classification framework.

datasets (calibration and evaluation) are available for each subject. Arrow indicators (left, right, up, or down) were displayed as visual indicators for the calibration data on a screen. The subject was instructed to conduct the arrow-guided MI task for 4 seconds [28]. These intervals were separated by 2 seconds of blank screen time and 2 seconds of a fixation cross visible in the centre of the screen. The complete marker information for these datasets is provided. The words "left," "right," and "foot" were used as cues for the MI tasks during the evaluation data for periods ranging in length from 1.5 to 8 seconds. The word "stop" is indicated at the end of the MI period. The time between intermittent intervals is ranged from 1.5 to 8 seconds, which is shown in Figure 3.2. In order to synchronize the MI-EEG command, it is necessary to segment EEG data into the appropriate size epochs. An appropriate time window selection would affect the quality of feature selection and classification results. In the present study, we have defined a segmented window with 3 seconds time length over the EEG signal and selected an epoch size of 3 seconds. This study used 80-hour evaluation datasets, one for testing and the other for training.

In the preprocessing stage, the raw EEG signals have been band-pass filtered with pass-band frequency (0.1–65 Hz) and sampled at 250 Hz. We have used 3 of the 22 available recorded EEG channels in our work: $C_3(8)$, $C_4(12)$, and $C_z(10)$ [2]. Here, we have first divided the epochs of each EEG signal into four MI-EEG classes. During these experiments, each MI task was measured in epochs, with each epoch

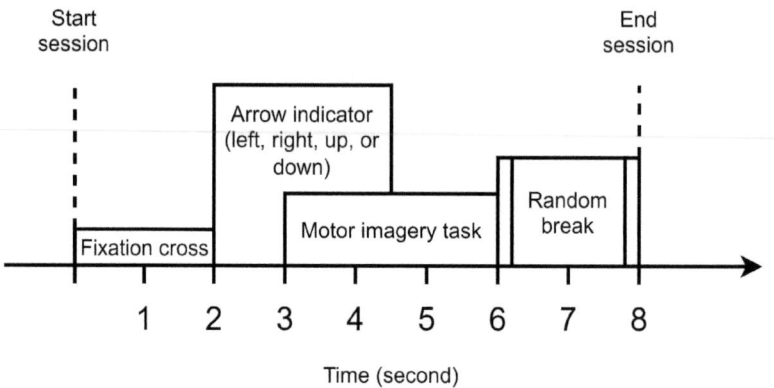

FIGURE 3.2   Separation of EEG segment with 3 seconds time window with a single session.

consisting of a series of 12 repetitions. Thus, for each MI task epoch, there are 12 × 22 = 244 trials, 240 of which are target trials. Out of the epochs of the 18 subjects, we have considered the smallest epoch size of 240 into consideration for each subject trial. The BCI Competition IV dataset 2A data was measured from nine subjects and consists of the training set of 27,878 trials (7,058 for LH task, 7,450 for RH task, 6,602 for BF task, and 6,767 for T task) and a test set of 30,932 trials (7,834 for LH task, 8,268 for RH task, 7,331 for BF task, and 7,499 for T task). For each subject, Table 3.1 shows the total number of epochs computed from the MI-EEG signals.

3.3.2 SWARM DECOMPOSITION METHOD

SWD is a meta-heuristic decomposition technique based on swarm intelligence motivated by flocking bird behaviour, individually and collectively [22]. It is a data-driven approach in which SWD can result in the globally ideal solution with a limited number of repetitions. Based on appropriately SWF parameterizing, SWD, an adaptive decomposition method, can iteratively extract the dominating OCs from the nonstationary multicomponent signals. The core idea of SWF is that the filter output has been produced by analyzing the input signal $x[n]$ with the swarm's trajectory, which represents the swarm prey's route for the hunting the prey. The SWD has been shown to be an effective decomposition technique in recent decades, and numerous researchers have carried out a wide range of experiments to investigate the potential of this method for various applications [22, 46, 47]. In SWD, it involves iteratively applying SwF on the residual of a multicomponent input signal to detect OCs. The OCs are extracted by selecting the frequency band with the highest amplitude peak by considering the energy spectral density (ESD) [13]. In every iteration, the remaining input is subtracted from the currently extracted notable OCs in order to process subsequent oscillatory modes. This process keeps going until there is no longer any observable spectral energy in the residual signal.

TABLE 3.1

Description of Obtained Epochs from BCI Competition IV Dataset 2A MI-EEG Signals for Four Classes LH, RH, BF, and T

Patient ID	A01-T	A02-T	A03-T	A04-T	A05-T	A06-T	A07-T	A08-T	A09-T	A01-E	A02-E	A03-E	A04-E	A05-E	A06-E	A07-E	A08-E
For C_3 channel	566	572	555	498	578	572	576	570	568	581	557	543	555	574	561	568	570
For C_4 channel	555	561	543	488	566	561	564	559	580	593	568	555	566	568	555	562	581
For C_z channel	549	555	538	483	572	566	559	576	557	564	528	526	538	568	559	551	576

The following steps are used to compute the fitness function value for every swarm member (SM) and extract dominant OCs [13].

1. Set the position and velocity vectors of every SM, $P_i[n]$ and $V_i[n]$, to uniformly distributed integer random values. Under initial condition, prey $P_i[0]$ and velocity $V_i[0]$ values are assigned with a random value with symmetrical position in the swarm-prey hunting mechanism. It is represented in the following equations:

$$P_i[0] = P_{prey}[0] + d_{cr}\left(i - 1 - \frac{M}{2}\right), \forall_i = 1,\ldots,Q \tag{3.1}$$

$$v_i[0] = 0, \forall_i = 1,\ldots,Q \tag{3.2}$$

Where Q represents the velocity value for each SM.

2. Find the driving force ($F_{Dr,i}^n$) and cohesive force ($F_{Coh,i}^n$) as follows:

$$F_{Dr,i}^n = P_{prey}[n] - p_i[n-1] \tag{3.3}$$

$$F_{Coh,i}^n = \frac{1}{M-1} \sum_{j=1, j \neq i}^{M} f\left(P_i[n-1] - P_j[n-1]\right) \tag{3.4}$$

$$f(d) = -\operatorname{sgn}(d).\ln\left(\frac{|d|}{dcr}\right) \tag{3.5}$$

Where $F_{Dr,i}^n$ is the attractive force that members experience as prey. The force that induces attraction and repulsion among SMs is $F_{Coh,i}^n$. The cohesion force between the jth member and the ith member with distance (d) is represented by the expression $\ln(|d|/d_{cr})$, and it is measured by the function $f(d)$ with range $d = (1, 4)$. The sgn represents the sign function. The minimum critical distance (d_{cr}) between members must be set at 0.5, 1, or 2 in order for it not to affect each member's performance individually. Here we have selected $d_{cr} = 1$.

3. In this stage, the parameters are updated in accordance with Equations 3.6 and 3.7. Each SM changes its current position $P_i[n]$ and velocity $V_i[n]$ in each cycle of swarm-prey hunting, as demonstrated in the following equations:

$$v_i[n] = v_i[n-1] + \delta\left(F_{Dr,i}^n + F_{Coh,i}^n\right) \tag{3.6}$$

$$p_i[n] = p_i[n-1] + \delta v_i[n], \tag{3.7}$$

where δ regulates the flexibility of SM in the hunting space with [0, 1].

4. The OCs of SWD are obtained by adding the positions of the SMs. The output $y(n)$ is defined as follows:

$$y(n) = \beta \sum_{i=1}^{M} p_i[n], \tag{3.8}$$

where β is a scaled factor.

Here, OC modes are measured using the positions of the SMs along the swarm's trajectory [22]. The obtained dominant OCs from the SwF filter locate the specific subband signals with compact bandwidth and centre frequencies.

5. Swarm response is optimized with each subsequent iteration using the following fitness function, which is defined as follows:

$$Y_{it+1}[n] = Y_{it}[n] - Y_{it}', \, if S_{x_{it}}'(\omega) < PS_{th} \tag{3.9}$$

Where, at i^{th} iteration, the extracted signal $Y'_{it}[n]$ is subtracted from the input signal $Y_{it}[n]$; $Y_{it+1}[n]$ is the resultant output signal. The iteration is stopped if the difference between two successive iterations is less than the PS_{th} value of 0.1; otherwise, the process is repeated.

Further, the SWD method has been applied to the segmented filtered EEG output to decompose the EEG signals into OCs. In our work, the MI-EEG signal experimentation is conducted by choosing the best SWD parameters from the values indicated in [22]. Also, we compare the effectiveness of the SWD method with EMD [41]. Figures 3.3–3.10 show the obtained bands (6–24 Hz, 16–24 Hz, and 8–12 Hz), using the SWD and EMD methods, for each MI class as the subject performs MI-EEG tasks. In order to compare the performance, here we have considered the EEG epoch sample from channel C_3 of the trial 2 of A02-T subject which corresponding to the four class MI-EEG movements. These bands can contribute spectral information, which can be useful for the detection of MI-EEG tasks [2]. These bands are computed from the obtained modes by the SWD and EMD methods. The decomposed modes obtained by the SWD and EMD have a similar spectral structure. However, the experiment shows that the first four modes are crucial for band selection and further feature extraction, which contain most of the MI-EEG task information [2]. But due to its mode mixing nature, EMD could not decompose properly and generate few relevant modes for respective bands. Therefore, the obtained bands by the EMD method contribute less MI-EEG task information, as shown in Figures 3.7–3.10. Another side, experimental results have been shown that the SWD has decomposed more relevant modes into separated bands, which gather all relevant MI-EEG task information for further classification.

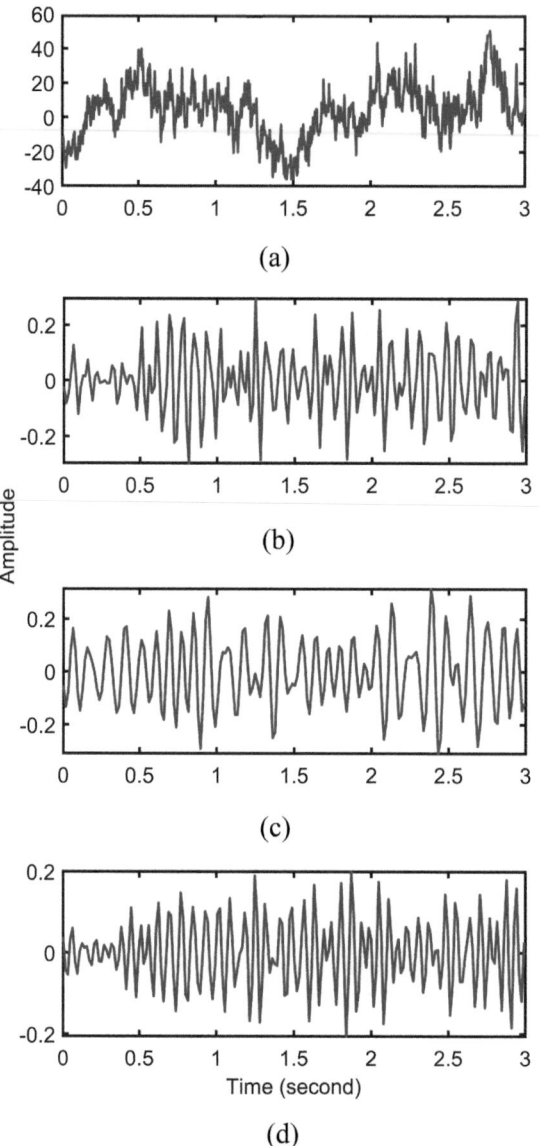

FIGURE 3.3 Plots of (a) the EEG epoch from C_3 of the trial 2 of A02-T subject corresponding to MI-EEG RH movement, (b) the obtained band 6–24 Hz, (c) the obtained band 16–24 Hz, and (d) the obtained band 8–12 Hz by using SWD for RH MI task.

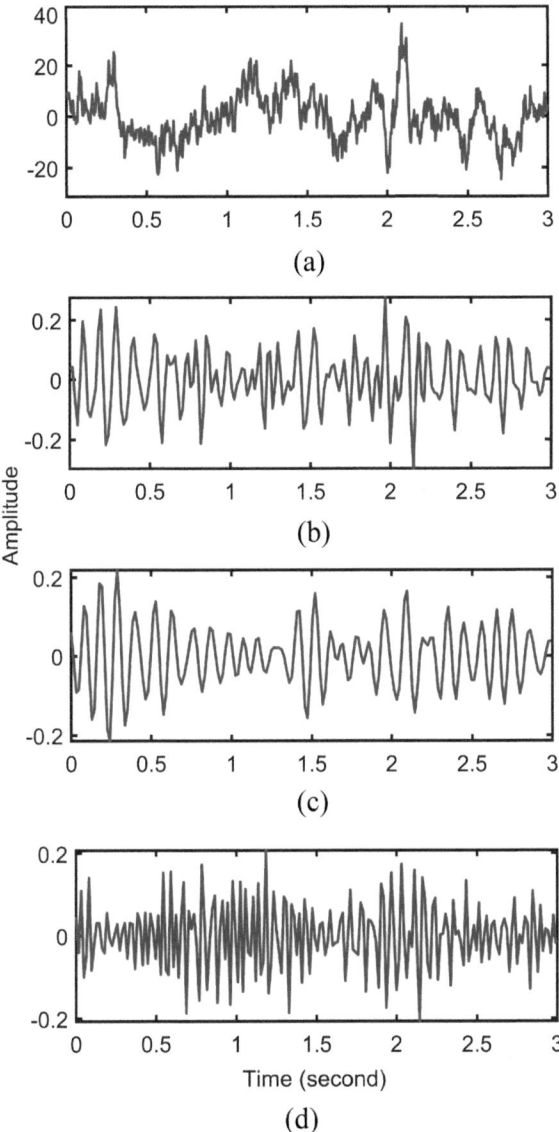

FIGURE 3.4 Plots of (a) the EEG epoch from C_3 of the trial 2 of A02-T subject corresponding to MI-EEG LH movement, (b) the obtained band 6–24 Hz, (c) the obtained band 16–24 Hz, and (d) the obtained band 8–12 Hz by using SWD for LH MI task.

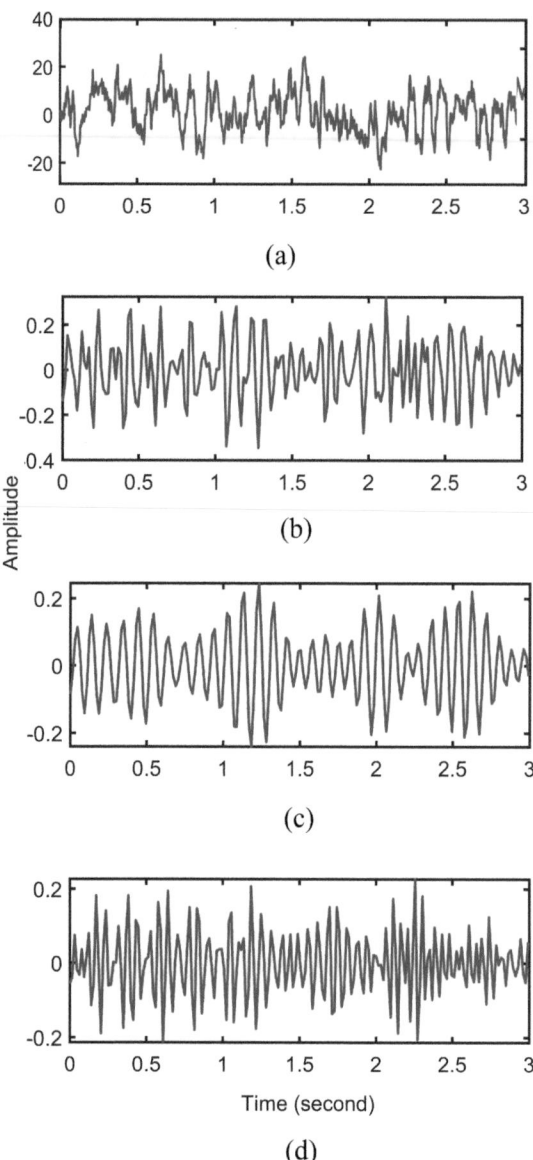

FIGURE 3.5   Plots of (a) the EEG epoch from C_3 of the trial 2 of A02-T subject correspond-ing to MI-EEG BF movement, (b) the obtained band 6–24 Hz, (c) the obtained band 16–24 Hz, and (d) the obtained band 8–12 Hz by using SWD for BF MI task.

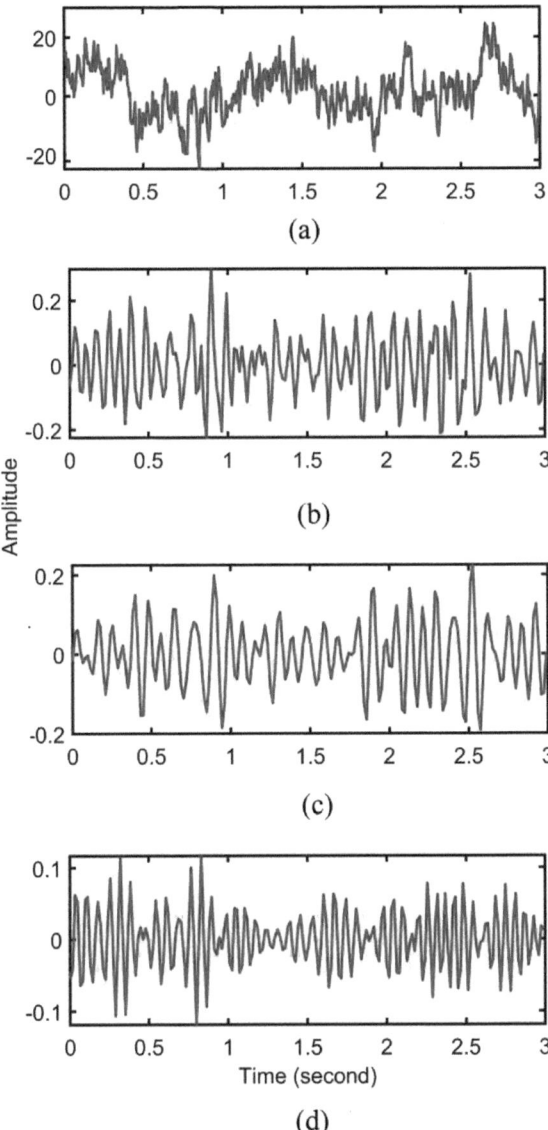

FIGURE 3.6 Plots of (a) the EEG epoch from C_3 of the trial 2 of A02-T subject corresponding to MI-EEG T movement, (b) the obtained band 6–24 Hz, (c) the obtained band 16–24 Hz, and (d) the obtained band 8–12 Hz by using SWD for T MI task.

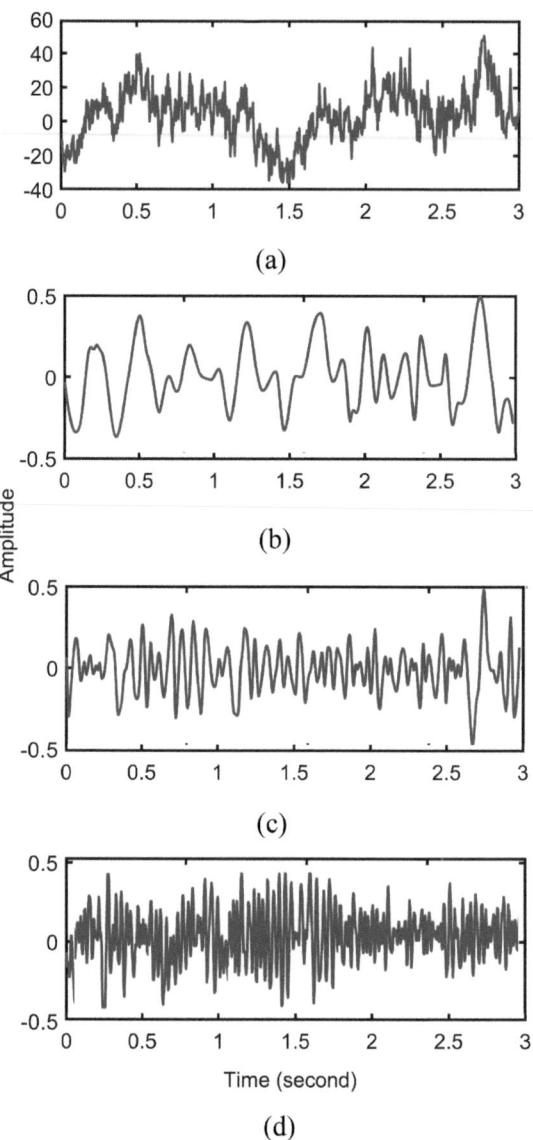

FIGURE 3.7   Plots of (a) the EEG epoch from C_3 of the trial 2 of A02-T subject correspond-ing to MI-EEG RH movement, (b) the obtained band 6–24 Hz, (c) the obtained band 16–24 Hz, and (d) the obtained band 8–12 Hz by using EMD for RH MI task.

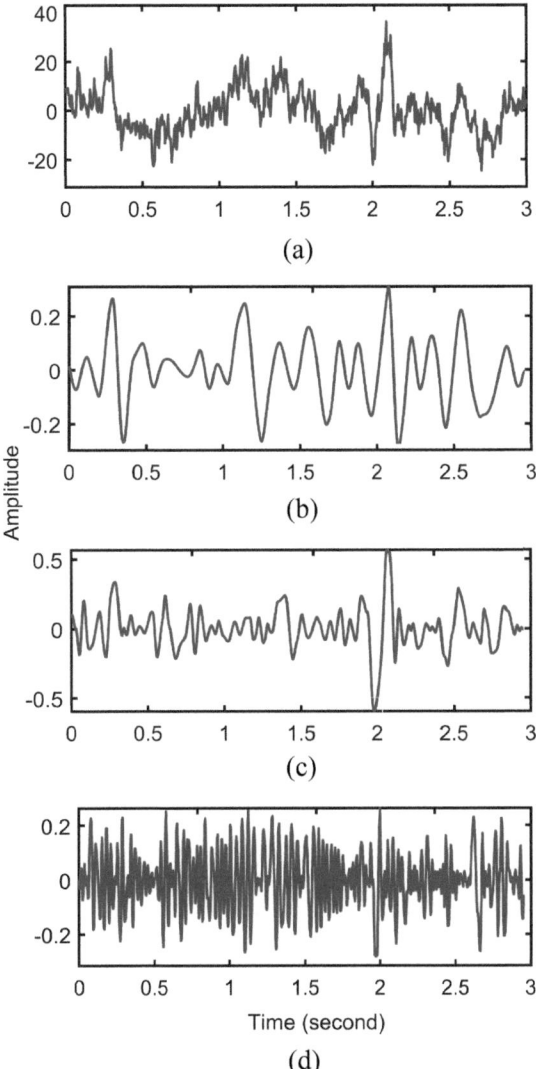

FIGURE 3.8 Plots of (a) the EEG epoch from C_3 of the trial 2 of A02-T subject corresponding to MI-EEG LH movement, (b) the obtained band 6–24 Hz, (c) the obtained band 16–24 Hz, and (d) the obtained band 8–12 Hz by using EMD for LH MI task.

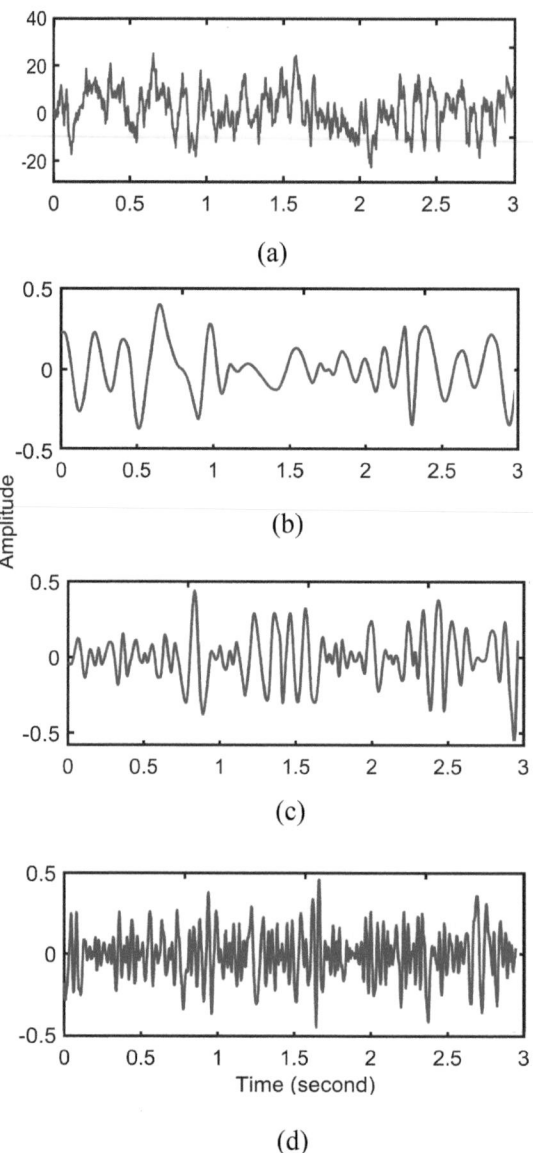

FIGURE 3.9   Plots of (a) the EEG epoch from C_3 of the trial 2 of A02-T subject corresponding to MI-EEG BF movement, (b) the obtained band 6–24 Hz, (c) the obtained band 16–24 Hz, and (d) the obtained band 8–12 Hz by using EMD for BF MI task.

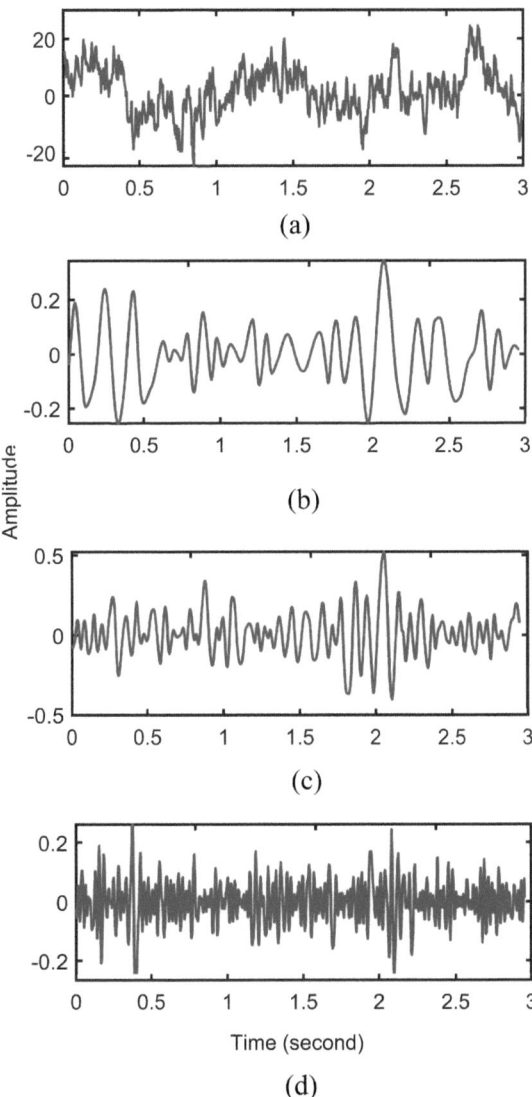

FIGURE 3.10   Plots of (a) the EEG epoch from C_3 of the trial 2 of A02-T subject corresponding to MI-EEG T movement, (b) the obtained band 6–24 Hz, (c) the obtained band 16–24 Hz, and (d) the obtained band 8–12 Hz by using EMD for T MI task.

3.3.3   CONSTRUCTION OF FEATURE DATABASE

MI task is a complex cognitive process carried about by neural activity in various brain regions. From the point of machine learning, adequate feature extraction is necessary to extract meaningful information from MI EEG data. In our work, we have used new HFs that were derived by combining statistical features, namely RASM, DASM, and DCF [17], and we have carefully experimented to choose the best combination of these features [23]. The HFs were extracted from the improved EEG data in order to detect MI-EEG signals. The variation characteristics of EEG signals show a strong association to MI tasks by the statistical parameters DCF, DASM, and RASM, which are defined in the time domain and frequency domain [2, 33, 34, 40].

Below is a discussion of the features that were extracted from the SWD-based rhythms.

1.Asymmetry feature

Since the MI-EEG signals are inherently nonstationary, we have used the HFs in the frequency domain to better represent these MI-EEG signals and find a strong correlation to the MI-EEG tasks. Here, we have taken the acquired OCs using the SWD and separated them into frequency subbands, including 6–24 Hz, 8–12 Hz, and 16–24 Hz, by considering the mean frequency criteria [40]. We then estimated the power values of these frequency subbands (P_{freq}) using the below equation [23]:

$$P_{\text{freq}} = \frac{1}{n} \sum_{j=1}^{n} p\left(S_j\right)^2 \tag{3.10}$$

Where p represents the power value and S_j is signal in the frequency domain for jth number of subbands. Since asymmetry ratios examine signal features from both the left and right hemispheres of the brain, it has been demonstrated that they considerably impact MI-EEG task classification. Some MI tasks target the right hemisphere of the brain, while others target the left. We have extracted RASM and DASM features from the power of frequency bands. In order to calculate the RASM and DASM features, we have used the 6–24 Hz, 8–12 Hz, and 16–24 Hz bands. The RASM and DASM features are derived from the following equations [23]:

$$\text{RASM} = \frac{P_{\text{left}}}{P_{\text{right}}} \tag{3.11}$$

$$\text{DASM} = P_{\text{left}} - P_{\text{right}} \tag{3.12}$$

Where P_{left} and P_{right} represent the power from the channel on the left and right hemispheres of the brain (e.g., C_3 and C_4). The three-channel pairs, C_3/C_z, C_4/C_z, and C_3/C_4, were used in our computation of these feature values. It is observed that the accuracy of MI-EEG task identification greatly improved when only the C_3/C_4 pair was used for the RASM and DASM features.

2.Deep CNN feature

Statistical features computed from EEG signals via conventional signal processing techniques are generally not sufficient to explore the nonrhythmic dynamic characteristics. On the other hand, compact CNN enhances the micro-scale dynamics by directly extracting deep features from time series through adaptive learning [48]. To improve the representation of MI-EEG task, it is preferable for EEG features to simultaneously explore the micro-scale characteristics. In our method, DCF is directly extracted from the obtained SWD-based OCs (from three channels, C_3, C_4, and C_z) using a CNN model and asymmetry feature (from cross-channel C_3/C_4) to enhance the categorization of MI-EEG tasks. According to Figure 3.1, there are six layers in the CNN model. There are three one-dimentional (1-D) CNN layers, one rectified linear unit (ReLU), one batch normalization (BNL), activation functions, and five max-pooling layers (MPLs) with three successive dense layers (DLs). Temporal information from features is computed through the 1-D CNN layer by selecting a small kernel size [33]. The parameter setting for the used architecture of the CNN is shown in Table 3.2. The band information is fed into the pooling layer and two DLs to obtained the final DCF of size 1,024.

In our study, 12 MI-EEG combinational feature types were calculated using the SWD method for three EEG bands (6–24 Hz, 8–12 Hz, and 16–24 Hz), which

TABLE 3.2

Parameter Setting of CNN Architecture

Layer	Layer type	Output shape	Kernel size	Strides
Layer-1	1-D CNN layer 1	16×1000	6	[1 1]
Layer-2	BNL	16×1000	-	[1 1]
Layer-3	MPL 1	16×500	2	[1 2]
Layer-4	1-D CNN layer 2	16×500	6	[1 1]
Layer-5	MPL 2	16×250	2	[1 2]
Layer-6	1-D CNN layer 3	64×250	6	[1 2]
Layer-7	MPL 3	32×250	2	[1 2]
Layer-8	MPL 4	32×125	2	[1 2]
Layer-9	MPL 5	32×64	2	[1 2]
Layer-10	Concatenate	2048	-	-
Layer-11	DL 1	2048	-	-
Layer-12	DL 2	1024	-	-
Layer-13	Concatenate (selected features)	1024	-	-
Layer-14	DL 3	512	-	-
Layer-15	Dense layer 4 (activation function)	4	-	-

correspond to three C3, C4, and Cz channels. Then, top 12 features (out of 78 features) were chosen using the feature ranking method, namely Student's t-test [48] with selection of lowest probability (p)-values. Similar to this, eight features for EMD have been chosen after taking into consideration criteria ($p < 0.01$). The significant features produced by the SWD approach exhibit better discriminative characteristics than those produced by the EMD method, according to the lowest p-values. Here, the used convention for features is denoted as channel-feature-band. For EMD and SWD methods, Table 3.3 presents the most significant features based on the Student's t-test ranking method. It should be noted that the features 12868, 13575, 12007, and 12393 have been chosen specifically for the LH, RH, BF, and T MI-EEG classes, respectively. Finally, the BiLSTM classifier has been employed to these significant features space to detect different classes corresponding to the LH, RH, BF, and T MI tasks [2, 49].

3.4   RESULTS AND DISCUSSIONS

In order to classify the LH, RH, BF, and T MI-EEG tasks, experiments have been conducted to evaluate the performance of the extracted HFs using the BiLSTM classifier. Table 3.4 shows architecture model for the BiLSTM classifier. The reason for choosing the BiLSTM classifier is that it has the ability to select the most discriminative features with suitable model parameters and overcome the overfitting issue with optimum performance with the optimizing hidden layers [27]. Additionally, the proposed model has also been evaluated in contrast to current state-of-the-art methodologies. Table 3.4 shows the classification results with selective HFs, which were obtained using the SWD-HF-BiLSTM and EMD-HF-BiLSTM frameworks from nine subjects. Here, we have evaluated the proposed SWD-based framework (SWD-HF-BiLSTM) with EMD-based framework (EMD-HF-BiLSTM) for obtained HFs for MI-EEG task recognition application. All classifiers are configured with 70% of the data used for network training and 30% used for network testing. Here we have considered training set has 45,753 feature samples with 750 sample lengths each for MI task events. The validation test set has 5,084 feature samples. To reduce complexity, feature smoothing has been done using the Student's t-test ranking method. To work with imbalanced EEG dataset, the highest accuracy may not mean an excellent performance of the MI-EEG tasks classification model but must deliver the performance on the subject-independent approach. Under the cross-validation scheme, the leave-one-leave-out approach is accommodated for subject-independent analysis in which a cross-subject dataset has been considered in the training and testing phase [42]. For each subject, this procedure was repeated, and average results were considered with a subject-independent approach. In our work, we have employed evaluation metrics to measure the model's performance, namely accuracy, recall, precision, and F1-score as expressed as follows:

$$\text{Recall} = \frac{Tp}{Tp + Fn} \tag{3.13}$$

TABLE 3.3

The Selected Most Significant 15 Features Using the SWD and EMD Methods

Features (channel-feature-band)	SWD method					EMD method				
	LH class (mean ± std)	RH class (mean ± std)	BF class (mean ± std)	T class (mean ± std)	p-value	LH class (mean ± std)	RH class (mean ± std)	BF class (mean ± std)	T class (mean ± std)	p-value
C_1C_2-(CDF-RASM-DASM)-B_1	1.45 ± 1.72	3.79 ± 2.56	6.73 ± 6.71	9.37 ± 8.35	1.98e-10	0.82 ± 1.64	2.89 ± 2.38	4.24 ± 5.76	7.95 ± 9.49	1.83e-8
C_1C_2-RASM-B_2	3.73 ± 3.98	6.07 ± 4.84	9.01 ± 8.99	11.65 ± 10.63	4.26e-10	4.55 ± 5.37	6.42 ± 6.03	8.65 ± 9.49	11.68 ± 10.14	5.56e-8
C_1C_2-(CDF-RASM-DASM)-B_3	4.84 ± 5.09	7.18 ± 5.95	10.12 ± 9.12	12.76 ± 11.74	5.37e-9	2.82 ± 3.64	4.13 ± 4.32	6.21 ± 7.76	9.95 ± 8.41	3.83e-8
C_1C_2-(CDF-RASM-DASM)-B_2	7.01 ± 7.26	9.35 ± 8.12	12.29 ± 11.27	14.93 ± 13.91	7.54e-9	3.82 ± 4.62	5.69 ± 5.23	7.92 ± 8.75	10.95 ± 9.49	4.83e-7
C_1C_2-(CDF-RASM)-B_2	3.42 ± 3.67	5.76 ± 4.53	8.70 ± 7.68	13.34 ± 12.32	6.95e-8	4.24 ± 5.56	6.11 ± 6.22	8.34 ± 9.18	11.37 ± 12.91	5.28e-7
C_1C_2-(CDF-DASM)-B_1	1.78 ± 2.03	4.12 ± 2.89	7.06 ± 6.04	9.70 ± 8.68	4.31e-8	2.60 ± 4.92	4.47 ± 5.52	6.70 ± 7.54	9.73 ± 8.63	3.61e-6
C_1C_2-CDF-B_2	0.38 ± 0.63	2.72 ± 1.49	5.66 ± 4.64	8.30 ± 7.28	0.91e-7	8.82 ± 11.19	10.69 ± 11.82	12.92 ± 13.76	15.95 ± 14.82	9.83e-6
C_1C_2-CDF-B_1	4.59 ± 4.23	6.93 ± 9.09	9.87 ± 11.89	12.51 ± 11.49	14.12e-7	5.31 ± 8.63	7.18 ± 9.29	9.41 ± 8.58	13.72 ± 12.63	6.32e-5
C_1C_2-DASM-B_1	4.12 ± 3.76	6.46 ± 8.62	9.40 ± 7.38	12.04 ± 11.02	13.65e-7	4.84 ± 8.16	6.71 ± 8.82	8.94 ± 8.11	11.97 ± 10.82	5.85e-5
C_1C_2-(CDF-RASM)-B_1	1.97 ± 1.61	4.31 ± 6.45	7.25 ± 5.23	9.89 ± 8.87	2.50e-6	2.69 ± 6.01	4.81 ± 6.63	6.79 ± 5.96	9.82 ± 8.72	3.70e-5
C_1C_2-RASM-B_1	5.04 ± 4.68	7.38 ± 9.52	10.32 ± 8.38	12.96 ± 11.94	9.57e-6	1.72 ± 5.06	3.84 ± 5.72	5.82 ± 4.99	8.85 ± 7.75	2.73e-4
C_1C_2-DASM-B_2	2.93 ± 2.57	5.27 ± 7.21	8.21 ± 11.23	10.85 ± 9.83	3.46e-5	3.65 ± 2.34	5.75 ± 2.93	7.75 ± 8.59	10.78 ± 9.68	4.66e-3
C_1-CDF-B_3	4.96 ± 4.6	7.30 ± 5.37	10.24 ± 13.26	12.88 ± 11.86	12.49e-4	6.72 ± 5.42	8.84 ± 6.83	10.82 ± 11.64	13.85 ± 12.75	7.78e-3
C_1-CDF-B_2	4.58 ± 4.22	6.92 ± 4.94	9.86 ± 12.88	12.50 ± 15.17	5.11e-4	2.78 ± 1.44	4.90 ± 2.19	6.88 ± 7.72	9.91 ± 8.81	**3.79e-2**
C_1-CDF-B_1	3.82 ± 3.46	6.17 ± 5.23	9.10 ± 10.12	11.74 ± 12.77	13.35e-3	6.72 ± 5.29	8.84 ± 6.89	10.82 ± 11.62	13.85 ± 12.75	7.73e-2

RASM: rational asymmetry, DASM: differential asymmetry, CDF: CNN deep features, B_1: 6–24 Hz band, B_2: 8–12 Hz band, and B_3: 16–24 Hz band

TABLE 3.4

Feature-Specific Performance Obtained Using the SWD and EMD Methods

Features	SWD method				EMD method			
	Average accuracy (%)	Precision	Recall	F1-score	Average accuracy (%)	Precision	Recall	F1-score
C_1-CDF-B_1	72.08	0.66	0.60	0.543	64.87	0.59	0.57	0.516
C_1-CDF-B_2	72.98	0.63	0.60	0.550	65.68	0.58	0.57	0.522
C_1-CDF-B_3	73.88	0.67	0.61	0.557	66.49	0.56	0.58	0.529
C_1C_2-CDF-B_1	**76.18**	0.69	**0.64**	0.597	71.26	**0.61**	**0.59**	0.567
C_1C_2-RASM-B_1	75.45	0.69	0.62	0.569	67.91	0.60	0.59	0.540
C_1C_2-DASM-B_1	76.12	0.70	0.63	**0.608**	68.92	0.59	0.58	0.548
C_1C_2-CDF-B_2	76.51	0.67	0.62	0.564	**71.46**	0.59	0.54	**0.568**
C_1C_2-RASM-B_2	78.38	0.68	0.62	0.591	70.54	0.54	0.57	0.561
C_1C_2-DASM-B_2	74.28	0.66	0.63	0.562	71.35	0.56	0.56	0.539
C_1C_2-(CDF-DASM)-B_1	76.60	0.69	0.63	0.570	68.04	0.59	0.59	0.541
C_1C_2-(CDF-RASM)-B_2	76.90	0.71	0.65	0.587	70.11	0.62	0.61	0.558
C_1C_2-(CDF-RASM)-B_1	75.82	**0.73**	0.66	0.602	71.84	**0.66**	**0.63**	0.571
C_1C_2-(CDF-RASM-DASM)-B_3	78.32	0.71	0.65	0.590	70.49	0.64	0.62	0.561
C_1C_2-(CDF-RASM-DASM)-B_2	77.54	0.71	0.64	0.584	69.78	0.64	0.61	0.504
C_1C_2-(CDF-RASM-DASM)-B_1	**78.62**	0.72	**0.66**	**0.684**	**72.89**	0.64	0.61	**0.582**

RASM: rational asymmetry, DASM: differential asymmetry, CDF: CNN deep features, B_1: 6–24 Hz band, B_2: 8–12 Hz band, and B_3: 16–24 Hz band. The best values for each performance parameter are shown in bold.

$$\text{Precision} = \frac{Tp}{Tp + Fp} \qquad (3.14)$$

$$\text{F1-score} = 2 \times \frac{\text{Recall} \times \text{Precision}}{\text{Recall} + \text{Precision}} \qquad (3.15)$$

$$\text{Accuracy} = \frac{Tp + Tn}{Tp + Tn + Fp + Fn} \qquad (3.16)$$

Where T_p (true positive), T_n (true negative), F_p (false positive), and F_n (false negative) are the correctly identified LH MI-EEG task, RH MI-EEG task, misidentified LH task, and misidentified RH task, respectively.Figures 3.12 and 3.13 show the box plots of HF and individual features are mentioned in Table 3.4. It compares the classification results with the two proposed classification models (SWD-HF-BiLSTM and EMD-HF-BiLSTM) with the experimentation of hybrid and individual features. It has proved the potential of HF over the individual feature in case of MI-EEG classification. With HF (C_1/C_2-(CDF-RASM-DASM)-B_1), the proposed SWD-HF-BiLSTM technique improves the recall rate, precision, and F1-score by 0.72, 0.66, and 0.684, respectively. The proposed model significantly improves the recall rate when HFs are used and enhances the F1-score compared to the EMD-HF-BiLSTM model. Overall, MI-EEG task detection has significantly improved in terms of classification performance.

Table 3.4 shows that the proposed SWD-HF-BiLSTM framework found suitability for MI-EEG task classification and delivered the overall best average classification accuracy of 78.62% as compared to the EMD-HF-BiLSTM framework. For the SWD-HF-BiLSTM framework, precision, recall, and F1-score of 0.72, 0.66, and 0.684 are obtained, respectively. Whereras EMD-HF-BiLSTM framework has achieved accuracy, precision, recall, and F1-score of 72.89%, 0.64, 0.61, and 0.582, respectively. However, as compared to the SWD-HF-BiLSTM, it performs marginally worse. From the detailed experimental analysis, it is found that the HFs significantly contribute to the MI-EEG classification problem. In the case of an individual feature, the highest classification accuracy using derived feature C1-CDF-B3 by SWD is 76.18%, whereas with HF (C_1/C_2-(CDF-RASM-DASM)-B_1), it achieved the highest classification accuracy of 78.62%. In contrast, EMD method-based features (C_1/C_2-RASM-B_1) and (C_1/C_2-(CDF-RASM-DASM)-B_1) have delivered 71.46% and 72.89% in the case of an individual and hybrid feature, respectively.

Figure 3.11 shows the normalized confusion matrices of the classification accuracies of MI-EEG tasks using the hybrid and individual features by the SWD and EMD methods. The proposed classification model has a training accuracy of 87.6%, a validation accuracy of 79.54%, and a testing accuracy of 78.62% during the 10-fold cross-validation. For the individual MI-EEG tasks classification, the accuracy was 82.6% for RH, 79.4% for LH, 80.1% for BF, and 75.2% for T in case of SWD-HF-BiLSTM framework with the HF parameter (C_1/C_2-(CDF-RASM-DASM)-B_1), which

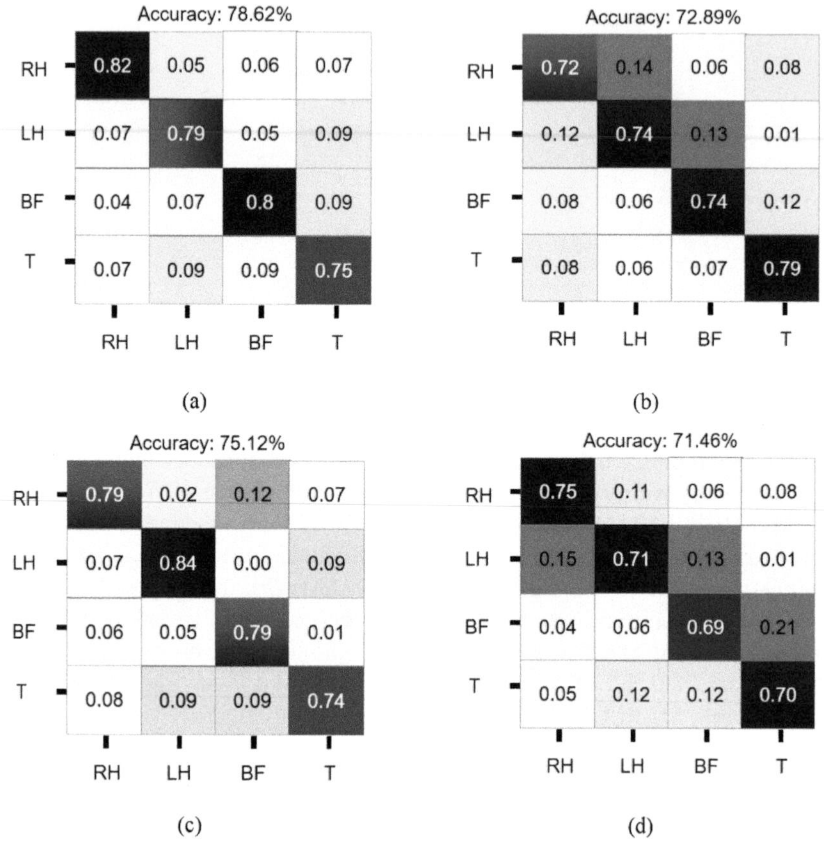

Accuracy: 78.62%

	RH	LH	BF	T
RH	0.82	0.05	0.06	0.07
LH	0.07	0.79	0.05	0.09
BF	0.04	0.07	0.8	0.09
T	0.07	0.09	0.09	0.75

(a)

Accuracy: 72.89%

	RH	LH	BF	T
RH	0.72	0.14	0.06	0.08
LH	0.12	0.74	0.13	0.01
BF	0.08	0.06	0.74	0.12
T	0.08	0.06	0.07	0.79

(b)

Accuracy: 75.12%

	RH	LH	BF	T
RH	0.79	0.02	0.12	0.07
LH	0.07	0.84	0.00	0.09
BF	0.06	0.05	0.79	0.01
T	0.08	0.09	0.09	0.74

(c)

Accuracy: 71.46%

	RH	LH	BF	T
RH	0.75	0.11	0.06	0.08
LH	0.15	0.71	0.13	0.01
BF	0.04	0.06	0.69	0.21
T	0.05	0.12	0.12	0.70

(d)

FIGURE 3.11 Normalized confusion matrices of the classification accuracies of MI-EEG tasks for (a) HF using the SWD, (b) HF using the EMD, (c) individual feature using the SWD, and (d) individual feature using the EMD.

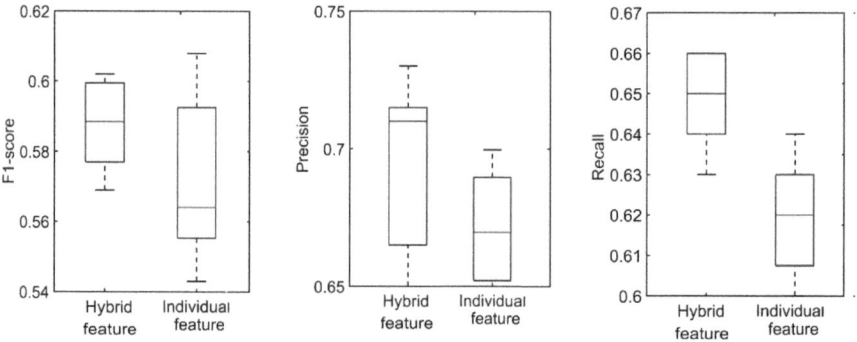

FIGURE 3.12 Comparison of the classification result obtained with the HF and individual features using the SWD method.

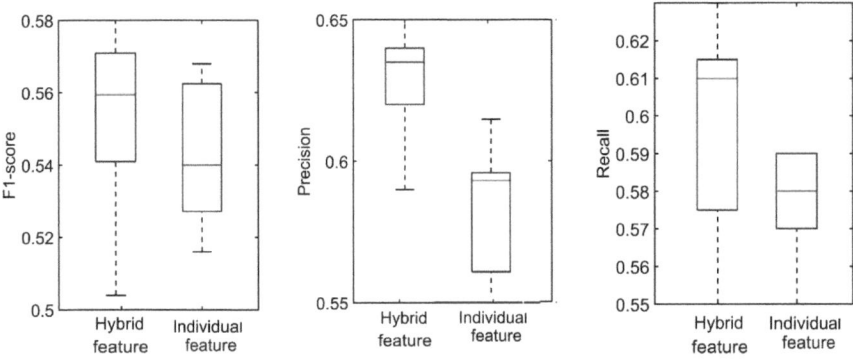

FIGURE 3.13 Comparison of the classification result obtained with the HF and individual features using the EMD method.

is shown in Figure 3.11(a). Whereas in EMD-HF-BiLSTM, the identified MI-EEG tasks are RH with 79.3%, LH with 84.2%, BF with 79.19%, and T with 74.7%, and it is achieved with 75.12% accuracy (see Figure 3.11(c)). Overall, the proposed SWD-HF-BiLSTM technique for classifying MI-EEG tasks with the HF (C_1/C_2-(CDF-RASM-DASM)-B_1) and BiLSTM classifiers has shown improved classification accuracy rates of 78.62% over the EMD-HF-BiLSTM.

Further, we have compared the proposed classification framework with some recent works where the same BCI Competition IV dataset 2A has been utilized to

TABLE 3.5
Architectures of the Implemented BiLSTM Model

Layer	Memory cell	Activation	Parameters
BatchNorm-0	-	-	4
RNN-1	256	Sigmoid	64
BatchNorm-1		-	1568
RNN-2	2016	Sigmoid	128
BatchNorm-2		-	5890
RNN-3	128	Sigmoid	256
BatchNorm-3		-	
RNN-4	64	Sigmoid	32760
BatchNorm-4		-	240
Dense layer 1		-	16920
RNN-5	16	Sigmoid	7640
BatchNorm-5		-	140
Dense layer 2		16	560
Dense layer 3	2	Softmax	56

classify MI-EEG tasks with different approaches. Table 3.6 shows the comparison of the proposed methodology with the existing work available in the literature. For classifying MI-EEG signals for BCI application, many research groups have utilized different signal decomposition methods with use of advanced feature extraction and machine learning techniques. Work mentioned by the research groups in [17, 28–32] has achieved higher accuracy of about 75% with the use of different signal decomposition techniques and machine learning model for MI-EEG classification. However, the proposed SWD-HF-BiLSTM model delivers a classification accuracy of 78.62%. Further, Gaur et al. [2] have used a MEMD-based filtering method and Riemannian geometry classifier for MI-EEG binary classification, which gives an accuracy of 79.93%. Whereas our method has achieved improved performance with multitask classification approach against to work [2]. Further in [18], the proposed methodology has employed a PSO optimizer on the CSP parameters using LightGBM classifier to find the best correlation matrix to discriminate artifacts-driven EEG channel information. It has shown superior performance to multiclass MI-EEG classification to all the existing reported work (see Table 3.6). Whereas our proposed method has utilized a subject-independent approach in performance validation for multitask MI-EEG classification, which was not found in work [18]. Thus, it is noted that the proposed approach gives significant enhancement in MI-EEG signal using SWD and computed new HFs for multitask MI-EEG classification. At every trial from all nine subjects, it has delivered a higher average classification performance in subject-independent approach. As a result, we can confidently claim that our method outperforms almost every method except [2, 18] with significant improvement.

TABLE 3.6

Performance Comparison of the Different Existing Methods with the Proposed Method

Method	MI-EEG tasks	Accuracy
Lotte et al. [3], 2010	LH and RH	78.01%
Ang et al. [28], 2012	LH, RH, BF, and T	68.0%
Raza et al. [29], 2016	LH and RH	73.84%
Gaur et al. [2], 2018	LH and RH	79.93%
Sakhavi et al. [30], 2018	LH, RH, BF, and T	74.4%
Lawhern et al. [31], 2018	LH, RH, BF, and T	69.0%
Amin et al. [32], 2019	LH, RH, BF, and T	75.7%
Amin et al. [17], 2019	LH, RH, BF, and T	73.80%
Abenna et al. [18], 2022	LH and RH	96%
Proposed work 1 (SWD-HF-BiLSTM)	LH, RH, BF, and T	**78.62%**
Proposed work 2 (EMD-HF-BiLSTM)	LH, RH, BF, and T	**72.89%**

3.5 CONCLUSIONS

The chapter introduced a novel SWD-based classification framework that significantly enhanced the classification of MI-EEG tasks. The proposed method has been applied to MI-EEG signals and achieved improved classification performance against the presented EMD-based approach. The proposed SWD-HF-BiLSTM and EMD-HF-BiLSTM classification frameworks are used for the MI-EEG multitask classification problem and tested using a 10-fold cross-validation approach for a subject-independent analysis. New SWD-based HFs have been formulated and tested with BiLSTM models using various testing configurations to determine the optimal performance for MI-EEG multitask classification. With an appropriate selection of HF, it is demonstrated that the BiLSTM classification network has delivered the best classification results with less computational complexity and overfitting issue. A comparison of the results shows that the proposed framework has delivered the best classification accuracy (78.62%), precision (0.72), recall (0.66), and F1-score (0.604). Therefore, the proposed SWD-HF-BiLSTM can be a powerful tool for the detection of MI tasks from MI-EEG signals. In the future, it would be interesting to improve the accuracy of the proposed MI-EEG classification model by adopting hybrid signal modalities and practical classifiers and making the proposed BCI framework compatible with the already-in-use clinical automated system.

REFERENCES

1. J. Kevric and A. Subasi, "Comparison of signal decomposition methods in classification of EEG signals for motor-imagery BCI system," *Biomed. Signal Process. Control*, vol. 31, pp. 398–406, 2017.
2. P. Gaur, R. B. Pachori, H. Wang and G. Prasad, "A multi-class EEG-based BCI classification using multivariate empirical mode decomposition based filtering and Riemannian geometry," *Expert Syst. Appl.*, vol. 95, pp. 201–211, 2018.
3. F. Lotte and C. Guan, "Regularizing common spatial patterns to improve BCI designs: Unified theory and new algorithms," *IEEE Trans. Biomed. Eng.*, vol. 58, no. 2, pp. 355–362, 2010.
4. L. Duan, H. Zhong, J. Miao, Z. Yang, W. Ma and X. Zhang, "A voting optimized strategy based on ELM for improving classification of motor imagery BCI data," *Cogn. Comput.*, vol. 6, no. 3, pp. 477–483, 2014.
5. R. Janapati, V. Dalal, R. Sengupta, U. Desai, P. V. Raja Shekar and S. Kollem, "Towards a more theory-driven BCI using source reconstructed dynamics of EEG time-series," *Nano LIFE*, vol. 12, no. 02, p. 2250005, 2022.
6. R. Janapati, V. Dalal, G. M. Kumar, P. Anuradha and P. R. Shekar, "Web interface applications controllers used by autonomous EEG-BCI technologies," in *AIP Conference Proceedings*, vol. 2418, no. 1, p. 030038, 2022.
7. R. Janapati, V. Dalal and R. Sengupta, "Advances in experimental paradigms for EEG-BCI," in Proceedings of the 2nd International Conference on Recent Trends in Machine Learning, IoT, Smart Cities and Applications, 2022, pp. 163–170.
8. R. Janapati, V. Dalal and R. Sengupta, "Advances in modern EEG-BCI signal processing: A review," *Mater. Today Proc.*, vol. 80, pp. 2563–2566, 2021.
9. R. Janapati, V. Dalal, R. Sengupta and R. S. PV, "Progression of EEG-BCI classification techniques: A study," *Inven. Syst. Control*, vol. 204, pp. 161–170, 2021.

10. R. Janapati, V. Dalal, N. Govardhan and R. Sengupta, "Signal processing algorithms based on evolutionary optimization techniques in the BCI: A review," *Comput. Vis. Bio-Inspired Comput.*, pp. 165–174, 2021.

11. R. Janapati, V. Dalal, N. Govardhan and R. S. Gupta, "Review on EEG-BCI classification techniques advancements," in *IOP Conference Series: Materials Science and Engineering*, vol. 981, no. 3, p. 032019, 2020.

12. R. Janapati, V. Dalal, P. Anuradha and P. R. Shekar, "Various signals used for device navigation in BCI production," in *IOP Conference Series: Materials Science and Engineering*, vol. 981, no. 3, p. 032003, 2020.

13. R. P. Kumar, S. S. Vandana, D. Tejaswi, K. Charan, R. Janapati and U. Desai, "Classification of SSVEP signals using neural networks for BCI applications," in 2022 International Conference on Intelligent Controller and Computing for Smart Power (ICICCSP), pp. 1–6, 2022.

14. C. Park, D. Looney, N. ur Rehman, A. Ahrabian and D. P. Mandic, "Classification of motor imagery BCI using multivariate empirical mode decomposition," *IEEE Trans. Neural Syst. Rehabil. Eng.*, vol. 21, no. 1, pp. 10–22, 2012.

15. R. Tang, Z. Li and X. Xie, "Motor imagery EEG signal classification using upper triangle filter bank auto-encode method," *Biomed. Signal Process. Control*, vol. 68, p. 102608, 2021.

16. W. Ma et al., "A novel multi-branch hybrid neural network for motor imagery EEG signal classification," *Biomed. Signal Process. Control*, vol. 77, p. 103718, Aug. 2022, doi: 10.1016/j.bspc.2022.103718.

17. S. U. Amin, M. Alsulaiman, G. Muhammad, M. A. Mekhtiche and M. S. Hossain, "Deep learning for EEG motor imagery classification based on multi-layer CNNs feature fusion," *Future Gener. Comput. Syst.*, vol. 101, pp. 542–554, Dec. 2019, doi: 10.1016/j.future.2019.06.027.

18. S. Abenna, M. Nahid and A. Bajit, "Motor imagery based brain-computer interface: Improving the EEG classification using Delta rhythm and LightGBM algorithm," *Biomed. Signal Process. Control*, vol. 71, p. 103102, Jan. 2022, doi: 10.1016/j.bspc.2021.103102.

19. A. Khorshidtalab, M. J. E. Salami and R. Akmeliawati, "Motor imagery task classification using transformation based features," *Biomed. Signal Process. Control*, vol. 33, pp. 213–219, Mar. 2017, doi: 10.1016/j.bspc.2016.12.006.

20. W. Huang, J. Zeng, Z. Wang and J. Liang, "Partial noise assisted multivariate EMD: An improved noise assisted method for multivariate signals decomposition," *Biomed. Signal Process. Control*, vol. 36, pp. 205–220, Jul. 2017, doi: 10.1016/j.bspc.2017.04.003.

21. M. Thilagaraj and M. P. Rajasekaran, "An empirical mode decomposition (EMD)-based scheme for alcoholism identification," *Pattern Recognit. Lett.*, vol. 125, pp. 133–139, 2019.

22. G. K. Apostolidis and L. J. Hadjileontiadis, "Swarm decomposition: A novel signal analysis using swarm intelligence," *Signal Process.*, vol. 132, pp. 40–50, Mar. 2017, doi: 10.1016/j.sigpro.2016.09.004.

23. M. Khateeb, S. M. Anwar and M. Alnowami, "Multi-domain feature fusion for emotion classification using DEAP dataset," *IEEE Access*, vol. 9, pp. 12134–12142, 2021.

24. A. Rényi, "On measures of entropy and information," in Proceedings of the Fourth Berkeley Symposium on Mathematical Statistics and Probability, vol. 1, 547–561, 1961.

25. B. Hjorth, "EEG analysis based on time domain properties," *Electroencephalogr. Clin. Neurophysiol.*, vol. 29, no. 3, pp. 306–310, Sep. 1970, doi: 10.1016/0013-4694(70)90143-4.

26. B. C. Biswas and S. V. Bhalerao, "A real time based wireless wearable EEG device for epilepsy seizure control," in 2015 International Conference on Communications and Signal Processing (ICCSP), pp. 0149–0153, 2015.

27. Y. Wang, Z. Xiao, S. Fang, W. Li, J. Wang and X. Zhao, "BI-Directional long short-term memory for automatic detection of sleep apnea events based on single channel EEG signal," *Comput. Biol. Med.*, vol. 142, p. 105211, 2022.

28. K. K. Ang, Z. Y. Chin, C. Wang, C. Guan and H. Zhang, "Filter bank common spatial pattern algorithm on BCI Competition IV datasets 2a and 2b," *Front. Neurosci.*, vol. 6, 2012, Accessed: Jul. 04, 2022. [Online]. Available: https://www.frontiersin.org/articles /10.3389/fnins.2012.00039

29. H. Raza, H. Cecotti, Y. Li and G. Prasad, "Adaptive learning with covariate shift-detection for motor imagery-based brain–computer interface," *Soft Comput.*, vol. 20, no. 8, pp. 3085–3096, 2016.

30. S. Sakhavi, C. Guan and S. Yan, "Learning temporal information for brain-computer interface using convolutional neural networks," *IEEE Trans. Neural Netw. Learn. Syst.*, vol. 29, no. 11, pp. 5619–5629, Nov. 2018, doi: 10.1109/TNNLS.2018.2789927.

31. V. J. Lawhern, A. J. Solon, N. R. Waytowich, S. M. Gordon, C. P. Hung and B. J. Lance, "EEGNet: A compact convolutional neural network for EEG-based brain–computer interfaces," *J. Neural Eng.*, vol. 15, no. 5, p. 056013, 2018.

32. S. U. Amin, M. Alsulaiman, G. Muhammad, M. A. Bencherif and M. S. Hossain, "Multilevel weighted feature fusion using convolutional neural networks for EEG motor imagery classification," *IEEE Access*, vol. 7, pp. 18940–18950, 2019.

33. S. Dutta, M. Singh and A. Kumar, "Automated classification of non-motor mental task in electroencephalogram based brain-computer interface using multivariate autoregressive model in the intrinsic mode function domain," *Biomed. Signal Process. Control*, vol. 43, pp. 174–182, May 2018, doi: 10.1016/j.bspc.2018.02.016.

34. A. Jiang, J. Shang, X. Liu, Y. Tang, H. K. Kwan and Y. Zhu, "Efficient CSP algorithm with spatio-temporal filtering for motor imagery classification," *IEEE Trans. Neural Syst. Rehabil. Eng.*, vol. 28, no. 4, pp. 1006–1016, Apr. 2020, doi: 10.1109/ TNSRE.2020.2979464.

35. Y. L. Siuly and P. Wen, "Modified CC-LR algorithm with three diverse feature sets for motor imagery tasks classification in EEG based brain–computer interface," *Comput. Methods Programs Biomed.*, vol. 113, no. 3, pp. 767–780, Mar. 2014, doi: 10.1016/j. cmpb.2013.12.020.

36. S. Jayalakshmy, J. K. Pragatheeswaran, D. Saraswathi and N. Poonguzhali, "Scattering convolutional network based predictive model for cognitive activity of brain using empirical wavelet decomposition," *Biomed. Signal Process. Control*, vol. 66, p. 102501, 2021.

37. J. Meng, G. Liu, G. Huang and X. Zhu, "Automated selecting subset of channels based on CSP in motor imagery brain-computer interface system," in 2009 IEEE International Conference on Robotics and Biomimetics (ROBIO), Dec. 2009, pp. 2290–2294. doi: 10.1109/ROBIO.2009.5420462.

38. D. Gutiérrez and R. Salazar-Varas, "Using eigenstructure decompositions of time-varying autoregressions in common spatial patterns-based EEG signal classification," *Biomed. Signal Process. Control*, vol. 7, no. 6, pp. 622–631, 2012.

39. A. Kamble, P. Ghare and V. Kumar, "Machine-learning-enabled adaptive signal decomposition for a brain-computer interface using EEG," *Biomed. Signal Process. Control*, vol. 74, p. 103526, 2022.

40. H. Mirvaziri and Z. S. Mobarakeh, "Improvement of EEG-based motor imagery classification using ring topology-based particle swarm optimization," *Biomed. Signal Process. Control*, vol. 32, pp. 69–75, Feb. 2017, doi: 10.1016/j.bspc.2016.10.015.

41. X. Zhao, H. Zhang, G. Zhu, F. You, S. Kuang and L. Sun, "A multi-branch 3D convolutional neural network for EEG-based motor imagery classification," *IEEE Trans. Neural Syst. Rehabil. Eng.*, vol. 27, no. 10, pp. 2164–2177, 2019.

42. S. Tiwari, S. Goel and A. Bhardwaj, "MIDNN-a classification approach for the EEG based motor imagery tasks using deep neural network," *Appl. Intell.*, vol. 52, no. 5, pp. 4824–4843, 2022.

43. C. Brunner, R. Leeb, G. R. Muller-Putz and A. Schlogl, "BCI competition 2008: Graz data set A," *Institute for Knowledge Discovery (Laboratory of Brain-Computer Interfaces), Graz University of Technology*, vol. 16, pp. 1–6, 2008.

44. B. Blankertz et al., "The BCI competition 2003: Progress and perspectives in detection and discrimination of EEG single trials," *IEEE Trans. Biomed. Eng.*, vol. 51, no. 6, pp. 1044–1051, Jun. 2004, doi: 10.1109/TBME.2004.826692.

45. R. Leeb and C. Brunner, "BCI competition 2008 { Graz data set B," *Undefined*, 2008, Accessed: Jul. 04, 2022. [Online]. Available: https://www.semanticscholar.org/paper /BCI-Competition-2008-%7B-Graz-data-set-B-Leeb-Brunner/9031aac7e9b6adae909 ce22fa35fa74ec52a52ec

46. S. V. Bhalerao and R. B. Pachori, "Sparse spectrum based swarm decomposition for robust nonstationary signal analysis with application to sleep apnea detection from EEG," *Biomed. Signal Process. Control*, vol. 77, p. 103792, Aug. 2022, doi: 10.1016/j. bspc.2022.103792.

47. Y. Miao, M. Zhao, V. Makis and J. Lin, "Optimal swarm decomposition with whale optimization algorithm for weak feature extraction from multicomponent modulation signal," *Mech. Syst. Signal Process.*, vol. 122, pp. 673–691, May 2019, doi: 10.1016/j. ymssp.2018.12.034.

48. L. Zhang, D. Chen, P. Chen, W. Li and X. Li, "Dual-CNN based multi-modal sleep scoring with temporal correlation driven fine-tuning," *Neurocomputing*, vol. 420, pp. 317–328, 2021.

49. P. Gaur, R. B. Pachori, H. Wang and G. Prasad, "An empirical mode decomposition based filtering method for classification of motor-imagery EEG signals for enhancing brain-computer interface," in 2015 International Joint Conference on Neural Networks (IJCNN), 2015, pp. 1–7.

4 Analysis of Mental Stress during Public Speaking Using Feature Optimization Techniques on EEG Signals

Dr. T. Ananda Babu, Dr. D. Krishna, G. Sridevi,
Y. Mamillu, Dr. S. Ravi, Dr. P. Rajesh Kumar

CONTENTS

4.1 INTRODUCTION

Psychological stress is a perception of emotional tension and mental pressure (Mental Health America 2018) and helps to improve performance if its presence is more diminutive. The cause of stress (the type of stressor) can be unpredictable (natural disasters like war, pandemic), major life events (marriage, death of loved one), hassles occurred in daily life (meeting deadlines at school or work, making decisions), or ambient (pollution, traffic, crowding). Stress may escalate the risk of cardiac arrest, ulcers, and mental disorders like schizophrenia and depression and can intensify the previous ailments (Hans 1974, Reisman 1997, Pastorino and Portillo 2009, Schlotzet al. 2011, Espinosa et al. 2017, Duman 2014). The anxieties about the COVID-19 pandemic have transformed into perceived stress that impacts faculty mental health and their profession. A substantial rise in emotional exhaustion and work-related stress or frustrations was found in a group of faculty (Course

DOI: 10.1201/9781003326830-4

Hero 2020). The source of stress is the struggle to adjust to online teaching and new learning modalities (Ethan 2020). Changes in class size and teaching mode, reliance on teaching assistants, worry about losing the job, Zoom fatigue, childcare, financial issues, world events, and social unrest are the additional stressors observed in them (Susan 2020, Greta 2020). The mental and emotional pressures the faculty are experiencing due to online teaching can be conveyed through deteriorated physical health.

Psychological stress can be categorized as acute stress, which is instantaneous, and perceived (chronic) stress, which is long-lasting. The effects of adverse mental health due to coronavirus are severe and long-lasting (American Psychological Association, 2020). The subjective assessment of stress is carried out through different questionnaires or surveys (Koh et al. 2000, Meng and Guan 2018). Stress can change the working of the hypothalamus by activating the pituitary and adrenaline glands (Herbert and Cohen 1993). The circulation of adrenaline, immunoglobulin A, and cortisol causes many physiological changes, which can be noticed with variations in heartbeats, respiration, blood pressure, electrodermal activity, muscular activity, pupil dilatation, and brain activity (Dieleman et al. 2015, Xu et al. 2015). Neuroimaging techniques such as positron emission tomography (PET), functional magnetic resonance imaging (fMRI), and transcranial Doppler sonography (TCD) are employed in the assessment of brain activation during stressful conditions (Parasuraman and Caggiano 2005). Electroencephalogram (EEG) is a suitable approach for an unobstructive and continuous measure of stress as it can measure the brain's response during cognitive workload (Jensen and Tesche 2002). Recent advances in communication and sensor technologies made wearable EEG sensors more portable and economical than neuroimaging techniques to understand brain activity.

K-means clustering with general linear regression neural network is employed in Xu et al. (2015) for evaluating stress. They used state-trait anxiety inventory (STAI) for evaluating stress levels from the physiological signals such as ECG, EEG, EMG, and GSR. The discrete cosine transform (DCT) is used to find the correlation of cognitive stress with EEG rhythms (bands) (Chee-Keong et al. 2015). The brain activity is separated into five EEG bands (rhythms): delta (1–4 Hz), theta (4–7 Hz), alpha (8–15 Hz), beta (16–31 Hz), and gamma (>32 Hz) (Robertson et al. 2012, Antoine et al. 2019). Increased activity of a band is correlated with a particular task; for instance, theta activity (Ishihara and Yoshii 1972, Ishii et al. 2014) and lower alpha activity (Boksem et al. 2005) can be found in the frontal midline when subjects perform complex arithmetic tasks. Wu et al. (2015) used heart rate variability and accelerometer features for evaluating the stress levels from eight subjects. The EEG activity was analyzed to predict different stress levels using nonlinear features where the Stroop test was used as the stressor (Hou et al. 2015). Luijcks et al. (2015) investigated the perceived stress effect on cortical activity using electric shockers as stimuli. The combination of the Stroop test and arithmetic test is the stressor used in the study (Jun and Smitha 2016).

Montreal imaging stress task (MIST) is another stressor for a stress detection system based on a machine learning technique (Subhani et al. 2017). Features extracted through FFT are selected using the correlation-based feature selection method and classified with SVM (Sanay et al. 2018). Deep learning techniques can also be employed for real-time detection of acute stress as directed by He et al. (2019) with the help of ECG. Chronic stress will alter the brain's hippocampus, which may harm the perspective memory (Chen et al. 2019). The long-term stress was evaluated, but the labeling was done with the help of a psychology expert in Saeed et al. (2020) for the first time in EEG-based studies. Jebelli et al. (2019) used online multitask learning techniques for the real-time recognition of stress from subjects under various stressors (working in different conditions). Recently, perceived mental stress has been detected during pre-activity and post-activity by classifying EEG signals with different machine learning techniques (Arsalan et al. 2019). Several research groups make an effort to detect chronic stress under different stressors. Literature is available for recognition of the occupational stress induced in faculty with the help of surveys or questionnaires (Meng et al. 2018, Adrian et al. 2014). This chapter proposes a framework for analyzing the stress induced during speech activity. The EEG recordings of 28 subjects are made before, during, and after the activity. Different statistical, wavelet, and temporal features are extracted and optimized with three optimization techniques. The optimized features are categorized with five machine learning algorithms for effective stress assessment.

4.2 MATERIAL AND METHODS

4.2.1 Database Description

A Muse headband comprises of four channels to record EEG from AF7, AF8, TP9, and TP10 positions on the scalp. The EEG signals analyzed in the work are the recordings of 28 faculty or students of the University of Engineering and Technology, Taxila (Arsalan et al. 2019). They were first informed to fill out the perceived stress scale (PSS) questionnaire designed by psychologists. The PSS scores are used to label the subject to a particular group: stressed or nonstressed. Later they are asked for preparing a presentation on a topic in which content is delivered at the same time. After the preparation, they presented in front of a 15–25 audience. EEG data was recorded in three phases (pre, post, and during activity) by providing relaxation between each stage.

The EEG signals have a sampling frequency of 256 Hz. An onboard digital signal processing module was used for the analysis. The module deployed a fast Fourier transform on the signals with 90% overlapping, and the window size was 256. As a result, the EEG signal from each sensor is divided into five EEG bands (brain waves), namely delta (0–4 Hz), theta (4–7 Hz), alpha (8–12 Hz), beta (12–30 Hz), and gamma (30–50 Hz) frequency bands. Figure 4.1 depicts the EEG bands of the TP9 sensor for

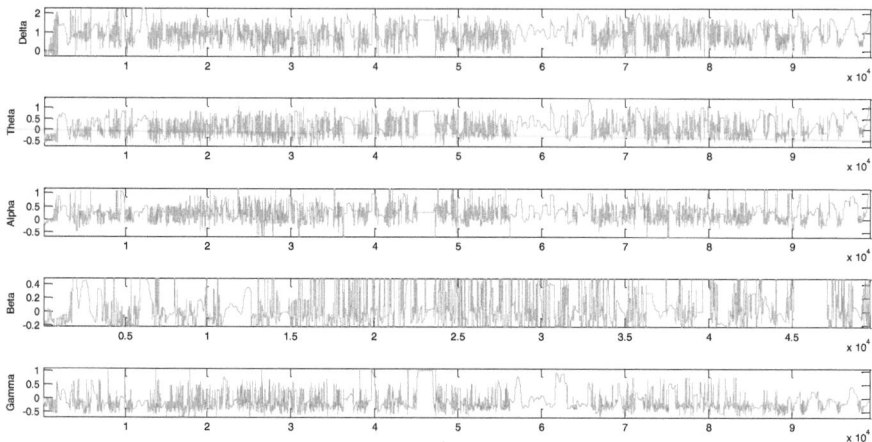

FIGURE 4.1   EEG bands of TP9 sensor for subject 1 during speech activity.

subject 1, where the x-axis comprises of time samples, and the y-axis resembles the magnitude in microvolts.

4.2.2   FEATURE ENGINEERING

The objective of the feature extraction is to analyze the raw signals to get informative and nonredundant parameters that can be used in the classification instead of the raw data. Feature selection methods are used to reduce the feature space in pattern recognition and classification. So, feature engineering is the most important and crucial step in medicinal diagnosis as these features can assist human interpretations in an emergency. The following features were extracted from the EEG signals: The time-domain and statistical features are mean, maximum, minimum, skewness, and kurtosis from the time samples.

Kurtosis

The tailedness of the probability density function around its mean is measured by kurtosis.

$$K = \frac{E\left[\left(x - \mu\right)^4\right]}{\sigma^4} \tag{4.1}$$

Skewness

The degree of deviation of a PDF from its mean value is known as coefficient of skewness.

$$S = \frac{E\left[(x-\mu)^3\right]}{\sigma^3} \tag{4.2}$$

For computing skewness and kurtosis, x is the EEG signal, μ is the mean, and σ is the standard deviation. The expectation operator is E [.]. The skewness and kurtosis are the third order and fourth order central moments, respectively.

Wavelet features are variance, energy, wavelength, standard deviation, and entropy. These features are extracted by applying 6-level DWT on EEG signals. The coif5 is used as mother wavelet. All the coefficients obtained in DWT are processed for features.

Energy
Wavelet energy at each level of decomposition ($i = 1, \ldots, l$) is determined as follows:

$$E_{D_i} = \sum_{j=1}^{n} |D_{ij}|^2 \quad i = 1,2,3,\ldots l \tag{4.3}$$

Where l = 7, level of decompositions

$$E_{A_i} = \sum_{j=1}^{n} |A_{ij}|^2 \quad i = 1,2,3\ldots .l \tag{4.4}$$

Entropy
To calculate wavelet entropy, the wavelet energy E_j of the signal is determined at scale j as follows:

$$E_j = \sum_{k=1}^{L_j} d(k)^2 \tag{4.5}$$

Where k is summation index and L_j is the no. of coefficients at scale j in a given epoch. The total energy of all scales is measured by:

$$E_{Total} = \sum_{j} E_j = \sum_{j} \sum_{k}^{L_j} d_j(k)^2 \tag{4.6}$$

Then the ratio of wavelet energy and total energy yields relative wavelet energy at scale j:

$$p_j = \frac{E_j}{E_{Total}} \tag{4.7}$$

Entropy is calculated as:

$$S = -\sum_j p_j \log p_j \tag{4.8}$$

Waveform Length

It is described as the cumulative length of waveform over a signal segment. It can be defined as:

$$WL = \sum_{i=1}^{N-1} \left| X_{i+1} - X_i \right| \tag{4.9}$$

Variance

Variance (VAR) is the average of the deviated square values of that variable. However, mean value of the EEG signal is nearly 0. Hence, variance of EEG signal is expressed as:

$$VAR = \frac{1}{N-1} \sum_{i=1}^{N} X_i^2 \tag{4.10}$$

Standard Deviation

Standard deviation denotes the quantity of change in frequency of a signal and determined using the following equation:

$$SD = \frac{1}{N-1} \sum_{i=1}^{N} \left(x_i - \mu \right)^2 \tag{4.11}$$

Where

μ—mean of the segment
x_i—an ith sample of EEG data in a segment
N—length of segment

The genetic algorithm (GA) is one feature optimization method based on the famous Darwin theory, survival of the fittest (Bazarghan et al. 2012). It mimics the natural evolution by forming a dynamic population, and members of it are known as chromosomes. The training dataset is used to enumerate the criterion function (fitness function). The mutation and crossover operations are executed during the evolution process. The method allows survival and reproduction of the fittest chromosomes, hence the fitness function is maximized or minimized effectively for subsequent generations. The process will stop when maximum number of generations is touched or the required fitness is attained. The parameters are chosen prudently to avert early convergence. The success of GA also depends on the selection of the initial population. The maximum number of generations is considered 100, and pressure and elite count parameters are taken as 6 and 2 for the work.

The PSO algorithm, developed by Eberhartand Kennedy (Kennedy and Eberhart 1995), is a population-based search technique based on the social behavior of birds or fishes. The individuals (particles) of the swarm learn from each other through exploration and exploitation. With their exploitation and exploration, the particle of the swarm hover through hyperspace and have two essential capabilities: the knowledge about local best (own best position) and memory of the global best (global or neighborhood's best). The PSO parameters used in work are inertia weight of 1, damping ratio of 0.99, and the personal learning coefficient of 1.5 and global learning coefficient of 2.0.

Another nature-inspired technique developed by Xin-She Yang (Yang 2009), firefly algorithm (FFA), is also used in the research. The algorithm be governed by the flashing characteristics of fireflies: (a) A firefly is attracted to by other irrespective of its sex. (b) A firefly with a lesser amount of brightness is attracted to the brighter firefly. The attractiveness of a butterfly is directly proportional to brightness, and these values decrease when they move away from each other. (c) The objective or fitness function determines the brightness of a firefly. The FFA is implemented by considering the step size of $\alpha = 0.5$, light absorption coefficient as 1 (γ), and 25 is the number of fireflies. The mutual information between the class labels (stressed or nonstressed) and the features is used as the objective function in all optimization techniques (Pohjajalainena et al. 2013).

4.2.3 CLASSIFICATION

The Naïve-Bayes algorithm has many applications, such as weather forecasting, medical diagnosis, and spam mail filtering. Using the Bayes theorem, it computes the feature's probability to categorize either stressed or nonstressed. The Naïve-Bayes technique has gotten attention nowadays as it's more straightforward and effective. It requires a lesser number of training samples to evaluate classification parameters (means and variances of the variables). In the K-nearest neighbor (KNN) classifier, the features are classified depending on the class of nearest neighbor. Samples with labels are used for training the classifier. For a given test sample (label unknown), that is represented by a vector of feature space, estimate the distances between test data and each point in train data. After arranging distances, the decision test sample class label will be made according to the label of nearest training samples. KNN classifier is based on Euclidean distance metric used in our research. The quadratic discriminant analysis (QDA) is engaged to discover the features combination that can discern the stressed and nonstressed signals. Though the primary application is for dimensionality reduction, it can be used as a classification algorithm, as the literature shows.

It has been proven that artificial neural networks (ANNs) are suitable for predicting stressed and nonstressed signals. Multilayer perceptron (MLP) is the neural network used in this work and is termed ANN throughout the chapter. The neural network bias values and weights are updated using scaled conjugate gradient back-propagation function. Support vector machine (SVM) is a robust classification algorithm that is applied to electrophysiological signals. Though it was implemented for two-class problems, it can also be used for multiple-class issues. The optimization condition is the breadth of the margin among the classes, i.e., the empty region around the separating hyperplane (decision boundary) demarcated by the distance to the nearest patterns. These patterns are support vectors that describe the classification function. In our work, a polynomial kernel was implemented for the classification. The features and the classification techniques are adapted as these have proven history, as evident from the literature and our previous work (Ananda and Rajesh 2020). Supervised classification is deployed in the work, and two-third of the feature space is a train set, and the residual is a test set. In order to predict stress, the classifier capability is tested by classifying stressed and nonstressed signals.

4.3   RESULTS AND DISCUSSION

Psychological tasks such as Stroop, imaginary, open/close eye, and arithmetic tasks are widely used to evaluate acute or perceived stress (Xu et al. 2015, Chee-Keong et al. 2015, Jun and Smitha 2016, Subhani et al. 2017, Sanay et al. 2018, Saeed et al. 2020). These tasks are suitable for assessing cognitive stress in laboratory settings. Representation of natural conditions in daily life is not simpler. The literature reported some realistic stressors such as electric shocks (Luijcks et al. 2015) and everyday duties at different heights (Jebelli et al. 2019). Some studies reported stress evaluation in the faculty (Meng and Guan 2018, Adrian et al. 2014), but they totally depend on surveys and questionnaires. To cope with the natural conditions, the paper process the EEG signals recorded from the subjects during speech activity (Arsalan et al. 2019).

The following metrics are used to scrutinize the classifier's performance: accuracy, specificity, precision, recall, F_score, and FPrate. These metrics are derived from true positive (TP), false positive (FP), true negative (TN), and false negative (FN) values taken from the confusion matrix of a classifier. Table 4.1 displays the accuracy values of different classifiers in each activity. The classification is done before any feature optimization method. The activity represents the stage of the EEG recordings: pre-activity means EEG signals taken before the start of the actual task,

TABLE 4.1
Accuracy Values (%) for Several Classifiers with Different Activities

Classifier	Pre-activity	Post-activity	During-activity
NB	42.7778	66.6667	37.2222
QDA	55.5556	53.3333	53.3333
ANN	58.3333	52.7778	52.7778
KNN	60	74.4444	85
SVM	63.3333	66.1111	72.2222

i.e., speech. Post-activity resembles the recording done after the completion, while during-activity is in the middle of the task.

Table 4.1 shows that for pre-activity, the SVM classifier is suitable for discerning the stressed signals with 63.3333% accuracy. The classifier's performance slightly increased for the signals recorded after the activity as the accuracy is 66.1111%, and also, the KNN classifier goes to 74.4444% against 60% for pre-activity. Moreover, KNN accuracy reached 85% when the signals recorded during the activity were fed. There is a lot of variation in the performance of the NB classifier, while the ANN classification metric is consistent among the activities. The ANN performance is decreased in other activities when compared with pre-activity. The KNN and SVM classifier performance varies similarly among the activities. All the features extracted using different signal processing techniques are employed for direct classification. Feature optimization techniques are deployed in subsequent steps to reduce the redundant features and increase the relevancy of the features. Table 4.2 represents the classifier accuracies for different activities after the features are optimized using GA, PSO, and FFA.

For the EEG signals recorded before the activity, the classifier performance increased with the feature optimization. The accuracy of the NB classifier is raised a small amount for the features optimized using FFA. The QDA accuracy is improved,

TABLE 4.2
Accuracy Values (%) of Classifiers after Feature Optimization with Different Activites

Classifier	Optimization	Pre-activity	Post-activity	During-activity
NB	No	42.7778	66.6667	37.2222
QDA	No	55.5556	53.3333	53.3333
ANN	No	58.3333	52.7778	52.7778
KNN	No	60	74.4444	85
SVM	No	63.3333	66.1111	72.2222
NB	GA	48.3333	63.8889	41.1111
QDA	GA	55.5556	58.8889	56.1111
ANN	GA	59.4444	52.7778	53.3333
KNN	GA	65.5556	78.3333	88.8889
SVM	GA	63.8889	78.8889	85.5556
NB	PSO	47.7778	67.7778	44.4444
QDA	PSO	58.3333	62.7778	55.5556
ANN	PSO	61.1111	56.6667	56.1111
KNN	PSO	87.2222	81.6667	90.5556
SVM	PSO	90	91.6667	93.8889
NB	FFA	48.3333	70.5556	45
QDA	FFA	60	63.8889	57.7778
ANN	FFA	59.4444	57.7778	56.1111
KNN	FFA	93.3333	85.5556	92.7778
SVM	FFA	94.4444	92.2222	94.4444

but the maximum value achieved is only 60%. The ANN classifier performance is almost constant for all the cases. The KNN performance is not mainly increased for GA optimized values, but for others, its value reached 90%, with the maximum of 93.3333% for FFA. The SVM classier performance is similar to KNN and achieved maximum accuracy of 94.4444%. For the signals recorded after the activity, the pattern of the classifier performance is repeated. The classifier accuracy increased slightly for GA optimization compared to the direct classification (without optimization). A significant improvement in accuracy can be observed for the PSO and FFA optimizations. The maximum value achieved is 92.2222% of the SVM classifier when the features are optimized with FFA.

The NB classifier's poor performance continued in the EEG signals' classification during the activity. It is not improved even with the feature optimization as the accuracy does not reach even fifties. The QDA and ANN accuracies are also not significant even after optimization. The maximum value among these classifiers is for QDA, with 57.7778% with FFA-optimized signals. The KNN accuracy shows a 7% increment between direct classification and FFA optimization, increasing from 85% to 92.7778%. A significant improvement in accuracy can be observed for the SVM classifier for FFA compared with the direct classification, which is increased from 72.2222% to 94.4444%, which translates to approximately 22% improvement. However, the maximum gain can be observed for the pre-activity as the KNN performance improvement is almost 33%, which is 60% to 93.3333%. The maximum accuracy value observed for the EEG signals recorded pre-activity and during-activity by the SVM classifier with FFA optimization (94.4444%).

Figure 4.2 displays the classification metrics accuracy, specificity, precision, recall, and F_score of the classifiers for different optimization techniques. In Figure 4.2,

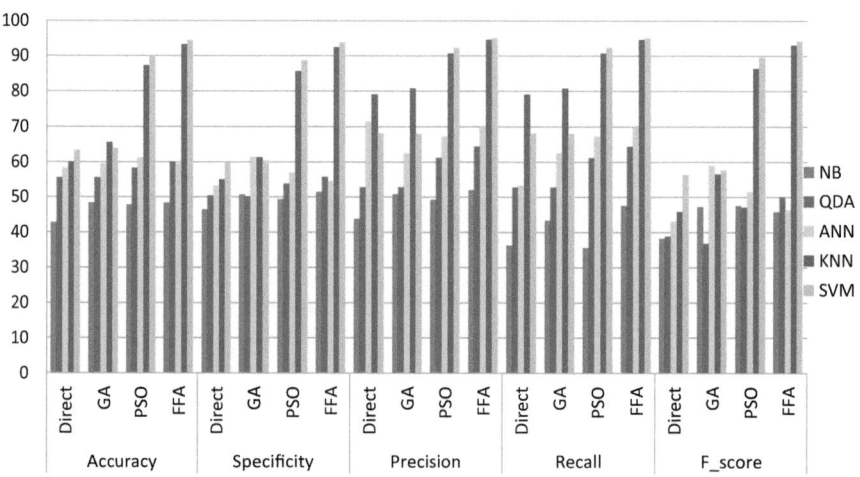

FIGURE 4.2 Classification metrics of the classifiers for different optimization techniques on EEG signals collected in pre-activity stage.

Direct represents the classification without optimization; the remaining are the classification metrics after optimization techniques GA, PSO, and FFA.

From Figure 4.2, it is evident that all parameters are increased with optimization techniques. The KNN and SVM classifiers' performance is way more than other classifiers. KNN precision and recall values are more for both direct and GA when compared with the values of the SVM classifier. However, the SVM classifier achieved superior performance among the classifiers for pre-activity. The same pattern is repeated for the EEG signals recorded post-activity and during-activity. FPrate significance is somewhat different among the performance metrics as it measures the misclassification rate. The FPrate is expected to reduce with the utilization of optimization techniques. Figure 4.3 is the plot of FPrate of several classifiers computed for different optimization techniques across the activities: pre-activity, post-activity, and during-activity, respectively.

FPrates of classifiers are decreased with the introduction of feature optimization, as observed in Figure 4.3. ANN classifier FPrate is increased for both PSO and FFA compared to GA. This behavior is observed in all the activities. KNN value is smaller than SVM if the FPrate is measured in direct classification without any optimization. With the optimization, KNN FPrate is more than that of the SVM classifier. The smaller FPrate of the KNN classifier is 7.1302% observed for the signals recorded during the speech activity, and the optimization is FFA. The SVM classifier achieves NillFPrate for during-activity with the FFA optimization. The optimized signals using PSO also generate a small FPrate of 0.5495%. From all the metrics, it can be concluded that the SVM is the best classifier for classifying EEG signals recorded during pre-activities whose features are optimized using FFA.

A Muse headband is used for recording the EEG signals from the scalp. It covers only the frontal and temporal lobes. The frontal lobe and temporal lobes deal with attention, memory processing, and sensory input processing; the processing of EEG signals from those areas is sufficient for stress assessment. The database used in the study is 28 EEG recordings collected from the pool of university faculty and students. The maximum accuracy of 94.4444% achieved by the SVM classifier worked on FFA-optimized features obtained in pre-activity and during-activity stages. For post-activity, maximum accuracy is achieved by SVM again for FFA optimization, and its value is 92.2222%. It is worth mentioning that the FPrates achieved by the SVM classifier for PSO and FFA techniques are 0.5495% and 0%, respectively. Each activity is analyzed separately, and cross-activity will be attempted in feature works. Recognizing the stress at the starting level and providing counseling will help improve the physical and mental health of the faculty and student. It is advantageous if they can recognize stress levels in real time and explore methods to enhance their presentation skills.

4.4   CONCLUSION

Stress induced in faculty or students escalates the threat of cardiac and mental diseases. The chapter proposed a new method to analyze stress in faculty and students during speech. The EEG signals recorded from 28 subjects during a presentation

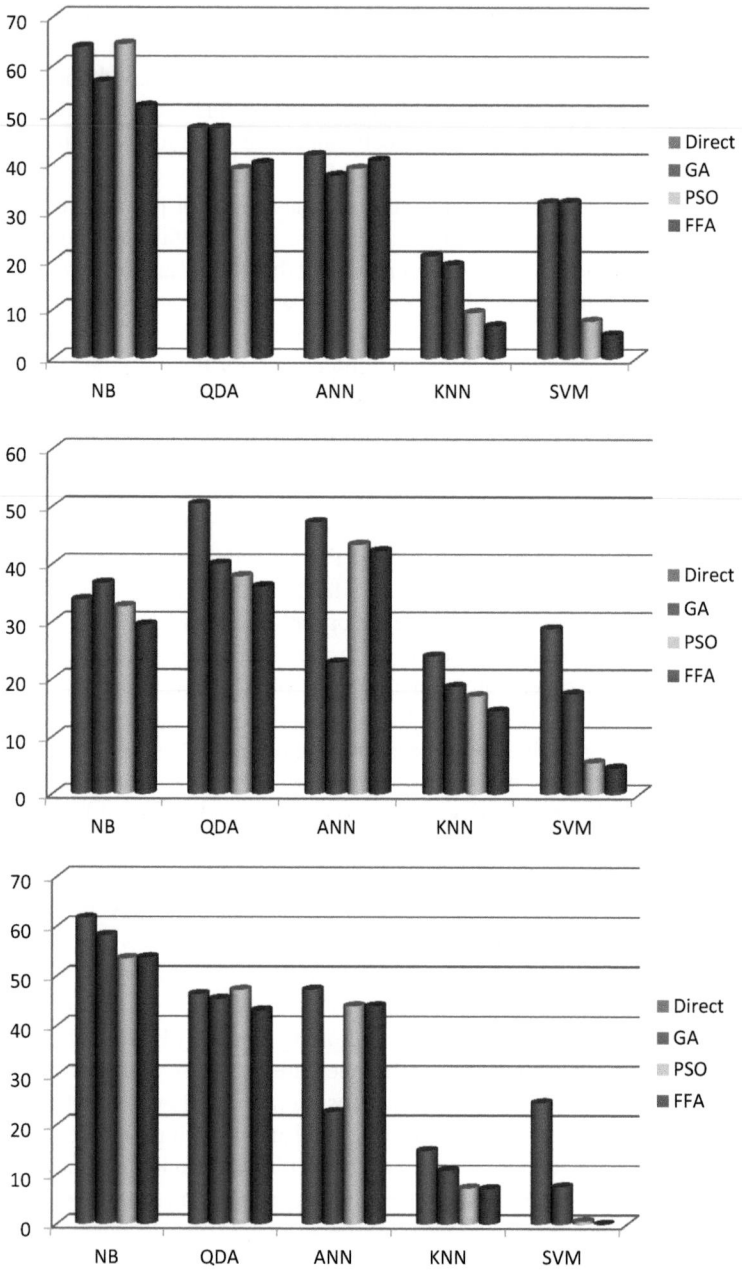

FIGURE 4.3   FPrates of different classifiers for pre-, post-, and during-activities, respectively.

are separated into stressed and nonstressed groups based on the score. Different signal processing techniques were adopted to extract the features from EEG signals. These features were selected using GA, PSO, and FFA. The optimized features are then fed to prominent classifiers for discerning the stressed and nonstressed subjects. Quadratic discriminant analysis, Naïve-Bayes, artificial neural networks, support vector machine (SVM), and K-nearest neighbor are used as classifiers. The stress levels are analyzed at three different stages: pre-activity, during-activity, and post-activity. The SVM classifier achieved an accuracy of 94.4444%, revealing that the proposed algorithm can assess stress more effectively than the existing techniques, either before or during the activity. The proposed techniques may be extended with more subjects and EEG devices with more sensors to make the most robust and reliable method.

REFERENCES

Adrian et al., 2014. Issues causing stress among business faculty members, *Journal of Academic Administration in Higher Education*, vol. 10, no. 1, pp. 41–42.

Ananda Babu T, Rajesh Kumar P, 2020. Optimized feature selection for the classification of uterinemagnetomyography signals for the detection of term delivery, *Biomedical Signal Processing and Control*, vol. 58, p. 101880, https://doi.org/10.1016/j.bspc.2020.101880.

Anderson G, 2020. Mental health needs rise with pandemic, *Inside Higher ED*, Sep 11, 2020, https://www.insidehighered.com/news/2020/09/11/students-great-need-mental-health-support-during-pandemic.

American Psychological Association, 2020. *Stress in the Time of COVID-19*, Vol 1, May 2020.

Arsalan A, Majid M, A. R. Butt and S. M. Anwar, 2019. Classification of perceived mental stress using a commercially available EEG headband, *IEEE Journal of Biomedical and Health Informatics*, vol. 23, no. 6, pp. 2257–2264, Nov. 2019, https://doi.org/10.1109/JBHI.2019.2926407.

Bazarghan M, Amandi R, Jaberi Y, Abedi M, Shahbazi A, Karami E, 2012. Automatic ECG beat arrhythmia detection. *CoRR*, https://doi.org/10.48550/arXiv.1209.0167 [cs.NE] 7 Nov 2012.

Blum SD, 2020. Why we're exhausted by zoom, *Inside Higher ED*, April 22, 2020, https://www.insidehighered.com/advice/2020/04/22/professor-explores-why-zoom-classes-deplete-her-energy-opinion.

Boksem MA, Meijman TF, Lorist MM, 2005. Effects of mental fatigue on attention: An ERP study, *Cognitive Brain Research*, vol. 25, no. 1, 107–16.

Chee-Keong AL, Chia WC, 2015. Analysis of single-electrode EEG rhythms using MATLAB to elicit correlation with cognitive stress, *International Journal of Computer Theory and Engineering*, vol. 7, no. 2, pp. 149–154. DOI: 10.7763/IJCTE.2015.V7.947.

Chen J, Wei Z, Han H et al., 2019. An effect of chronic stress on prospective memory via alteration of resting-state hippocampal subregion functional connectivity, *Scientific Report*, vol. 9, p. 19698. https://doi.org/10.1038/s41598-019-56111-9.

Course Hero, 2020. *Faculty Wellness and Careers, Course Hero*, November 18, 2020, https://www.coursehero.com/blog/faculty-wellness-research.

Dieleman, et al. 2015, Alterations in HPA-axis and autonomic nervous system functioning in childhood anxiety disorders point to a chronic stress hypothesis, *Psychoneuroendocrinology*, vol. 51, pp. 135–150.

Duman RS, 2014. Neurobiology of stress, depression, and rapid acting antidepressants: Remodeling synaptic connections, *Depression Anxiety*, vol. 31, no. 4, pp. 291–296.

Espinosa-Garcia C et al., 2017. Stress primes microglial polarization after global ischemia: Therapeutic potential of progesterone, *Brain, Behavior, Immunity*, vol. 66, pp. 177–192.

Ethan Kross, 2020, *Chatter: The Voice in Our Head, Why It Matters, and How to HarnessIt*, Penguin Random House.

Gaume A, Dreyfus G, Vialatte F-B, 2019. A cognitive brain–computer interface monitoring sustained attentional variations during a continuous task, *Cognitive Neurodynamics*, vol. 13, pp. 257–269, https://doi.org/10.1007/s11571-019-09521-4.

Hans Selye, 1974. *Stress Without Distress*, Philadelphia: J.B. Lippincott Company, p. 171, 1974.

He J, Li K, X. Liao, P. Zhang, N. Jiang, 2019. Real-time detection of acute cognitive stress using a convolutional neural network from electrocardiographic signal, *IEEE Access*, vol. 7, pp. 42710–42717, https://doi.org/10.1109/ACCESS.2019.2907076.

Herbert TB, Cohen S, 1993, Stress and immunity in humans: A meta-analytic review, *Psychosomatic Medicine*, vol. 55, no. 4, pp. 364–379, CiteSeerX 10.1.1.125.6544, https://doi.org/10.1097/00006842-199307000-00004. PMID 8416086. S2CID 2025176.

Hou X, Liu Y, Sourina O, Tan YRE, Wang L, Mueller-Wittig W, 2015. EEG based stress monitoring, in 2015 IEEE International Conference on Systems, Man, and Cybernetics, Hong Kong, China, pp. 3110–3115, https://doi.org/10.1109/SMC.2015.540.

Ishihara T, Yoshii N, 1972. Multivariate analytic study of EEG and mental activity in Juvenile delinquents, *Electroencephalography Clin. Neurophysiol.*, vol. 33, no. 1, pp. 71–80.

Ishii et al., 2014. Frontal midline theta rhythm and gamma power changes during focussed attention on mental calculation: An MEG beam former analysis, *Frontiers in Human Neuroscience*, 07/2014, https://doi.org/10.3389.fnhum.2014.00406.

Jebelli H, M. Mahdi Khalili, S. Lee, 2019. A continuously updated, computationally efficient stress recognition framework using electroencephalogram (EEG) by applying online multitask learning algorithms (OMTL), *IEEE Journal of Biomedical and Health Informatics*, vol. 23, no. 5, pp. 1928–1939, Sept. 2019, https://doi.org/10.1109/JBHI.2018.2870963.

Jensen O, Tesche CD, 2002. Frontal theta activity in humans increases with memory load in a working memory task, *Eur. J. Neurosci.*, vol. 15, no. 8, pp. 1395–1399.

Jun G, Smitha KG, 2016. EEG based stress level identification, in 2016 IEEE International Conference on Systems, Man, and Cybernetics (SMC), Budapest, Hungary, pp. 003270–003274, https://doi.org/10.1109/SMC.2016.7844738.

Kennedy J, Eberhart RC, 1995. Particle swarm optimization. In: Proceedings of IEEE International Conference on Neural Networks IV, pp. 1942–1948.

Koh KB, Park JK, Kim CH, 2000. Development of the stress response inventory, *J. Korean Neuropsychiatric Assoc.*, vol. 39, no. 4, pp. 707–719.

Luijcks R, Vossen CJ, Hermens HJ, van Os J, Lousberg R, 2015. The influence of perceived stress on cortical reactivity: A proof-of-principle study, *PLoS ONE*, vol. 10, no. 6, p. e0129220, https://doi.org/10.1371/journal.pone.0129220.

Meng Q, Wang G, 2018. A research on sources of university faculty occupational stress: A Chinese case study. *Psychology Research and Behavior Management*, vol. 11, pp. 597–605. 7 Dec. 2018, https://doi.org/10.2147/PRBM.S187295.

Mental Health America, 2018. *Stress*, 2013-11-18. https://www.mhanational.org/conditions/stress.

Parasuraman R, Caggiano G, 2005. Neural and genetic assays of mental workload, In D. McBride and D. Schmorrow (Ed.) *Quantifying Human Information Processing* (pp 123–155). Lanham: Rowman and Littlefield.

Pastorino E, Doyle-Portillo S, 2009. *What is Psychology?*, 2nd Ed. Belmont, CA: Thompson Higher Education.

Pohjajalainena J, Rasanen O, Kadioglub S, 2013. Feature selection methods and their combinations in high-dimensional classification of speaker likability, intelligibility and personality traits, *Computer Speech & Language*, November 2013, https://doi.org/10.1016/j.csl.2013.11.004.

Reisman S, 1997. Measurement of physiological stress, in Proc. IEEE 23rd Northeast Bioeng. Conf., pp. 21–23.

Robertson et al., *2012 Primer on the Autonomic Nervous System*, 3rd ed. Amsterdam, The Netherlands: Elsevier.

Saeed SMU, Anwar SM, Majid M, Awais M, Alnowami M, 2018. Selection of neural oscillatory features for human stress classification with single channel EEG headset, *BioMed Research International*, vol. 2018, Article ID 1049257, 8 pages. https://doi.org/10.1155/2018/1049257.

Saeed SMU, Anwar SM, Khalid H, Majid M, Bagci AU, 2020. EEG based classification of long-term stress using psychological labeling, *Sensors*, vol. 20, no. 7, p. 1886, https://doi.org/10.3390/s20071886.

Schlotz W, Yim IS, Zoccola PM, Jansen L, Schulz P, 2011. The perceived stress reactivity scale: Measurement invariance, stability, and validity in three countries, *Psychol Assess.*, pp. 80–94.

Subhani AR, Mumtaz WMN, Saad BM, Kamel N, Malik AS, 2017. Machine learning framework for the detection of mental stress at multiple levels, *IEEE Access*, vol. 5, pp. 13545–13556, https://doi.org/10.1109/ACCESS.2017.2723622.

Wu M, Cao H, Nguyen HL, Surmacz K, Hargrove C, 2015. Modeling perceived stress via HRV and accelerometer sensor streams, in Annu Int Conf IEEE Eng Med Biol Soc. 2015 Aug, pp. 1625–8. doi: 10.1109/EMBC.2015.7318686. PMID: 26736586.

Xu TLN, Guan C, 2015. Cluster-based analysis for personalized stress evaluation using physiological signals, *IEEE J. Biomed. Health Inform.*, vol. 19, no. 1, pp. 275–281, Jan. 2015.

Yang XS, 2009. Firefly algorithms for multimodal optimization, in: *Stochastic Algorithms: Foundations and Applications, SAGA 2009, Lecture Notes in Computer Sciences*, vol. 5792, pp. 169–178.

5 Classification of Human Emotions Based on Electroencephalogram (EEG) Using Two-Layer Gated Recurrent Units (GRUs)

*Ch. Rajendra Prasad, Sreedhar Kollem,
Ravichander Janapati, Srinivas Samala,
and Sandip Bhattacharya*

CONTENTS

5.1   INTRODUCTION

Computer vision is a key study area in state-of-the-art human–computer interface for improving the experience of video game engagement, virtual reality, educational systems, and the detection of mental disorders [1]. A plethora of paradigms is being thoroughly examined in order to get insight into psychological condition and its relevance to emotion identification. Initial phases of this research area, facial expression, voice, and gesture were selected because they were interpretable and palpable signs suiting general humans' cognizance, and the data was pervasive and possibly will be

TABLE 5.1
Frequency Domain Features Frequency Bands and Their Ranges

Frequency band	Δ	θ	α	β	Υ
Range (Hz)	1–4	4–8	8–14	14–31	31–50

gathered on a wide scale in the laboratories [2, 3]. In spite of these positives, external data has a flaw since individuals have a propensity to conceal their emotions, and their facial expressions may not reflect the genuine feelings. Computer vision researchers are progressively focused on the internal data that can't be faked. The internal data includes blood pressure, heart rate, brain signal, and skin conductance, which possess the true emissions. In the past, internal signal measuring equipment was cumbersome, difficult to operate, and in broad deployment [4].

At present, internal signal measuring equipment is handy, and some kinds of information can be collected by smart devices like smartwatches. Among several types of physiological signs, brainwave signals have been drawing more attention due to their objectivity and reliability. The EEG device consists of electrode pads, which are deployed on the scalp [5]. The electrode pads could noninvasively record the neural activities that reflect several brain conditions such as emotional state, concentration level, sleep disorders, and mental illness. Brainwave-based emotion classification study is still fascinating, since the methods for extracting the features and techniques for the classification are substantially diversified [6, 7].

A traditional EEG-based emotion detection method comprises of four basic stages, i.e., raw data processing, feature extraction, feature selection, and prediction [8]. Out of these, feature extraction is critical to the effectiveness of emotion identification. Brainwave signals may be explored in several domains (time, frequency, time-frequency etc.), and every domain has numerous matching properties. Time domain feature can be named as statistical feature, principal component analysis (PCA), Hjorth feature, fractal dimension feature, high-order crossing feature, etc. For frequency domain features, the time domain raw EEG signals are applied to autoregressive algorithms and fast Fourier transform to produce five primary frequency bands [9]. The frequency bands are namely alpha (α), beta (β), theta (θ), delta (Δ), and Gamma (Υ), and ranges are tabulated in Table 5.1.

Whereas for the time-frequency domain features, the time domain raw EEG signals are applied to wave packet decomposition or wavelet transform. After getting the requisite features, diverse machine and deep learning models were proposed to predict the human emotions based on the features [10].

5.2   RELATED WORK

At present, most of the researchers are developing deep learning techniques for EEG-based emotion classification instead of conventional methods. From the last five years, several convolutional neural network (CNN)- and recurrent neural network (RNN)-based network models have been proposed [11]. Deep learning-based algorithms outperform earlier ones and enable new ways to evaluate EEG data. These deep learning algorithms transform brainwave signals' one-, two-, or three-dimensional matrices based on their features. This information can be useful for the researchers to analyze and generate novel models [12, 13].

The algorithms for EEG-based emotion classification were depend on the selection of features and classifiers. A few studies applied manual feature extraction before moving on to the next phase, while others employed networks to interpret raw EEG data automatically and predict emotions. In [14], the authors presented DenseNet model to study nonlinear time-domain features from the brainwave signals. Eventually, the learned features are employed for emotion classification. In [15], authors divided the brainwave signals into several overlapping sectors and from each sector features were extracted. The probability vectors of all sectors corresponding to the identical classes were then computed and given to the support vector machine (SVM) for classification. In [16], to extract the time-frequency domain features, wavelet transform is applied on raw EEG signals and then symmetric positive definite matrices were constructed based on the extracted features. On the Riemannian manifold, a supervised matrix factorization approach was constructed to minimize the high dimensionality of the symmetric positive definite matrices. Finally, the data was translated to tangent space and classified using the random forest (RF), K-nearest neighbors (KNN), and SVM models.

In [17], the authors presented a network model that is fine-grained emotion, which is employed for feature extraction in spatial-temporal domain. Each stage of the fine-grained emotion network is connected to the raw EEG data in order to prevent overfitting and generate spatial domain features. Also, all of the scale features from the previous layers are added to the current layer of spatial features to improve the scale features in the spatial component. Also, long short-term memory (LSTM) is used to pull out temporal features based on spatial components and put fine-grained emotions into categories. The suggested technique outperformed representational approaches in emotion classification and methods with a comparable structural framework in cross-session and subject-dependent experiments. In [18], "ATtention-based LSTM with Domain Discriminator (ATDD-LSTM)," a deep learning algorithm, is presented. This deep learning algorithm has two characteristics: (1) An attention method selects EEG channels for emotion recognition from LSTM feature vectors. This drive the trained network focus on emotion-based channels. (2) It employs a domain discriminator for changing the representation of data and creating invariable domain features. This model exhibits superior subject-dependent, subject-independent, and cross-session performance.

5.3 METHODS AND MATERIALS

The proposed model architecture is shown in Figure 5.1. In this architecture the raw data is collected with four electrodes such as AF7, AF8, TP9, and TP10. The collected data is preprocessed using feature extraction. The main function of feature extraction is to remove redundant and unwanted features from the raw dataset.

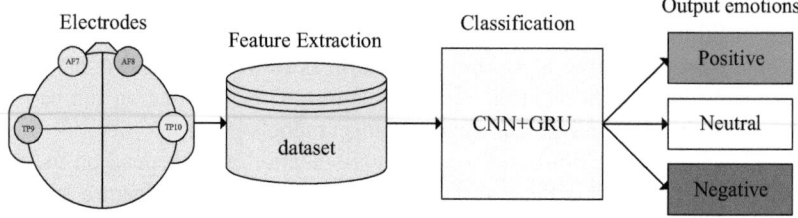

FIGURE 5.1   Proposed model architecture.

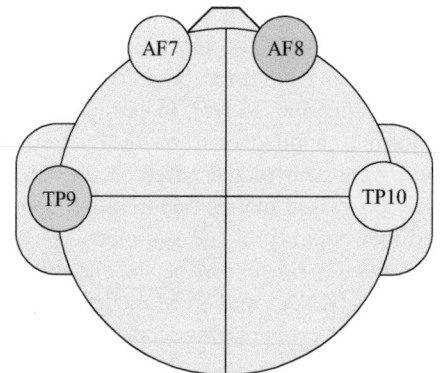

FIGURE 5.2   The placement of headset with four electrodes on the head.

Then, the deep neural network consists of CNN + GRU layers that are used to classify the data into three emotions, i.e., positive, neutral, and negative emotions. Further, the comprehensive explanation of the model is presented in the subsequent sections.

5.3.1   DATASET AND PREPROCESSING

For the purpose of this investigation, an EEG dataset that is freely accessible to the public and can be accessed from Kaggle [19] was utilized. The dataset was generated by employing a Muse EEG headset by means of a resolution with four different electrodes (AF7, AF8, TP9, and TP10) [20]. The placement of the electrodes on the head is illustrated in Figure 5.2.

Two subjects, one male, one female, aged from 20 to 22, were employed to create the data. Positive emotional state and negative emotional state are brought on by employing film clips by means of a clear valence. In addition, the neutral emotional state data was collected without any stimulus. The duration for every session was about 60 seconds. Table 5.2 shows the film clips employed as stimulus for creation of EEG data.

TABLE 5.2

Stimulus Used for Creation of EEG Data [21]

S. no.	Emotion	Stimulus
1	Positive	La La Land (2016)
2	Positive	Funny Dogs (2015)
3	Positive	Slow Life (2014)
4	Negative	Up (2009)
5	Negative	Marley & Me (2008)
6	Negative	My Girl (1991)
7	Neutral	Relaxed mode

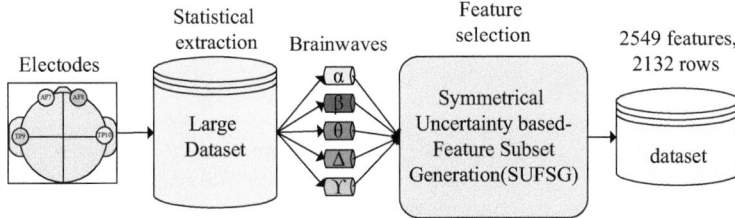

FIGURE 5.3 Dataset preprocessing and feature selection.

5.3.2 FEATURE EXTRACTION

Figure 5.3 illustrates dataset preprocessing and feature selection. The $\alpha, \beta, \theta, \Delta$, and Υ brainwaves are extracted statistically to generate a large-scale dataset. Symmetrical uncertainty-based feature subset generation (SUFSG) model is used to predict system uncertainty and to remove redundant feature map.

Symmetrical uncertainty-based feature subset generation (SUFSG): Symmetrical uncertainty is a statistical technique, which can be employed for dimensionality reduction [22]. In [23], authors presented feature selection model called fast correlation-based filter (FCBF). This model extracts the features by discarding the irrelevant and redundant features. In this model to identify the redundancy and irrelevant features, symmetrical uncertainty (SU) is employed and is defined as [24]:

$$SU(X,Y) = 2 \times \left[\frac{IG(X/Y)}{E(X) + E(Y)} \right] \qquad (5.1)$$

where $E(X)$ is the entropy of feature X, $E(Y)$ is the entropy of feature Y, and $IG(X/Y)$ is the information gain and is given as:

$$IG(X/Y) = E(X) - E(Y) \qquad (5.2)$$

C-correlation: denoted by $SU_{m,c}$ and it is defined as SU among any feature F_m and the class C.

F-correlation: denoted by $SU_{m,n}$ and it is defined as SU among any pair of features F_m and F_n $(m \neq n)$.

The step-by-step working of SUFSG model is given in the following algorithm:

SUFSG model

ASSUMPTIONS:

```
Let       F_m and F_n (m ≠ n), two relevant features
          σ is the SU threshold
if        SU_m,c ≥ σ, SU_n,c ≥ σ and SU_m,n ≥ (SU_m,c, SU_n,c)
then      F_m is the significant feature
          F_n will be the redundant feature of F_m
```

STEPS:

```
Step 1: Candidate list creation
        F_1, F_2, . . ., F_m are features and C is the class
                    Arrange features in descending order
                    based on SU_m,c
        if (SU_m,c < σ)
                    feature F_m is removed
        else
                    F_m is placed in the candidate (C) list.
Step 2: Analysis redundancy
        features in C-list and named as C_i
        If SU_m,n ≥ min(SU_m,c, SU_n,c)
                    C_m is placed in the significant feature (SF)
                    list.
                    C_n is placed in the noncandidate (NC) list.
                    do this for complete C-list.
Step 3: For the element C_i of C-list
        Check the first element (NC_1) of NC-list
        if (C-correlation of C_i > NC_1)
                    C_i is removed from C-list.
        else
                    continue with next element.
        Initialize the NC-list.
Step 4: Repeat above two steps until C-list empty.
```

5.3.3   CLASSIFICATION

Two-layer GRU model is employed for the emotion classification using EEG signals. The architectural two-layer GRU (2L-GRU) model is employed in this work for human emotion prediction illustrated in Figure 5.4. The 2L-GRU architecture consists of two GRU layers, the first layer with 128 units and the second layer with 64 GRU units. These two layers are followed by a dropout layer of probability $p = 0.25$. Then this dropout layer is followed by two dense fully connected (FC) layers with 16 neurons and 3 neurons. The output dense FC layer classifies the required emotions by using 'tanh' activation function [25].

Figure 5.5 shows that a GRU is an LSTM modified version with a simpler structure. This structure has two gates, namely reset and update. The update gate can determine how much previous data (past time steps) must be forwarded to the next stage. The output of update gate is denoted by z_t and is given in Equation 5.3. The reset gate can determine how much previous data (past time steps) is desirable to

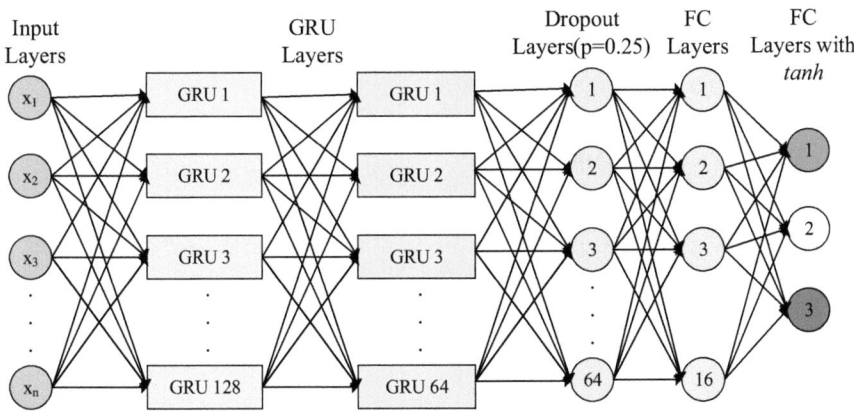

FIGURE 5.4   Two-layer GRU model for emotion classification.

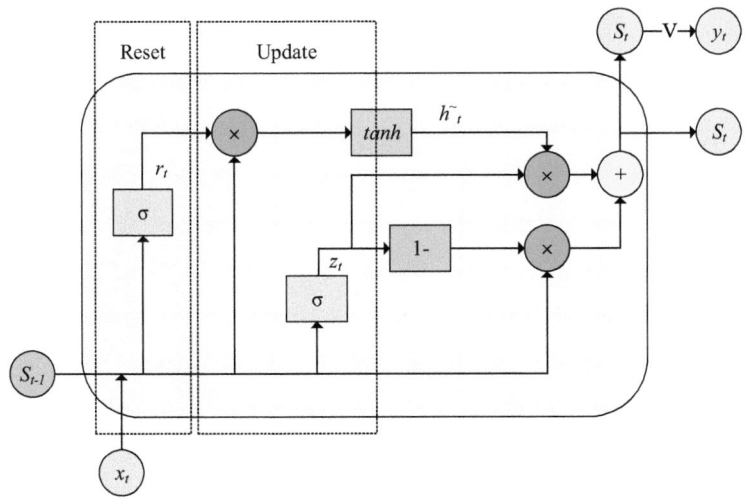

FIGURE 5.5   GRU layer internal logic diagram.

be neglected. The output of reset gate is denoted by r_t and is given in Equation 5.4. Equations 5.5 and 5.6 demonstrate detailed working of the GRU unit.

$$z_t = \sigma \times \left[(x_t, s_{t-1}) \times W_z + b_z \right] \tag{5.3}$$

$$r_t = \sigma \times \left[(x_t, s_{t-1}) \times W_r + b_r \right] \tag{5.4}$$

$$\hat{s_t} = tanh \times \left[(x_t, r_{t-1}) \times W_s + b_s \right] \tag{5.5}$$

$$s_t = z_t + (1 - z_t) \times s_{t-1} + \hat{s_t} \tag{5.6}$$

where x_t is the input vector, t is the time step, Wz, Wr, Ws are gate weight matrices of update, reset gates, and memory, bz, br, bs are connection bias at t of update, reset gates, and memory.

5.4 RESULTS AND DISCUSSION

To conduct this research, we utilized a publicly available electroencephalogram (EEG) brainwave-based emotion recognition dataset [19]. There are three classes in this dataset: positive, negative, and neutral. Figure 5.6 illustrates the 716 emotions under neural class, 708 emotions under negative class, and 708 emotions under positive class. After feature extraction the entire dataset consists of 2,549 features and

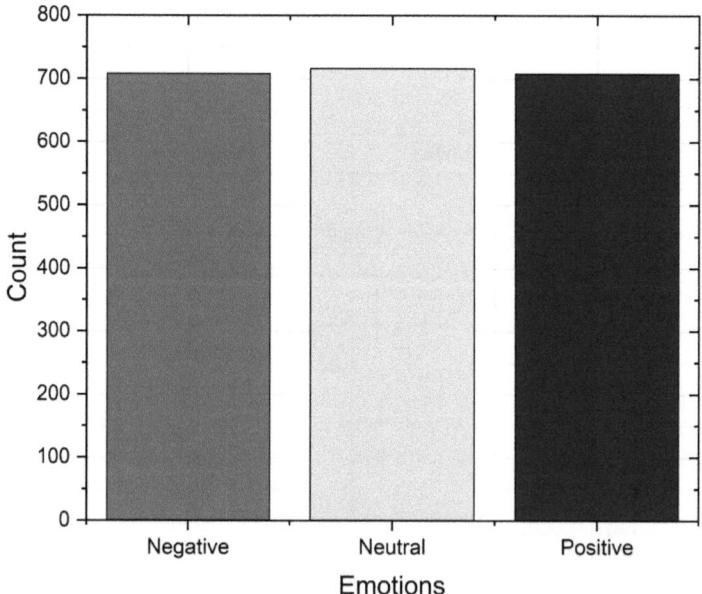

FIGURE 5.6 EEG brainwave dataset of three classes (negative, neutral and positive).

dimension of 2,132 rows. The features of data samples at location 0 and 2131 are show in Figures 5.7 and 5.8, respectively.

To compile the proposed model Adam optimizer [26] is employed; it is among the most widely used optimizers. The hyper parameters employed in the model are β1

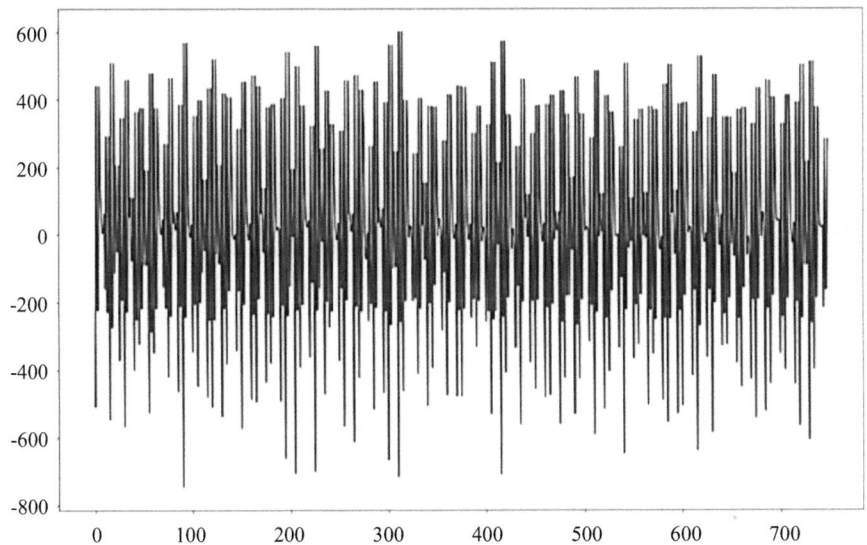

FIGURE 5.7   Sample data at location 0.

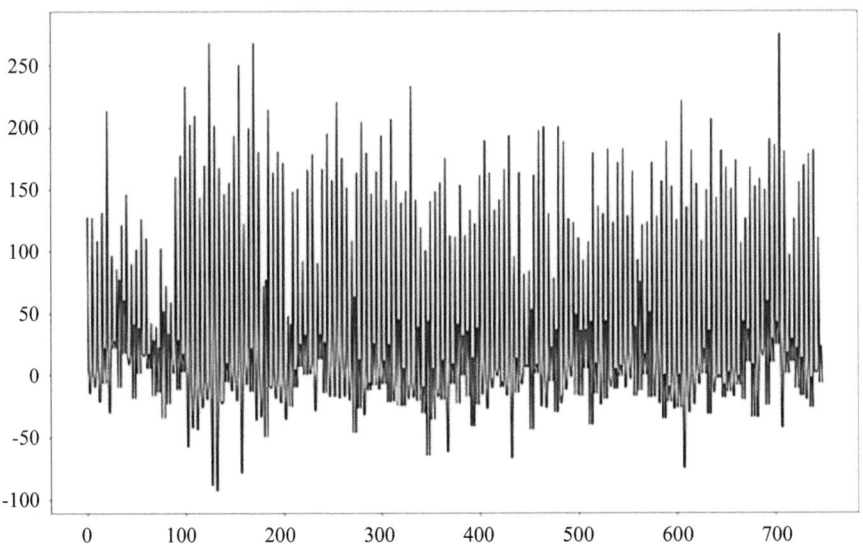

FIGURE 5.8   Sample data at location 2131.

= 0.9, β2 = 0.98, batch size of 128, and a learning rate of $1e^{-4}$. To achieve optimum accuracy that is 96%, the proposed model is trained for 200 epochs with sparse categorical cross-entropy. The accuracy curve of proposed model is illustrated in Figure 5.9. The loss curve of proposed model is illustrated in Figure 5.10. The confusion matrix of proposed model is illustrated in Figure 5.11.

The proposed model is compared with five recently proposed models by Duan et al., [27], Zhang et al. [28], Usman et al. [29], Aslam [30], and Xu [31]. The performance evaluation of the proposed model with aforementioned reference models is shown in Table 5.3. Three evaluation parameters, sensitivity, accuracy, and specificity, were employed to compare the performance of proposed model reference models. The evaluation parameter are defined as:

$$\text{Sensitivity} = \frac{\text{True Positives}}{\text{True Positives} + \text{False Negatives}} \tag{5.7}$$

$$\text{Accuracy} = \frac{\text{True Positives} + \text{True Negatives}}{\text{True Positives} + \text{True Negatives} + \text{False Positives} + \text{False Negatives}} \tag{5.8}$$

$$\text{Specificity} = \frac{\text{True Negatives}}{\text{True Negatives} + \text{False Positives}} \tag{5.9}$$

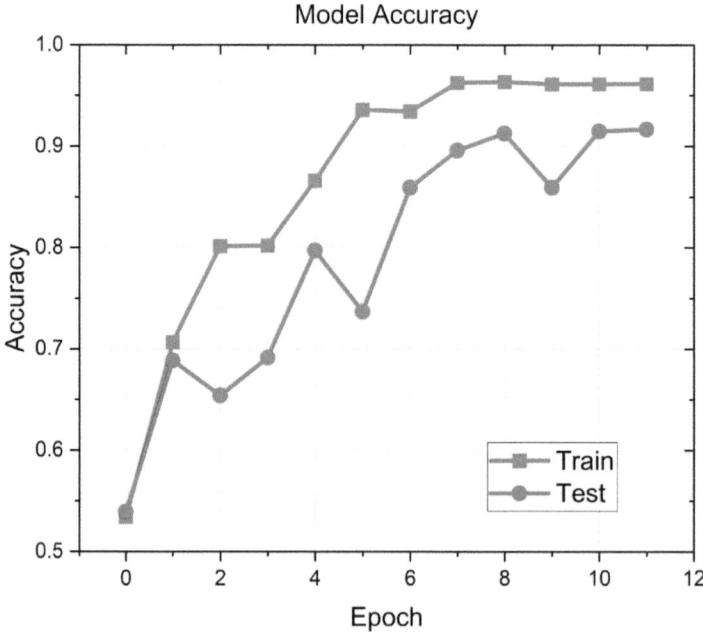

FIGURE 5.9 Accuracy of proposed model with respect to epochs.

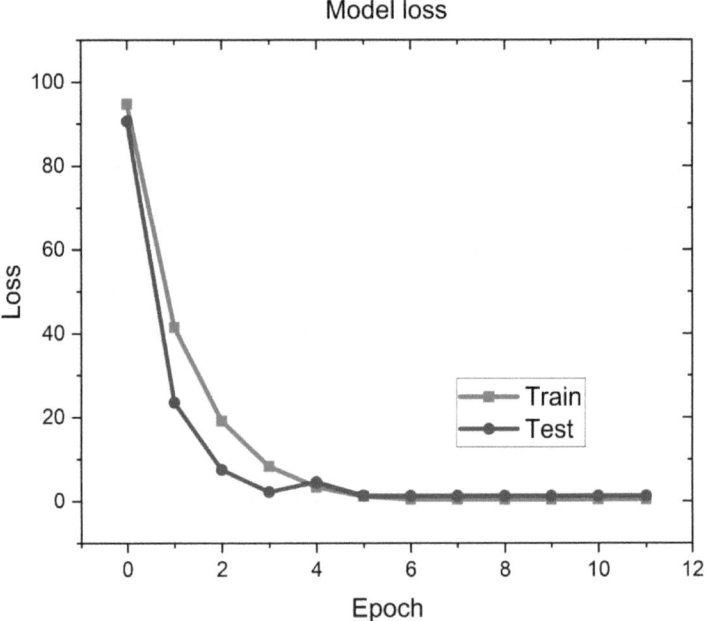

FIGURE 5.10   Loss of proposed model with respect to epochs.

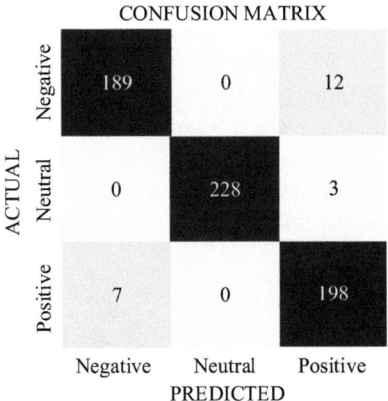

FIGURE 5.11   Confusion matrix of proposed deep learning model.

After the thorough analysis the proposed 2L-GRU model offers improved prediction outcomes for emotion classification. The simulation results of proposed model show that it is most suitable for real-time classification of EEG-based emotions in comparison with the recently proposed models.

TABLE 5.3

Proposed Model versus Conventional Models' Performance

Authors	Feature extraction technique	Classification model	Evaluation parameters
Proposed model	**Symmetrical uncertainty-based feature subset generation**	**2L-GRU**	**Specificity = 95.5%** **Accuracy = 96.00%**
Duan et al. [27]	Spectral subband using CNN	Bi-GRU	Sensitivity = 91.7%, Accuracy = 94.8%,
Zhang et al. [28]	Pearson correlation coefficient-based synchronization	CNN	Accuracy = 89.98%
Usman et al. [29]	Empirical mode and CNN decomposition	LSTM	Specificity = 92.5% Sensitivity = 93%
Aslam et al. [30]	Automated features using CNN	LSTM	Specificity = 91.2% Sensitivity = 93.8% Accuracy = 94%
Xu et al. [31]	Parallel convolution	Attention-based bidirectional LSTM	Accuracy = 90.02%

5.5   CONCLUSION

This chapter proposed deep learning model for classification of human emotions based on EEG using two-layer GRUs. The proposed model is implemented in two phases, namely feature extraction and emotion classification. In feature extraction phase SUFSG is applied to extract the most significant from the raw brainwave data. In emotion classification phase two-layer GRUs are employed to classify the human emotions such as neutral, positive, and negative. The proposed model is tested using the EEG brainwave signal dataset. Finally, the proposed model achieves better accuracy and faster tuning time in comparison with the classical classification models.

REFERENCES

1. Li, S., & Deng, W. (2020). Deep facial expression recognition: A survey. *IEEE Transactions on Affective Computing*, 13(3), 1195–1215.
2. Fahad, M. S., Ranjan, A., Yadav, J., & Deepak, A. (2021). A survey of speech emotion recognition in natural environment. *Digital Signal Processing*, 110, 102951.
3. Anbarjafari, G., Noroozi, F., Corneanu, C. A., Kamińska, D., Sapiński, T., & Escalera, S. (2018). *Survey on emotional body gesture recognition. IEEE Transactions on Affective Computing, 12*(2), 505–523.
4. Fu, Y., Wu, X., Li, X., Pan, Z., & Luo, D. (2020). Semantic neighborhood-aware deep facial expression recognition. *IEEE Transactions on Image Processing*, 29, 6535–6548.
5. Phan, T. D. T., Kim, S. H., Yang, H. J., & Lee, G. S. (2021). EEG-based emotion recognition by convolutional neural network with multi-scale Kernels. *Sensors*, 21(15), 5092.

6. Janapati, R., Dalal, V., & Sengupta, R. (2022). Advances in experimental paradigms for EEG-BCI. In Proceedings of the 2nd International Conference on Recent Trends in Machine Learning, IoT, Smart Cities and Applications. Springer, Singapore, pp. 163–170.

7. Wu, Y., Xia, M., Cai, H., Nie, L., & Zhang, Y. (2021). Exploring multiple scales and bi-hemispheric asymmetric EEG features for emotion recognition. *arXiv preprint arXiv:2110.06462.*

8. Yan, J., Chen, S., & Deng, S. (2019). A EEG-based emotion recognition model with rhythm and time characteristics. *Brain informatics*, 6(1), 1–8.

9. You, Y., Zhong, X., Liu, G., & Yang, Z. (2022). Automatic sleep stage classification: A light and efficient deep neural network model based on time, frequency and fractional Fourier transform domain features. *Artificial Intelligence in Medicine*, 127, 102279.

10. Janapati, R., Dalal, V., Govardhan, N., & Sengupta, R. (2021). Signal processing algorithms based on evolutionary optimization techniques in the BCI: A review. *Computational Vision and Bio-Inspired Computing, ICCVBIC 2020,* 165–174.

11. Torres, E. P., Torres, E. A., Hernández-Álvarez, M., &Yoo, S. G. (2020). EEG-based BCI emotion recognition: A survey. *Sensors*, 20(18), 5083.

12. Wang, Z., Wang, Y., Zhang, J., Hu, C., Yin, Z., & Song, Y. (2022). Spatial-temporal feature fusion neural network for EEG-based emotion recognition. *IEEE Transactions on Instrumentation and Measurement*, 71, 1–12.

13. Li, C., Wang, B., Zhang, S., Liu, Y., Song, R., Cheng, J., & Chen, X. (2022). Emotion recognition from EEG based on multi-task learning with capsule network and attention mechanism. *Computers in Biology and Medicine*, 143, 105303.

14. Dang, W. D., Lv, D. M., Li, R. M., Rui, L. G., Yang, Z. Y., Ma, C., & Gao, Z. K. (2022). Multilayer network-based CNN model for emotion recognition. *International Journal of Bifurcation and Chaos*, 32(01), 2250011.

15. Xing, X., Li, Z., Xu, T., Shu, L., Hu, B., & Xu, X. (2019). SAE+ LSTM: A new framework for emotion recognition from multi-channel EEG. *Frontiers in Neurorobotics*, 13, 37.

16. Gao, Y., Sun, X., Meng, M., & Zhang, Y. (2022). EEG emotion recognition based on enhanced SPD matrix and manifold dimensionality reduction. *Computers in Biology and Medicine*, 146, 105606.

17. Janapati, R., Dalal, V., Sengupta, R., Desai, U., Raja Shekar, P. V., &Kollem, S. (2022). Towards a more theory-driven BCI using source reconstructed dynamics of EEG time-series. *Nano LIFE*, 12(02), 2250005.

18. Lotfian, R., & Busso, C. (2017). Building naturalistic emotionally balanced speech corpus by retrieving emotional speech from existing podcast recordings. *IEEE Transactions on Affective Computing*, 10(4), 471–483.

19. Kaggle, https://www.kaggle.com/birdy654/eeg-brainwave-dataset-feeling-emotions, last accessed 2021/06/20.

20. Github repository, https://github.com/NeuroTechX/eeg-101/tree/master/EEG101, last accessed 2021/06/20.

21. Bird, J. J., Manso, L. J., Ribeiro, E. P., Ekárt, A., &Faria, D. R. (2018, September). A study on mental state classification using EEG-based brain-machine interface. In 2018 International Conference on Intelligent Systems (IS). IEEE, pp. 795–800.

22. Singh, B., Kushwaha, N., & Vyas, O. P. (2014). A feature subset selection technique for high dimensional data using symmetric uncertainty. *Journal of Data Analysis and Information Processing*, 2(04), 95.

23. Kumar, R. P., Vandana, S. S., Tejaswi, D., Charan, K., Janapati, R., & Desai, U. (2022, July). Classification of SSVEP signals using neural networks for BCI applications. In 2022 International Conference on Intelligent Controller and Computing for Smart Power (ICICCSP). IEEE, pp. 1–6.

24. Piao, M., Piao, Y., & Lee, J. Y. (2019). Symmetrical uncertainty-based feature subset generation and ensemble learning for electricity customer classification. *Symmetry*, 11(4), 498.

25. Choi, W., Kim, M. J., Yum, M. S., &Jeong, D. H. (2022). Deep convolutional gated recurrent unit combined with attention mechanism to classify pre-ictal from interictal EEG with minimized number of channels. *Journal of Personalized Medicine*, 12(5), 763.

26. Janapati, R., Dalal, V., Sengupta, R., & PV, R. S. (2021). Progression of EEG-BCI classification techniques: A study. *Inventive Systems and Control: Proceedings of ICISC 2021*, 161–170.

27. Duan, L., Hou, J., Qiao, Y., Miao, J. (2019). Epileptic seizure prediction based on convolutional recurrent neural network with multi-timescale. *Intelligence Science and Big Data Engineering. Big Data and Machine Learning*. Springer, Berlin, pp. 139–150.

28. Zhang, S., Chen, D., Ranjan, R., Ke, H., Tang, Y., & Zomaya, A. Y. (2021). A lightweight solution to epileptic seizure prediction based on EEG synchronization measurement. *The Journal of Supercomputing*, 77(4), 3914–3932.

29. Usman, S. M., Khalid, S., & Bashir, Z. (2021). Epileptic seizure prediction using scalp electroencephalogram signals. *Biocybernetics and Biomedical Engineering*, 41(1), 211–220.

30. Aslam, M. H., Usman, S. M., Khalid, S., Anwar, A., Alroobaea, R., Hussain, S., ... & Yasin, A. (2022). Classification of EEG signals for prediction of epileptic seizures. *Applied Sciences*, 12(14), 7251.

31. Xu, Y., Su, H., Ma, G., & Liu, X. (2022). A novel dual-modal emotion recognition algorithm with fusing hybrid features of audio signal and speech context. *Complex & Intelligent Systems*, 9(1), 1–13.

6 Epileptic Seizure Classification Using Machine Learning Algorithms with the Techniques of PCA Feature Reduction

Dr. B. Shadaksharappa and P. Ramkumar

CONTENTS

6.1 INTRODUCTION

A seizure, often called an epileptic seizure, is a fleeting neurological condition of the brain that can be brought on by a sudden increase in the activity level of nerve cells in the brain [1, 2]. It is a prevalent neurological disorder that can strike anybody at any point in their life. This sickness affects one person in every one hundred across the globe [3].

There are a variety of factors that can lead to epilepsy, such as vascular abnormalities, brain infections, brain tumors, dietary deficiencies, pyridoxine deficiency, and abnormalities of calcium metabolism. Research is required to fully comprehend

DOI: 10.1201/9781003326830-6

the factors that contribute to the development of epileptic diseases in order to arrive at a correct diagnosis of epilepsy.

The electroencephalogram (EEG) offers a series of physiological and pathological facts that are effective in the treatment of epileptic cases. For example, the EEG can be used to determine the epileptogenic zone for presurgical evaluations [5]. At the moment, the most important part of an EEG diagnosis is for neurologists to physically examine EEG recordings. The visual scoring of long-term EEG is a tedious process that takes a lot of time. As a consequence of this, the automated identification system is beneficial to neurologists while they are performing an analysis of EEG recordings or information.

In the past 20 years, significant advancements in the realm of artificial intelligence (AI) have been made specifically in the subfield known as machine learning (ML). ML makes use of principles from mathematics and computer science in addition to algorithms in order to uncover the underlying characteristics of data and the natural connections between them. In the domain of disease diagnostics, it is currently utilized in a significant way. These days, approaches from the field of ML are applied in the process of forecasting or diagnosing a variety of potentially fatal diseases, including seizures, tumor, insulin, heart problems, hepatitis, and so on. A person's life can be saved by the diagnosis and treatment of disorders like epilepsy at an early stage.

However, making an accurate diagnosis of probable seizures in advance can be a difficult task. Unanticipated convulsions account for the vast majority of epileptic episodes, and it has been difficult for researchers to identify methods that may reliably diagnose seizures in patients before they take place. The method described in this chapter will be of use in determining whether or not an individual is having a seizure at the given time.

The primary objective of this research is to identify the most effective classification method for neurological problems by making use of the principal components analysis (PCA) technique for the purpose of reducing the number of features present in the dataset. In this chapter, the proposed system applied the random forest (RF), artificial neural network (ANN), K-nearest neighbors (KNN), support vector machine (SVM), and decision tree (DT) algorithms to the dataset in order to envisage epilepsy. The effectiveness of the classifier is analyzed both with and without the use of the PCA method.

Diagnostic methods such as ultrasound, positron emission tomography (PET), magnetic resonance imaging (MRI), computed tomography (CT), and EEG are also accessible. However, diagnostic procedures are costly and cannot be utilized for long-term monitoring. On the other hand, EEG is a test that does not cost very much but can be utilized for long-term exposure. As a consequence of this, EEG is the most reliable technique for making a diagnosis of epilepsy [4].

6.2   RELATED STUDY

The indention of this part is dedicated to a number of researchers who have written about topics associated with epilepsy and have experimented with different types of

ML in order to forecast epileptic episodes. A variety of recent research on the identification of epileptic seizures utilizing EEG data is discussed in this chapter.

SVM classifier was used by Nandy et al. for the classification of epileptic seizures, and they applied a Bayesian optimization approach for the optimization of the hyperparameters of the SVM. In addition to this, they compared the results of the SVM classifier with certain classifiers in their study [4]. The results showed that the SVM classifier had an accuracy of 97.05%.

The discrete wavelet transform (DWT) approach was utilized by Hamad et al. in order to extract features, and then these features were utilized in order to train the SVM with the radial basis function (RBF) kernel function. To select the key attribute selection and the proper SVM parameters [6] an another optimizer of gray wolf in order to accomplish an efficient EEG classification. This was done so that the results could be used.

A DWT was used in 2016 by Sharmila and Geethanjali primarily in order to divide EEG data into different subbands and then obtain statistical features from those subbands. DWT-derived statistical features are utilized during the training process of the classifier. After that, the signals are put through two different classifiers in order to determine whether or not they are epileptic. In this particular research endeavor, two different kinds of classifiers, namely KNN and Naive Bayes, are utilized. In this study, the researchers examine the effectiveness of 14 distinct combinations of methods. The findings of the studies demonstrated that the classifier of Naïve bayes gets the maximum accuracy with less processing time required for the majority of dataset combinations [7]. This was found to be the case while diagnosing epileptic episodes.

6.3   MATERIAL AND METHODS

Through the utilization of the PCA feature reduction strategy in the dataset, the primary objective of this research is to identify the most effective classification algorithm for epilepsy.

The phases are going to be covered in the subsequent sections. The entire flow diagram of the developed method is illustrated in Figure 6.1.

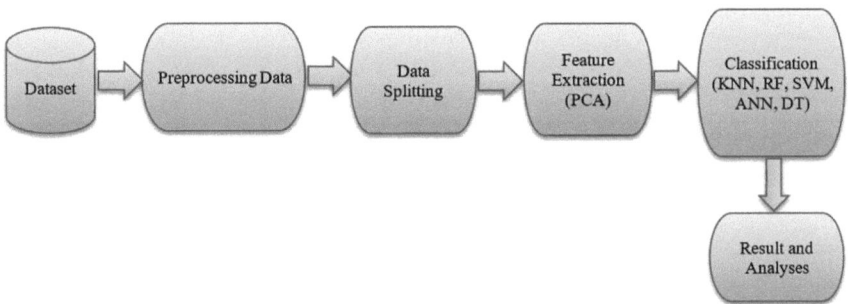

FIGURE 6.1   Proposed system.

6.3.1 DATASET DESCRIPTION

The dataset of this syndrome used in this work hasbeen collected from the website for the UCI Machine Repository [8]. This dataset has five classes, numbered 1 through 5, each consisting of 100 signals and having a duration of certain seconds. The various classes of data were obtained from five healthy human beings, two of whom had their eyes closed while the other two had their eyes open. Five epileptic patients provided data for the remaining three classes (3, 2, and 1) of the classification system. When there was no epileptic seizure present, both class 3 and class 2 seizures were recorded. In addition, the class 1 was recorded while the patient was having seizures.

The dataset contains 11,500 different samples, each with its own unique collection of 178 characteristics, and these characteristics are evenly dispersed across the samples. There has never been a single instance of epileptic seizure in any of the patients who fall into classifications 2, 3, 4, or 5. Seizures caused by epilepsy were only reported in class 1 patients [8]. As a consequence of this, this study will be a categorical structure for both instances of epileptic seizure and cases of nonepileptic seizure, with classes 2, 3, 4, and 5. Table 6.1 illustrates the number of instances for each of the groups that were used, and we can see that there is an equal amount of samples for each of the classes. The epilepsy episode dataset is displayed in an example view in Figure 6.2, which may be seen here.

TABLE 6.1
Comparison of Different Classifiers Using the PCA Feature Reduction Techniques

Techniques of classification	Accurateness	Precision	Recall	F1score	Train time(S)	Test time (S)
KNN	95%	95%	95%	95%	0.011	0.31
RF	**97%**	**97%**	**97%**	**97%**	0.92	0.05
SVM	89%	90%	89%	89%	2.21	0.42
ANN	87%	87%	87%	87%	5.60	0.003
DT	92%	92%	92%	92%	**0.052**	**0.0008**

	Unnamed: 0	X1	X2	X3	X4	X5	X6	X7	X8	X9	...	X170	X171	X172	X173	X174	X175	X176	X177	X178	y
11495	X22.V1.114	-22	-22	-23	-26	-36	-42	-45	-42	-45	...	15	16	12	5	-1	-18	-37	-47	-48	2
11496	X19.V1.354	-47	-11	28	77	141	211	246	240	193	...	-65	-33	-7	14	27	48	77	117	170	1
11497	X8.V1.28	14	6	-13	-16	10	26	27	-9	4	...	-65	-48	-61	-62	-67	-30	-2	-1	-8	5
11498	X10.V1.932	-40	-25	-9	-12	-2	12	7	19	22	...	121	135	148	143	116	86	68	59	55	3
11499	X16.V1.210	29	41	57	72	74	62	54	43	31	...	-59	-25	-4	2	5	4	-2	2	20	4

FIGURE 6.2 Sample view of epileptic seizure data.

6.3.2   DATA PREPROCESSING

When applying ML algorithms to a dataset, it is vital to apply this technique so that the resulting predictions are dependable, accurate, and successful [9]. Data preprocessing is a practice that entails transforming unprocessed and uncooked data into a format that is appropriate for the process of classification.

The data that comes from the actual world is often incorrect, deficient in specific behaviors or patterns, insufficient, and/or unreliable, and it also commonly contains a variety of inaccuracies. The data should be preprocessed before attempting to answer problems of this nature. The first processing of raw data prepares it for the future processing steps.

There are no unaccounted for data points in our dataset (NAN). In the bitwise group, there is an issue with an uneven class distribution, as seen in Table 6.1; in order to prevent this from happening, we make use of undersampling tactics.

Undersampling is a word that is in reference to a string of events of tactics for rebalancing the classification performance in a classification algorithm that has a skewed class dispersion. This term is used in the context of a classification dataset.

6.3.3   FEATURE EXTRACTION

During the phase known as "feature extraction," the goal is to reduce the total number of characteristics by deriving new ones from the preexisting ones in the dataset. This lowest expected feature set ought to be able to summarize the vast most of the data and characteristics contained inside the initial dataset. When the first set is combined with the subsequent sets, a condensed form of the primary characteristics can be produced [10].

In this dataset, there are 178 features, and the training process would take a very long time if it uses all of them. Because of this, PCA is employed in our research to identify and limit the number of features in our dataset (PCA).

6.3.4   PRINCIPAL COMPONENTS ANALYSIS

An approach to information extraction known as PCA involves mapping data obtained from a strong field onto a subspace with a lower order of dimension. It tries to save the critical pieces of the data that have the largest variation and eliminate the nonessential sections that have the lowest variation [11].

By reducing the number of dimensions in the original data, which is our goal when using PCA, it is hoped to find a set of input features that can best describe the data distribution that was first collected. PCA has the potential to accomplish this by increasing the variances and minimizing the prediction error by monitoring the fragmented distances. The majority of our information is projection onto a group of orthogonal axes, and the PCA assigns a weighting of importance to each of the axes.

When PCA is performed, the vast bulk of the variation that exists in the data is condensed into the first few parts. As a direct result of this, we only keep the parts of the system that differ significantly from one another, ignoring the rest.

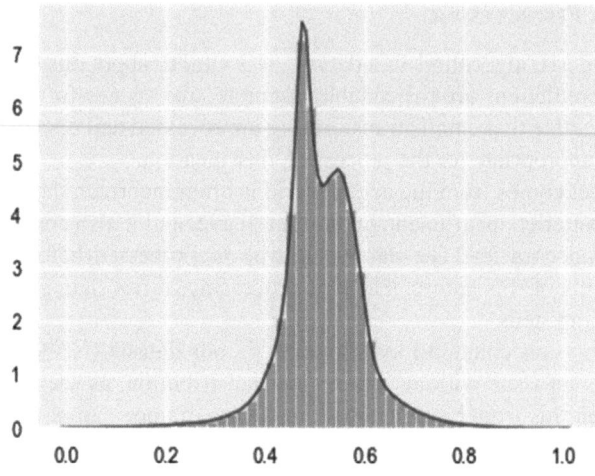

FIGURE 6.3   The distribution curve of training data without PCA.

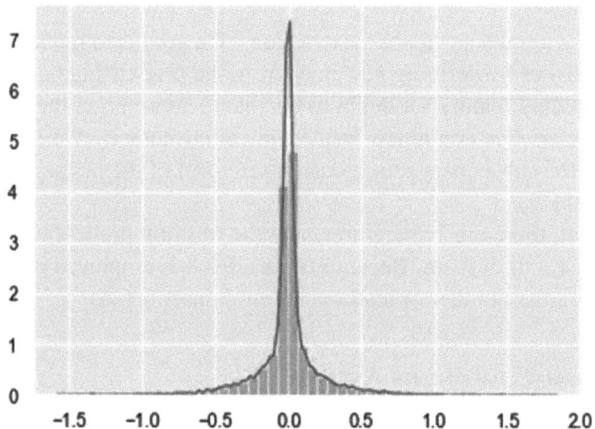

FIGURE 6.4   With PCA of the distributed training data.

After computing the n-dimensional mean vector m, an R matrix is next computed and assembled in declining order of its eigenvalues. Following the sorting process, the eigenvalues with the greatest magnitude are selected. It should be noted that noise is the additional dimension that exists. After performing some preprocessing, the data can be obtained by applying the equation below, provided that an n-by-n matrix A with n eigenvectors in each column has been created.

$$X' = At(x - \mu)$$
(6.1)

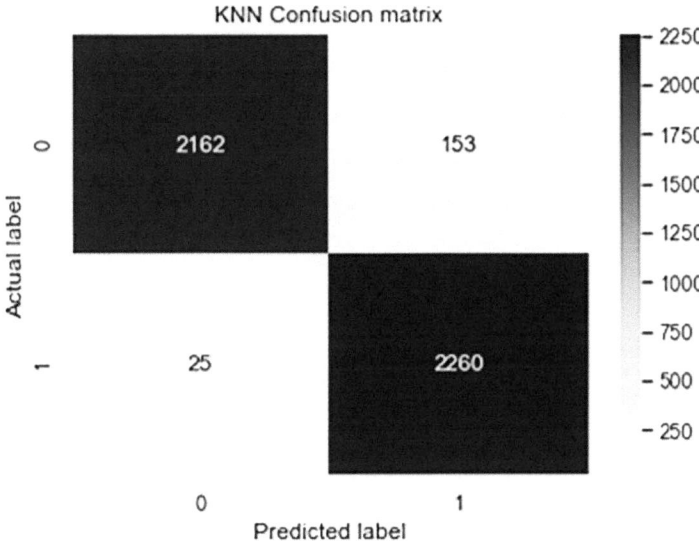

FIGURE 6.5 KNN algorithm confusion matrix with reduction of PCA features.

6.3.4.1 Classification Algorithms

The following are some of the classification and estimation algorithms as well as real ML methods that are applied all through this chapter:

6.3.4.1.1 K-Nearest Neighbor

In the field of classification research, one of the supervised learning strategies that is used extensively is called KNN. The KNN algorithm gets its name not from the number of neighbors, which is represented by the letter k, but from the requirement that the data to be classified must be located in close proximity to those neighbors. The problems associated with classification and regression are the focus of the KNN classification method, the goal of which is to provide a solution to these problems [12]. One of the algorithms that needs a considerable amount of time to be trained is the KNN method.

The KNN learning method uses computation to determine the sample distance as its foundation. When the algorithm comes across a new test dataset, the distance is computed for each sample that is included in the new data. After this calculation has been finished, the name of the class are grouped by locating k closest peers from priori knowledge data samples, comparing those samples with instances enclosed within the datasets in the new example, and evaluating the extent to which they are similar [13].

The following figure depicts the performance medium for the KNN algorithm when the value k=1 is chosen, and the PCA feature reduction method is applied in order to make a prediction regarding epilepsy. We can see in Figure 6.6 that the optimal value is one for k because it demonstrates the maximum accuracy both during preparation and validation.

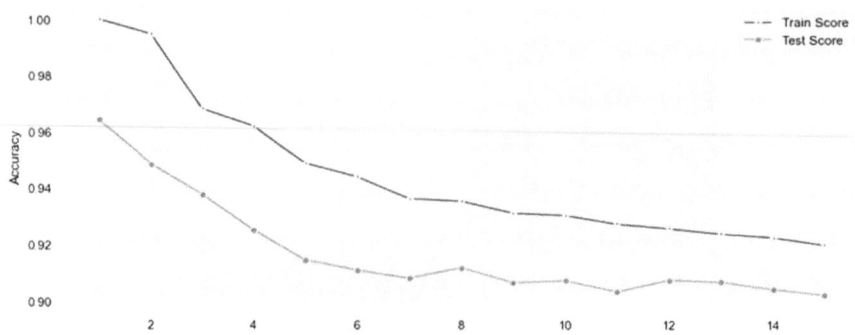

FIGURE 6.6   KNN algorithm with various k values of score value of test and train data. KNN algorithm with various values of score value for test and train data.

6.3.4.1.2   Random Forest

The classifier of RF, which was developed in the latest years, offers an advantage over the Velocity [14] and Classifier [15] techniques, which are recognized as efficient methodologies in group of learning, in the aspects of both rapidity and high precision. This advantage can be seen when comparing the RF classifier to the Velocity [14] and Classifier [15] methods. In comparison to other learning methods, the RF classifier's training phase is completed far more quickly, especially when compared to the Acceleration method. Because of its effectiveness and precision, it is an extremely helpful classifier [16].

The RF algorithm is a simple ML technique that frequently produces impressive outcomes even when its meta parameters are left same. Due to the ease with which it can be implemented and deployed, this ML method is among the most widely used of those that fall under the categories of "Classification" and "Regression." Using this approach, a forest would be generated at random. The constructed "forest" is actually a band known as "Decision Trees." Undoubtedly, this method is capable of dealing with very large datasets.

Figure 6.7 depicts about the metrics of RF values. Leo Bremen was the one who initially conceptualized RF. It then constructs a DT for each of the samples that it has randomly selected from the dataset after doing so. Each DT is given a score based on its ability to make predictions. After that, the model that receives the maximum votes is evaluated for the final forecast [17].

6.3.4.1.3   SVM

The SVM was initially introduced to the public by Vapnik and Chervonenkis. The SVM looks for a perfect hyperplane that can separate the instances of any given class. If there is a way to split the groups linearly, then maximum margin hyperplanes could be used to determine who belongs to which group. On the other hand, if the data cannot be divided linearly, it could be moved to a larger space in order to be divided linearly (i.e., feature space). The name given to this type of conversion

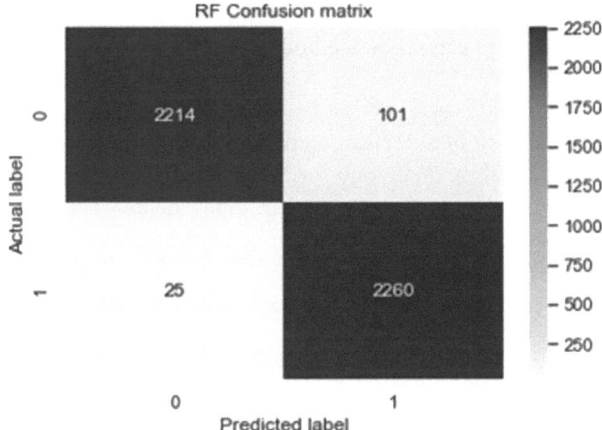

FIGURE 6.7 PCA feature reduction technique of confusion matrix for random forests algorithm.

is the kernel function. This classifier identifies the hyperplane that separates the points to place the greatest number of points that belong to the same class on the same side while simultaneously expanding the range of every group to such a hyperplane. The locations that are closest together along the hyperplane make up the support vectors. The distance between a class and a hyperplane is measured as the distance that is the smallest among all of them and the points that fall within that group [12].

In addition, the hyperplane can be used for the purposes of grouping or regression. The SVM classifies instances into the appropriate categories and can also identify the components of a sample that are not supported by the data. The hyperplane that carries out the division to the closest training place for any group brings the detachment to a successful conclusion.

In Figure 6.8, the metrics of the SVM algorithm that use the PCA feature extraction method to predict seizures are illustrated.

6.3.4.1.4 Artificial Neural Network

The human brain has served as inspiration for the development of ANNs, which are computational structures. The ANN is constructed out of a large number of linked unit operations that work together to process data. As a consequence of this, they frequently produce results that are to their advantage. In fact, the ANN is composed of subnets and network tasks, with the network layers consisting of the input layer, the hidden layer, and the outputlayer, respectively. The input neurons of the data mining model are responsible for determining the values of all the input attributes [18].

Utilizing the ANNs to perform tasks such as recognition of voice, recognition of image, and automatic system, the domain of AI has made a significant number of new advancements in recent years.

In Figure 6.9. In this example, the performance medium for the ANN algorithm that uses the PCA feature extraction method to predict the presence of epilepsy

6.3.4.1.5 Decision Tree

The use of tree modeling in a decision-making tool is denoted by the abbreviation "DT." DT is an essential methodology for both classification and regression analysis. The logistic regression is a useful tool in many fields, including research and operations. A DT is utilized whenever there is a categorical value associated with the result attribute.

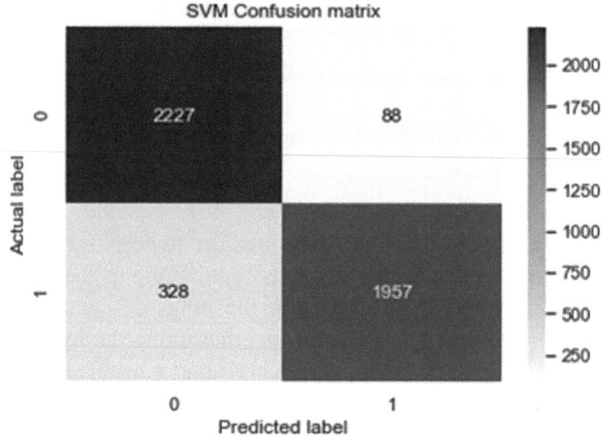

FIGURE 6.8 PCA feature reduction technique of confusion matrix for support vector machine.

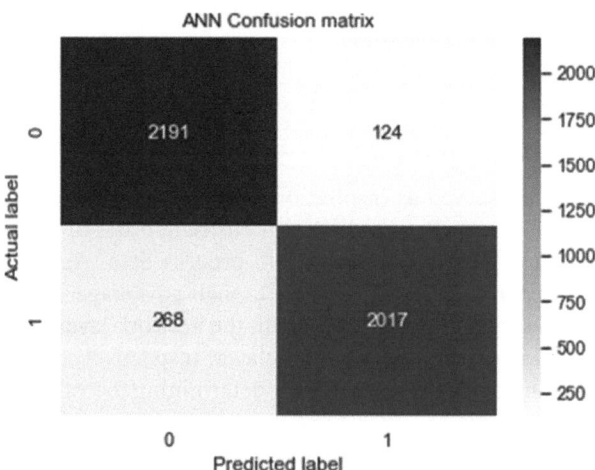

FIGURE 6.9 Performance of ANN algorithm with extracting features.

The DT graph can be broken down into its constituent parts: the root node, branches, and leaves. The categorization process occurs on the child node, whereas the results of every occurrence are recorded on the branches. When developing rules for categorization, it is important to take into consideration the pathways that lead us from parent node to child node [19].

The DT algorithm offers features that are both numeric and numerical in nature. It is possible for it to tolerate noise as well as values that are unreliable. This approach

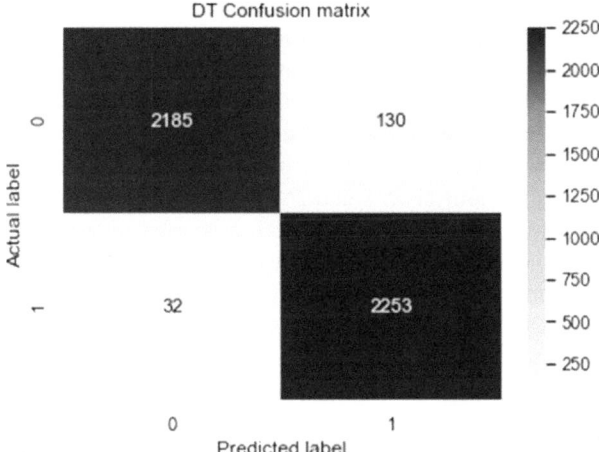

FIGURE 6.10   Confusion matrix of decision tree using PCA feature extraction.

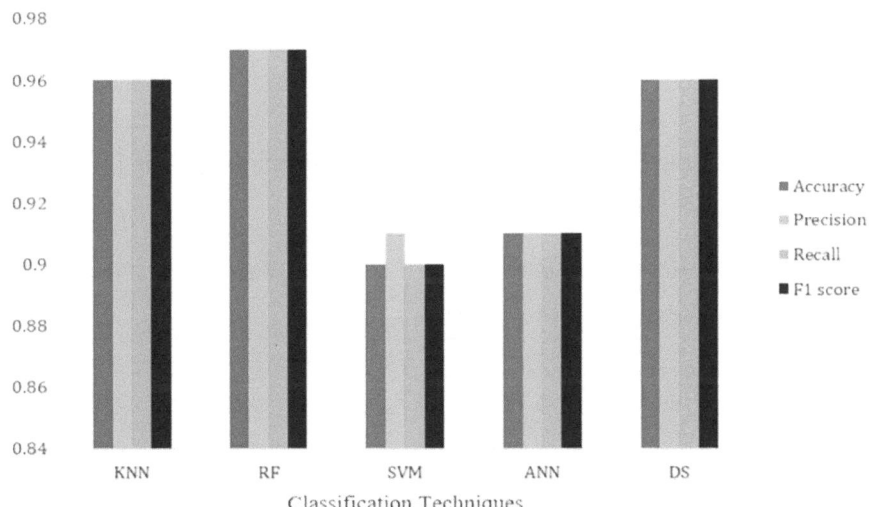

FIGURE 6.11   Comparative analysis of different classifiers using the PCA feature reduction techniques.

categorizes the data from the bottom-up. Each node in the graph represents a differ-ent possible value for the test attribute of the corresponding instance, and together, these possible values make up the set of possible scope for that point.

A DT can easily transform a given collection of examples with meaningful pat-terns, working its way down from the root node all the way to the node representing the attack class.

6.4 RESULTS AND DISCUSSIONS

In this section, it examines the results obtained from classifying the epilepsy dataset using multiple different classifiers as well as the routine of the categorization when using PCA feature reduction and when not using PCA feature reduction.

Table 6.1 and Diagram 11 present an evaluation of the various categorization methods by employing PCA characteristic diminution, dividing the dataset into training and testing portions of 75% and 25%, respectively, and then displaying the results of this analysis. According to Table 6.1, it is possible to perceive categoriza-tion algorithm. The desirable result is displayed by a model that has a reliability of 97% and a low computational time (including training and test time) cost.

The comparative analysis of various classification methods using the PCA attri-bute lessening method and set 75% of the dataset is reserved for the testing phase. In this instance, we find that the RF classifier demonstrates the best result, with an accuracy of 97%.

As per Table 6.2, it exhibited the classifier of RF achieves the top results when utilizing the PCA feature reduction technique, as it achieves a score of 97%. At the same time, without utilizing the PCA feature reduction technique, the KNN, or RF, it demonstrates the most accurate results possible with 99% precision.

TABLE 6.2
Calculation of Time Evaluation of Classification Techniques Using PCA and without Using the PCA

Classification techniques	Accuracy	Precision	Recall	F1score	Train time(S)	Test time(S)
KNN	99%	99%	99%	99%	2.00	14.86
RF	99%	99%	99%	99%	5.32	0.061
SVM	96%	96%	96%	96%	11.13	1.132
ANN	91%	91%	91%	91%	20.10	0.003
DT	94%	94%	94%	94%	2.12	0.007

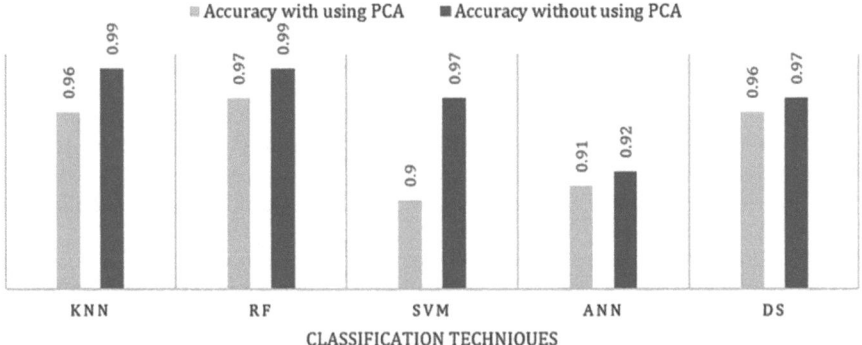

FIGURE 6.12 Comparative analysis of accuracy in classification techniques using the PCA and without using the PCA.

6.5 CONCLUSION

Seizures caused by epilepsy are currently one of the leading causes of morbidity and mortality around the world. Because epileptic seizures are becoming more common around the world and have a significant impact on people's lives, receiving an accurate diagnosis in a timely manner is more important than it has ever been.

By utilizing the PCA feature reduction strategy on the dataset, the primary objective for this study is to identify the most effective classification algorithm for epileptic seizures.

In this chapter, this proposed system applied KNN, RF, SVM, ANN, and DT algorithms by using the PCA feature reduction technique in the dataset to predict epilepsy. The performance of classifiers is analyzed with and without using the PCA technique.

It has been demonstrated that the RF classifier with an accuracy of 97% and with low computational times (training time and test time) shows the best result by utilizing the PCA feature reduction in the dataset. In addition, the KNN and RF classifiers with a 99% accuracy without using PCA feature reduction in the dataset show the best result.

It is important to emphasize the fact that the PCA feature reduction technique shortens the amount of time required for computation. The accuracy of the classifiers suffers as a direct consequence of this development as well. After applying the RF classifier, the KNN classifier has shown good results both when using PCA and when not using PCA. With 96% accuracy when using PCA and 99% accuracy when not using PCA, the KNN classifier has shown that it is possible to achieve good results.

REFERENCES

1. Fisher, R., Acevedo, C., Arzimanoglou, A., Bogacz, A., Cross, J., Elger, C., et al. (2014). ILAE official report: A practical clinical definition of epilepsy, *Epilepsia*, 55, 4, 475–482.
2. Ramgopal, S., Thome-Souza, S., Jackson, M., Kadish, N. E., Sánchez Fernández, I., Klehm, J., Bosl, W., Reinsberger, C., Schachter, S., &Loddenkemper, T. (2014). Seizure detection, seizure prediction, and closed- loop warning systems in epilepsy. *Epilepsy & Behavior: E&B*, 37, 291–307.
3. Lehnertz, K., Mormann, F., Kreuz, T., Andrzejak, R. G., Rieke, C., David, P., &Elger, C. E. (2003). Seizure prediction by nonlinear EEG analysis. *IEEE Engineering in Medicine and Biology Magazine: The Quarterly Magazine of the Engineering in Medicine & Biology Society*, 22, 1, 57–63.
4. Nandy, A., Alahe, M. A., Nasim Uddin, S. M., Alam, S., Nahid, A. A., &Awal, M. A. (2019). Feature extraction and classification of EEG signals for seizure detection. 2019 International Conference on Robotics, Electrical and Signal Processing Techniques (ICREST).
5. Almustafa, K. M. (2020). Classification of epileptic seizure dataset using different machine learning algorithms. *Informatics in Medicine Unlocked*, 21, 100444.
6. Hamad, A., Houssein, E. H., Hassanien, A. E., &Fahmy, A. A. (2017). A hybrid EEG signals classification approach based on grey wolf optimizer enhanced SVMs for epileptic detection. Proceedings of the International Conference on Advanced Intelligent Systems and Informatics, 2, 108–117.
7. Sharmila, A., &Geethanjali, P. (2016). DWT based detection of epileptic seizure from EEG signals using naive Bayes and k-NN classifiers. *IEEE Access*, 4, 7716–7727.
8. Andrzejak, R. G., Lehnertz, K., Mormann, F., Rieke, C., David, P., &Elger, C. E. (2001). Indications of nonlinear deterministic and finite-dimensional structures in time series of brain electrical activity: Dependence on recording region and brain state. *Physical Review E*, 64, 6.
9. Soni, M., &Varma, S. (2020). Diabetes prediction using machine learning techniques. *International Journal of Engineering Research & Technology (IJERT)*, 9, 142–151.
10. Ippolito, P. P. (2019). Feature extraction techniques: Towards data science. Retrieved, from   https://towardsdatascience.com/feature-extraction-techniques-d619b56e31be (Date of access: December 27, 2020).
11. Qiu, J., Wang, H., Lu, J., Zhang, B., &Du, K. L. (2012). Neural network implementations for PCA and its extensions. *ISRN Artificial Intelligence*, 2012, 1–19.
12. Rodrigues, J. D. C., Filho, P. P. R., Peixoto, E. N. A. K., &de Albuquerque, V. H. C. (2019). Classification of EEG signals to detect alcoholism using machine learning techniques. *Pattern Recognition Letters*, 125, 140–149.
13. Mitchell, T. M. (1997). Does machine learning really work?*AI Magazine*, 18, 3, 11.
14. Freund, Y., &Schapire, R. E.1996. Experiments with a new boosting algorithm. In: Machine Learning. Proceedings of the Thirteenth International Conference, 148–156.
15. Breiman, L. (1996). Bagging predictors. *Machine Learning*, 24, 2, 123–140.
16. Gislason, P. O., Benediktsson, J. A., &Sveinsson, J. R. (2006). Random forests for land cover classification. *Pattern Recognition Letters*, 27, 4, 294–300.

17. Fawagreh, K., Gaber, M. M., &Elyan, E. (2014). Random forests: from early developments to recent advancements. *Systems Science & Control Engineering*, 2, 1, 602–609.
18. Walczak, S., &Cerpa, N. (2003). Artificial neural networks. *Encyclopedia of Physical Science and Technology*, 3, 631–645.
19. Geetha, A., &Nasira, G. M. (2014). Data mining for meteorological applications: Decision trees for modeling rainfall prediction. 2014 IEEE International Conference on Computational Intelligence and Computing Research.

7 Nanotechnology Advancements in HMI

Vivek P. Chavda*, Mital Patel, Ritu Patel,
Kunjal Thakkar, Disha Patel, Pankti C. Balar,
Piyush Patel, and Usha Desai

CONTENTS

DOI: 10.1201/9781003326830-7

7.1   INTRODUCTION

Human–machine interaction (HMI) and human–machine interface (HMI) means how humans interact with any machines and devices, where machine can be described as "any mechanical or electrical device that transmits or modifies energy to perform or assist in the performance of human tasks." There are several ranges of HMI devices such as computer gadgets, kitchen appliances, medical machines, entertainment devices, manufacturing machinery, etc. At present, many new HMI devices appear in the market, like a whole range of various joysticks, mobile devices, graphics tablets, single-handed keyboards, and touch displays (Olmo and Domingo 2020; Wang, Lao, and Zhang 2017). Initially, the purpose of the HMI is to exclude manual interference from the manufacturing process, and in place of them with the help of artificial intelligence (AI), autonomous machine and gadgets are used as solutions (Saniuk, Grabowska, and Straka 2022). Industry 4.0, Operator 4.0, Web 4.0, and Machine tools 4.0 concepts have a significant effect on HMI usage (Bissoli et al. 2019). Today, practically, there is a machine for everything, from a small machine like a coffee maker to big vehicles, airplane engines, etc. (Adir et al. 2020). Devices that consist of HMI include energy supply devices, logic circuits, sensors, and data storage systems that need to be stretchable, flexible, imperceptible, self-healable, and biocompatible medical devices (Hong Wang, Ma, and Hao 2017). Stretchable, transparent sensor for sensing human skin also contains HMI and is made up of carbon nanotubes (Roh et al. 2015). Since self-powered sensors generate electrical impulses directly as opposed to older resistive or capacitive sensors, their use in HMI systems can significantly reduce energy consumption (Kwon et al. 2020). Even in case of supervisory control and data acquisition (SCADA) system HMI is added as a local supervisor, which is integrated with other devices also and forms a distributed architecture (Mendoza et al. 2021). Recently, the HMI has received more attention as it would provide an artificial media for humans to interact with the environments to promote people's lifestyles, for example, smart gloves made up of polydimethylsiloxane-carbon black (PDMS-CB) strain sensor is used to detect gesture of finger. Flexible and stretchable electronics such as wearable HMI are used to control the motion of the robot finger (Esposito et al. 2021).

7.2 EVOLUTION OF HMI

The human evolution started a few million years ago as chimps, eventually evolved into primates and then homo sapiens and eventually homo sapiens sapiens and finally they move to the computer era. A revolutionary HMI-based perception on a typewriter was created in the middle of the 20th century, which led to the creation of the pioneering keypads. A visual user interface is not utilized on earlier computers, which are mostly text based. The necessity for a mouse didn't arise until computers started to employ more graphics. The 1980s saw the beginning of the computer industry's "letter boom," during which a wide variety of new HMI devices—including graphic tablets, ATMs, and other devices—began to develop. Nowadays, everyone uses HMI devices in the industry for controlling automated processes. The evolution of HMI is shown in Figure 7.1.

Applying brain–computer interface technology to the HMI is a new concept in diagnosis. For decades, to get information regarding the ongoing activity of the brain electroencephalogram (EEG) was used, but the use of EEG has several drawbacks such as limited dimensional resolution and susceptibility to artifact sources—which are the main factors that limit current research in EEG. Other than EEG, one can

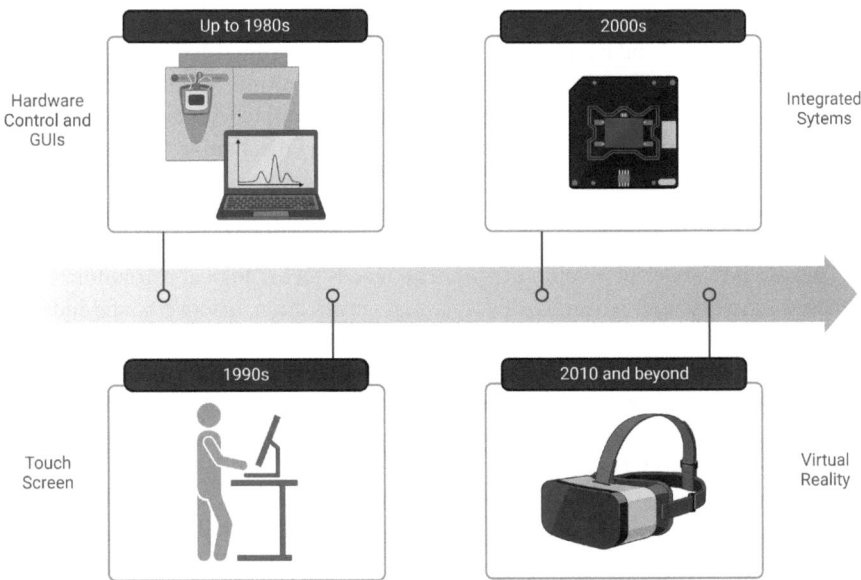

FIGURE 7.1 Development of technology over time period. (Created with Biorender.Com)

use functional magnetic resonance imaging (fMRI), functional near-infrared spectrography (fNIRS), and magnetoencephalography (MEG), but these measures are expensive and have the limitation of a little resolution because they concentrate on the blood-oxygen-level-dependent (BOLD) component (Malerba 1999a; Janapati et al. 2022). Therefore, real-time brain signal decoding (RBSD) is a preferred method. RBSD is the preferred method to attain information of dynamic state of mind (Janapati et al. 2022). Application provided by RBSD keeps record on user action and state over time, encompassing intents, circumstance perceptions, and feelings. It is termed as cognitive monitoring as it encodes human perception (Zander and Kothe 2011). Recent sensory substitution technology presents new opportunities for the development of systems that compensate the sensory loss (Chavda et al. 2021; Chavda 2019). In sensory substitution technology (e.g., artificial sight and vestibular function), artificial receptors collect the information and it reaches the brain via HMI; now brain port vision device containing HMI has an advanced feature that the blind person can visualize things as the normal person and identify the things and can do the tracking as well (Bach-y-Rita and W Kercel 2003).

7.3 INDUSTRY 4.0/INDUSTRY 5.0 AND HMI

The most noticeable aspect of global growth now is the industrialization. Earlier in the 18th century, the expansion picked up. Industries must deal with problems including supply chain management's demand forecasting and ambiguous economic situations (Vinitha et al. 2020). The uprising of the industry was distinguished by radical changes that put extensive pressure on companies. Each level of industrial history suggested several challenges, which had to be understood, examined, and lastly get better off (Zambon et al. 2019).

Global growth of industry started in the 18th century. First industrial revolution time span was up to 1840 from 1760. The first industrial revolution used water and steam power to mechanize production. Because mechanized versions realize eight times more production than conventional methods, the mechanized version of spinning wheels achieved eight times the volume. For the industrial revolution's growth, steam power and water power played a very important role. It was useful to make railways, roads, waterways, canals, and other developments (Vinitha et al. 2020).

The second industrial revolution is also known as technological revolution. It was a phase of rapid scientific standardization, mass production, discovery, and industrialization from the late 19th century to the early 20th century (Swathi et al. 2022). It was a period when electricity, petroleum, and steel production caused a series of innovations that changed society. The light bulb, the telephone, the automobile, the airplane, and synthetic dyes were the most important inventions of the second industrial revolution (Vinitha et al. 2020; Chavda et al. 2021).

The third industrial revolution designed through partial automation using memory programmable controls and computers. Humans are able to automate an entire production process without human assistance. The third industrial revolution begin in 1970. It produced astonishing upliftment in the area of engineering. Nano, bio/IT technologies, artificial intelligence, robotics, and 3D printing are the most important driver of the third industrial revolution. Automation is useful to decrease the dependence and accelerate productivity of the industries (Vinitha et al. 2020).

The fourth industrial revolution is the way of describing the boundaries between the digital, physical, and biological worlds. Industry 4.0 initiated in the 21st century. Industry 4.0 involves a fusion of advances in AI, the Internet of Things (IoT), 3D printing, genetic engineering, and quantum computing (Vinitha et al. 2020, 1). Related importance of industry 1.0 to 4.0 is described in Figure 7.2.

The main purpose of industry 4.0 is AI, where machines, thus supporting the technocentric paradigm, have visibly driven decision-making. It has its prominence in interconnecting the existing technology through the IoT (Houssein, Hammad, and Ali 2022), and it gives access to real-time data. Business owners understand every aspect of their operations and to better controls, as well as make full use of instant

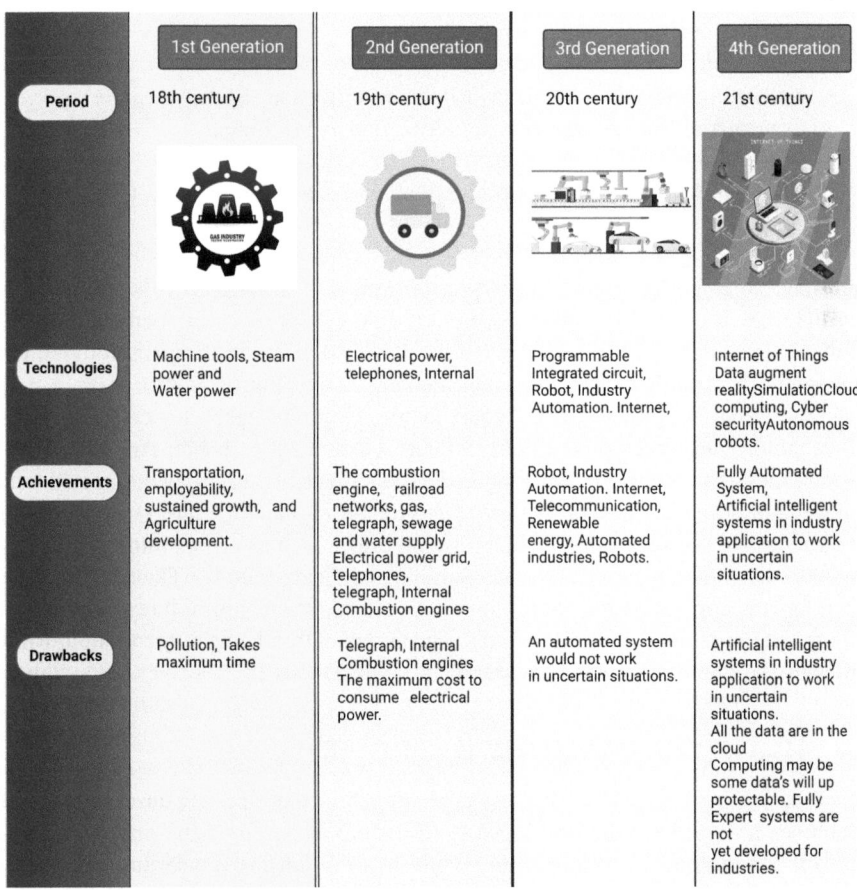

	1st Generation	2nd Generation	3rd Generation	4th Generation
Period	18th century	19th century	20th century	21st century
Technologies	Machine tools, Steam power and Water power	Electrical power, telephones, Internal	Programmable Integrated circuit, Robot, Industry Automation. Internet,	Internet of Things Data augment realitySimulationCloud computing, Cyber securityAutonomous robots.
Achievements	Transportation, employability, sustained growth, and Agriculture development.	The combustion engine, railroad networks, gas, telegraph, sewage and water supply Electrical power grid, telephones, telegraph, Internal Combustion engines	Robot, Industry Automation. Internet, Telecommunication, Renewable energy, Automated industries, Robots.	Fully Automated System, Artificial intelligent systems in industry application to work in uncertain situations.
Drawbacks	Pollution, Takes maximum time	Telegraph, Internal Combustion engines The maximum cost to consume electrical power.	An automated system would not work in uncertain situations.	Artificial intelligent systems in industry application to work in uncertain situations. All the data are in the cloud Computing may be some data's will up protectable. Fully Expert systems are not yet developed for industries.

FIGURE 7.2 Basic difference in different industrial revolutions such as first, second, third, and fourth. It describes the period of revolution, technologies employed, achievements, and their respective drawbacks. (Created with Biorender.Com)

data to improve the process, drive growth, and boost productivity (Grabowska, Saniuk, and Gajdzik 2022, 0).

To increase their profit efficiency and service administration with the environmental interaction, conventional machines must be transformed into self-aware and self-learning machines, which is a requirement for industry 4.0. Construction of an open, intelligent production console for an industrial digitally enabled implementation with actual data monitoring, tracking of the status and locations of the good or service as well as characteristics, and guidelines to control manufacturing processes is the primary goal of industry 4.0 (Vaidya, Ambad, and Bhosle 2018). Industry 4.0 technologies are useful for transform teaching and learning through online laboratories. The up-to-date manufacturing pattern looks to encourage optimization (Janapati, Dalal, and Sengupta 2021). The potential in production and organization purposed as below shown technologies:

- Additive and smart manufacturing.
- Cloud computing and big data.
- Networking (Business-to-Business communication, SCADA Supervisory Control and Data Acquisition, and Machine to Machine).
- Modern industrial automation.
- Bright system recording, domination, and decision-making. (Elbestawi et al. 2018)

Industry 5.0 is going to serve a very important role in society. It is a forthcoming industry. It will adjunct the existing industry 4.0 approach by focusing back on humans. The concept itself was defined in July 2020. Industry targets through interlinkage in the middle of machines and humans. Today, humans' efforts apace with machines, and they are secure to smart factories by intelligent devices. The world of technology, advanced production, and mass customization is breaking with transformation. Automatons are flattering and extra salient because they connected with the mind of humans via advances in AI and the brain–machine interface. Machines are interwined with the brain of humans and effort even as co-operators, so it does not work as a challenger (Grabowska, Saniuk, and Gajdzik 2022). The term industry 5.0 refers to people working alongside robots and smart machines. It is all about how robots help humans for working better and faster with advanced technologies like big data and the IoT (Saniuk, Grabowska, and Straka 2022).

7.4   HMI IN PHARMACEUTICAL INDUSTRY

Machine learning and AI have great applications in the pharmaceutical field also. Machines are used for the development to the clinical trial of the pharmaceutical product and systems as well. Starting from the drug design; pre-formulation to the actual production of any pharmaceutical system or drug. For QSAR study, solubility and dissolution study, and also in manufacturing study machine imparts great role in pharma industry (Damiati 2020). These computers or machine systems assess gestures, emotions, and body language like human-related activities. One can take an example of these all like the motion of an arm, leg, hand, or head gesture, emotions and body language like happy, sad, and worried at the time of doing any work or before it. In today's time, researchers are excited to work on the things that are applicable for performing a task for humans easily and conveniently. For example, wheelchairs that can be controlled by the head gesture or eye blinking and tracking devices and car parking gadgets. Also, included are devices that sense the touch and noise around them (Gao, Wang, and Xie 2022).

There are many applications available for controlling computers or TV by using optimum voice recognition technologies. Numerous applications exist for managing your TV, automating machines, telling the home to switch on the fans, and controlling robotics. All of these have been tried before, but only in the past 10 years has precision increased sufficiently to make them more practical. Voice-controlled home appliances, voice-controlled wheelchairs, or even voice-enabled in-car entertainment systems are some typical applications.

EEG is one of the widely used techniques. EEG signals are based on neuronal activity. HMI, also known as a brain–computer interface (BCI), is the central component of this theory. A system can acquire an EEG signal, extract features from it, "understand" the user's purpose, and control electronic devices like a PC, a robot, or a wheelchair (Ferreira et al. 2008; Feyissa and Tatum 2019).

Since electromyographic (EMG) HMI is a very practical interface, several applications have been created. Most significantly as a means of control for an artificial arm or leg. The human assisted limb (HAL) exoskeleton is one example of a full-body exoskeleton garment that can increase a wearer's strength (Sankai 2010). EMG gesture recognition is a similar technology to camera gesture recognition. Both stethoscopes and digital accelerometers record the mechanical transmission of signals via the skin, but they cannot be a continuous recording device, which is the major drawback of them. A soft, conformal device class is designed specifically for mechano-acoustic skin recording that can be applied to almost any area of the body. These devices can be utilized in multimodal configurations, such as electrophysiological recording, and they can be made to optimize detectable signals. It is well established that mechano-acoustic signals hold crucial data for applications in clinical diagnosis and health care. In particular, mechanical waves that travel through bodily tissues and fluids because of normal physiological activity show distinctive signatures of individual events, like the closing of cardiac valves, the contraction and relaxation of skeletal muscles, the vibration of vocal folds, and movement in the gastrointestinal tract (Liu et al. 2016).

Electrooculography (EOG) is the study of eye fluctuation by inserting electrodes above the eyes (Belkhiria and Peysakhovich 2020). For ocular tracking, it may be applied to projects like an EOG-based TV remote control. It is a highly practical machinery to help people with impairments who are unable to use conventional HMI devices. EOG is one of the very much helpful devices that records the movement of the eyes like the potential of cornea and retina generated from the hyperpolarization and depolarization together known as electrooculogram. A robotic wheelchair is one of the machines that works on this principle. The EOG-based device works by various types of guidance like automatic or semiautomatic, direct access guidance, and eye command strategy. Elderly people and handicapped personnel can use this device with great convenience and ease.

Tactile technology in the pharmaceutical world gives great devices like keyboard that is used until today very widely. Using various keys or buttons on the keyboard various commands are given to the machine by the humans (Handelzalts et al. 2021).

7.5 ADVANCED NANOPHOTONIC READOUT CIRCUITS

Nanophotonics, also known as nano-optics, is a branch of nanotechnology that studies light behavior on nanometer scales in addition to the interactions of nanometer-sized entities with light. Some of the advanced nanophotonic readout circuits include:

7.5.1 Wearable Triboelectric–Human–Machine Interface

The new advanced technology added to the human–machine interface is the use of wearable triboelectric technology, the combination of both is known as triboelectric–human–machine interface (THMI). There are various types of THMIs such as stretchable skin sensors. Earlier THMIs use conventional electronic readout but the major drawback of these is that they give unstable and missed transfer of information due to transient charge flow. Wearable electronics are widely developed in personalized health care and robotics (Saniuk, Grabowska, and Straka 2022). To overcome the obstacle of lossy transfer of interaction, THMIs can be used with robust nanophotonic aluminium nitride (AIN), which works in open circuit with minor charge (Dong et al. 2020).

7.5.2 Diamond-Based Quantum Emitters with Nanophotonic Circuits

The photonic circuit in a quantum system is advanced in the world. The nanophotonic circuit is introduced, and it is integrated with hundreds of optical elements and detectors that provide an excellent solution for quantum counting, but the limitation is that the number of independent sources generate a single photon. So, combining nanophotonic with the diamond-based emitter supplies a single photon into the nanophotonic network with the independent waveguide channel for excitation and emission (Bissoli et al. 2019). A study by Schrinner et al. uses material tantalum pentoxide (Ta_2O_5) because of its property of having least self-fluorescence and low loss of transmission, which favors the accurate reading. They incorporate lithographic positioning, which leads to the yield of 70%, which can be further improved by techniques such as super-resolution in frequency domain (Schrinner et al. 2020).

7.5.3 OCTANE: Optical Coherence Tomography Advanced Nanophotonic Engine

The optical coherence tomography advanced nanophotonic engine (OCTANE) is a turn-key spectrometer with a small size, low cost, single-mode fiber input, thermal and vibrational sensitivity, and many other features. It has a center wavelength of 400–1,100 nm and a spectral range of up to 350 nm (Adir et al. 2020). The first product line of chip-based spectrometers using silicon photonics for spectral domain optical coherence tomography (SD-OCT). It is well suited for low-cost, high-volume applications in both the medical and industrial fields (Hong Wang, Ma, and Hao 2017).

7.5.4 Nanophotonic Detector in Thermal Infrared Vision

Nanophotonic includes the study of interaction of light with nanoscale molecules. Silicon nanophotonic is used as nanophotonic for more than even a decade (Jalali and Fathpour 2006). Chinmay Khandekar et al. presented a new nanophotonic detector array instead of a traditional microbolometer as a potential minimalist approach to enable direct thermal IR vision. Their proposed design generates an infrared image as a spatially modulated wavefront of visible light reflected by the detector array, which is viewable with the human eye. This design eliminates the need for additional components in standard thermal IR cameras for electronic readout, A/D conversion, image processing, and information display. The proposed design may be useful for achieving the same functionality as infrared cameras at a lower cost, form factor, and power consumption (Roh et al. 2015).

7.5.5 Diamond Nanophotonic Circuits Functionalized by Dip-Pen Nanolithography

A potential manifesto in quantum optics, nonlinear optics, and optomechanics has been identified as diamond-based photonic components. Due to its outstanding material qualities, including its broadband optical transparency, great mechanical stability and toughness, high heat management, and good alchemical steadiness, diamond is employed in a wide impact on both innovation and industry. Due to its enticing optical qualities and biocompatibility, diamond is a desirable platform for biophotonic applications. Chemical vapor deposition (CVD) can now produce high-caliber diamond thin films on huge substrates, enabling the fabrication of functional devices employing wafer-scale processing methods. As a result, homogenous large-scale diamond thin films can develop. In terms of consequence, a sensitive readout platform for biotic indication may be successfully realized using the robust toolbox that integrated optics has to offer. It is necessary to establish the proper linkages to the required analytes before using integrated photonic devices for biosensing. Functional and biofunctional designs are deposited via a variety of lithography methods. A potential method for outstretching spatial resolution commensurate by the help of the normal length scales occupied by nanophotonic devices has recently been identified as dip-pen nanolithography (DPN). DPN applies "inks" such as tiny organic molecules, polymers, biomolecules, or nanoparticles to a rigid element using an atomic force microscope (AFM) tip as a "pen." Using biomimetic membrane stacks in immunological tests, for instance, the site-specific and multiplexed functionalization of surfaces is made possible by the use of phospholipids and lipid mixtures in DPN, sometimes referred to as lipid-DPN (L-DPN) (Kwon et al. 2020).

7.5.6   InP Photonic Integrated Circuits

InP PIC technology's capability is progressed in a way that can impact every field imaginable. This improvement is primarily accomplished by monolithically integrating and successfully demonstrating a wide range of photonic functions on the same chip. In addition to the development of high-performance next-generation telecom/datacom transceivers, the scientific community is focusing on the realization of complex and useful photonic systems based on InP PICs, which can enable a wide range of real-world applications such as sensing, imaging, and high-speed signal processing (Mendoza et al. 2021).

7.5.7   Nanophotonic: Shrinking Light-Based Technology

When light is shrinking in all three dimensions it interacts with the matter and illuminates an individual molecule with an individual nonparticulate radiation and its function is possible by the optical antennas, which are nanophotonic elements. The research of light has developed into a thriving sector to study (Nasrollahzadeh et al. 2019). The light is remarkably handled in novel ways employing geologic and insulative nanoparticle accurately refined into the 2D and 3D nano-architectures, which are impractical to produce from natural source as well as traditional geometries. A diverse variety of useful applications, including new areas for integrated electronics, optical computing, photovoltaic, and diagnostic products, as well as high hopes for several unique discoveries in the near future, have been made possible by this microscopic management of illumination (Esposito et al. 2021).

7.5.8   Helmet Identification

Helmets are an essential piece of safety equipment to protect employees during usage and inspection. Given that certain employees might not always abide by the law, video scrutiny that cover the entire facility and organizers is required to keep an eye on whether employees are donning helmets or not. Even with several monitoring displays, it might be tedious to detect any violation at once that may escort to catastrophic misfortune. An intelligent vision-based approach to helmet identification focuses on identifying the color of helmets while also determining whether workers are wearing helmets or not wearing helmets. After this, a hierarchical support vector machine (H-SVM) is built to categorize all features into four groups (red helmet, yellow helmet, blue helmet, and non-helmet) (Malerba 1999b).

7.5.9   Adaptable HMI for Mobile Devices

Currently, there are two types of user interfaces: static and dynamic. Static interface develops at the time of designing and its function is restricted. However, dynamic annexation is created, for smoothing daily work it requires human–machine interface generate at the run time. The HMI is required to produce a robust interface for the smartphones. HMI engine has two components: HMI definer and HMI builder, wherein the former displays the final interface on the device while the latter forms the main format of the juncture based on the work performed by the user and user preferences (Zander and Kothe 2011).

7.5.10 A REAL-TIME APPLICABLE 3D GESTURE RECOGNITION SYSTEM FOR AUTOMOBILE HMI

The 3D hand gesture recognition systems are used in automobiles, and these systems require an HMI device for recognition. These systems consist of two time-of-flight (ToF) sensors, and both are fixed to the center console. And the camera is placed in the center of the console. Data coming from two ToF sensors is fused by the device (Bach-y-Rita and W Kercel 2003). It is useful in various sectors such as gesture control, measurement of level of liquid, hurdle detection of robotics, autofocus camera assistant, and many more (Hongman Wang et al. 2021).

7.5.11 AVIONICS HUMAN-MACHINE INTERFACES AND INTERACTIONS FOR MANNED AND UNMANNED AIRCRAFT

HMI concept is used in civil and military aircraft. The use of HMI in aircraft both manned and unmanned (also referred to as remotely piloted aircraft) is to aviate, navigate, and communicate with the team, and air traffic management and its main function is safety (Lim et al. 2018). HMI collects the information from the sensor and displays the collected information to the team. HMI adaptiveness has significant potential to enhance the operator's effectiveness (Vinitha et al. 2020).

7.5.12 PERSONALIZED HMI INTERACTION IN ADVANCED DRIVER-ASSISTED SYSTEMS

In recent eras research in providing driving safety is flourishing. And for this advanced driver-assisted systems (ADASs) are used, which utilize various types of HMI devices. The addition of HMI in the automobile is difficult, so HMI is incorporated into the ADAS. ADAS provides information about the environment condition, driver state, and the probability of an accident (Kala 2016). This system gets information from the sensor, satellite, camera, ultrasound, and LiDAR. HMI uses input from the gesture and poses of the driver and gives output in form of an audiovisual (display on the screen). HMI in personalization also works on the driver's needs and driver characteristics. The framework of the system is based on a customizable and extensible set of personalization and adaption rules provided by the HMI expert and evaluated according to the driver, vehicle, and environmental parameters that activate HMI and produce adaptation decisions (Javaid et al. 2022).

7.5.13 YOLO v2

With the use of the robotics arm, the working efficiency of people who are physically challenged and constrain on a power wheelchair is increasing. Then automated wheelchairs come into existence and for people with major upper limb disabilities,

HMI with an artificial robotic arm is used for easy functioning. The concept behind the working of all systems is that camera present by the end of arm recognizes several elements with which the user may mesh, and the video stream produced by the camera is analyzed by you only look once (YOLO) v2 convolutional neural network, which functions as an object recognizer; these analyzed videos and images of an object are shown to the user (Sang et al. 2018). This information is passed through the target setting block and then the target setting block to track the object area, the letter message is sent to the logic system which evaluate the movement that the robot should assist to reach the object destination, root actuator drives the robot arm (Grabowska, Saniuk, and Gajdzik 2022).

7.5.14   Human–Machine Interfaces in Upper-Limb Prosthesis Control

Prostheses devices play an important role in individuals with the aim to regain some of the lost functions of their sever and cut limbs. HMIs are used for controlling these prosthetic devices, which play a critical role in users' experiences with prostheses (Vaidya, Ambad, and Bhosle 2018). Selection of prosthetic device is done on the basis of type of control required and can be divided into three types: myoelectric prosthetics, prosthetic controlled by buttons, and hybrid prostheses (Legrand et al. 2018). These all demonstrate the active form of prosthetics whereas for the passive form, aesthetic and functional prostheses are used (Ribeiro et al. 2019).

7.5.15   A Model-Driven Mobile HMI Framework (MMHF) for Industrial Control Systems

Industrial control systems grated over a period of time, which incorporate human machine interface to control and monitor the complex system. In industry, the function of HMI is to convert complex data into human-readable form. Tools such as ICONICS GraphWorX support the HMI. HMI that generates real-time data is costly and required special training, so to provide a generic, automated, easy-to-use, fast, and high-level mobile HMI development a MMHF is proposed. A concept like unified modeling language (UML) and model-to-text transformation are incorporated in MMHF.

7.6   MULTICLASS AND ADAPTABLE HUMAN–MACHINE INTERFACES

Human–machine interfaces for cooperative supervision and control by several human users, in either control rooms or in-group meetings, are dealt with. The information flow between the different human users and their overlapping information needs are explained (Schrinner et al. 2020).

A revitalized curiosity in developing intuitive and immersive interaction is sparked by recent developments in HMI. In order to build a glove-based multidimensional HMI, we created a simple-structured, high-resolution bending angle triboelectric sensor called the bending-angle triboelectric nanogenerator (BA-TENG) (Sun, Zhu, and Lee 2021).

Also, it is based on multichannel forehead bioelectric signals acquired by placing three pairs of electrodes (physical channels) on the frontalis and temporalis facial muscles. The acquired signals are passed through a parallel filter bank to explore three different subbands related to facial electromyogram, electrooculogram, and an electroencephalogram (Johannsen, Rijnsdorp, and Sage 1983).

It is used in driving emotion too. Emotions are having an opposing influence on driving. Feelings like anger, fear, and sadness hurt driver response time and may lead to deadly accidents. The spin-point emotions that may influence the driver are monitored. The method fuses both LBP and facial indicator detects emotion whereby support vector machine (SVM) is used for the classification of different emotions (Sukhavasi et al. 2022).

An alternative and natural kind of human–computer interaction (HCI), radar technology plays a vital role in the contactless identification of hand gestures or motions. Air writing refers to spoken characters or words written in open space using hand gestures. An air-writing system that refers to the interconnection of millimeter wave radars is suggested in the current study. The extraction and identification of handwritten motions is done in two steps.

Multimodal adaptive cognitive technological systems are being developed to direct, support, and monitor human workers in contexts with complex manual assembly. The desire for extremely flexible building infrastructure contradicts prolonged training and preparation phases for human workers. Nonspecialized industrial standby workers in a manufacturing task can be accurately allowed to execute the next processing step without any prior knowledge by supplying context-aware construction instructions over retina displays, text-to-speech commands, or acoustical signals (Wallhoff et al. 2007). For managing an application using hand gestures, a more conventional, intuitive, user-friendly, and less disturbing human–computer interface is offered. An effective hand gesture recognition system based on vision has been built for this use, and a new database has been made to test it. The three phases of the system are detection, tracking, and recognition. The detection step uses a binary SVM classifier and local binary patterns as a feature to scan through each frame of a video series for suitable hand poses (Arsalan and Santra 2019).

HMIs, which enable more natural human connection with the digital environment, have seen a booming development in recent years as we enter the 5G and into IoT age. A camera, microphone, inertial measurement unit (IMU), or other sophisticated sensing technology may be based on complementary metal-oxide-semiconductor (CMOS) or microelectromechanical system (MEMS) solutions (Schrinner et al. 2020).

Because of advancements in peculiar printing processes and delicate elements, wearable are changing to delicate form that allow comfortable synchronization with the epidermis. The innovation of majorly conductive nanomaterials and annealing techniques of printed inks, such as Ag9,10, Cu11, and carbon nanotube materials (CNT)12, enable low skin-to-electrode contact susceptibility in electrophysiological investigations during dynamic body movement. Several printed wearable systems have been demonstrated using these advances, but they are limited to passive electrodes and rely on rigid printed circuit boards fabricated using existing approaches (i.e., photolithography, spin coating, and high-vacuum deposition) for the active components. The incremental process should be able to efficiently print a multitude of ink materials with a broad range of apparent viscosity, as well as perfect layer alignment, as body-wearable devices often incorporate sensor components and an electronics module. The publications may come into direct touch with the skin over several days, hence biocompatibility tests should be done (Chavda et al. 2022; Bezbaruah et al. 2022). The manufacturing technique for the all-printed, nanomembrane hybrid electronics (referred to as "p-NHE") described in this study is based on in-depth fieldwork on nanostructured materials synthesis, material science, and printing optimization. To fully illustrate the viability of the all-printed EMG devices in advancing wearable healthcare and health monitoring, they incorporated multiple HMI circumstances, including hand gesture-controlled wireless target controls, such as drones and computer software. For remote monitoring of a robotic hand, deep learning algorithms can categorize individual finger movement in real time using synchronized multidevice EMG data. They showed how the suggested device integration, materials optimization, and EMG-based HMIs would change the objective printed electronics combined with brittle materials are employed to enhance individual intelligence and health care (Kwon et al. 2020).

HCI, such as using an assistive interface or for enjoyment, has ingrained itself into our daily lives. With the development of computer vision, hand gesture identification has become a common way for people to communicate with machines without actually touching any hardware. Consider smart home robotics, which enables users to operate numerous apps and gadgets such as TVs, music players, lights, and Wi-Fi by using hand gestures. Another option involves hand gestures being used to operate desktop applications, such as VLC, music players, or 2D games, as well as the creation of gesture-controlled holographic mice and keyboards. Therefore, users may manage programs in both scenarios without having to interact with them physically, which is highly helpful for the elderly and physically challenged. Wearable gloves-based hand gesture recognition and vision-based hand gesture recognition are the two main categories. The first technique has the disadvantages of being costly, requiring wearing on the hand to identify movements, and being unstable in specific circumstances. The second approach is based on image processing, and it uses a pipeline that includes webcam picture capture, segmentation, feature extraction, and gesture categorization. We are aware that, even when using diverse motions to comprehend various commands, our hand may not always be steady enough to control the mouse cursor's movement. Therefore, in order to address these problems, researchers initially suggested a vision-based hand gesture detection system utilizing CNN models, after which they continued their work to create a real-time gesture-controlled HMI. They have also created a virtual mouse using motion instructions to render this system quite practical and user-friendly. This chapter presents HCI based on a real-time hand gesture detection system. There are six steps in the system: hand detection, gesture segmentation, usage of six which was trained before CNN models utilizing the transfer learning approach, creation of a dynamic human–machine interface, and creation of a gesture-controlled holographic mouse are just a few of the techniques used. Using a Kalman filter to predict the hand location will enhance the pointer's motion's accuracy. To assess the model performances, three multiclass datasets—two publicly available and one custom—have been employed. The gesture recognition technology has been enhanced and is now utilized to operate multimedia apps (such as the 2D Super Mario Bros. game, audio player, file manager, and VLC player) with various custom gesture instructions in real-time settings. This system's 35 frames per second (fps) average speed satisfies the criteria for a real-time scenario (Sen, Mishra, and Dash 2022).

7.7 TRIBOELECTRIC BENDING SENSOR-BASED SMART GLOVE TOWARD INTUITIVE MULTIDIMENSIONAL HMI

Recent developments in HMI have rekindled interest in developing straightforward and realistic interaction. Here, a glove-based multidimensional HMI is built using a simple-structured, high-resolution twisting degree triboelectric sensor called the BA-TENG. The glove-based HMI demonstrates great susceptibility and minimal distortion in real-time finger motion sensors due to bespoke print circuit board (PCB). A 19.36 dB improvement in the signal-to-noise ratio (SNR) is made. A keyboard with identification capabilities, enhanced robotic control, and intuitive multidimensional HMIs for smart homes was all made possible by methodically collecting and processing the multifaceted signal properties of the BA-TENG. By utilizing the cutting-edge machine learning approach, the categorization efficiency of the virtual keyboard for seven users was increased to 93.1%. The suggested BA-TENG-based smart glove demonstrates its capabilities as a solution for multidimensional HMI with an intuitive minimalist design that has promise in a variety of applications (Greentree et al. 2008; Chavda et al. 2021).

HUMAN-MACHINE INTERFACE USING HYBRID SURFACE ELECTROMYOGRAPHY/A-MODE ULTRASOUND SENSING:

HMI research has been firmly centered on noninvasive muscle-based HMI. To enhance the functionality of muscular HMI, electroencephalographic and morphological alterations in muscle contraction must be obtained. However, a method for simultaneously displaying electrical and morphological data about the same muscle is currently lacking. A-mode ultrasound (AUS) can surveil the linguistic changes in muscle, and surface electromyography (sEMG) might portray the impulses of operational muscle function. Both techniques are noninvasive. A mobile hybrid sEMG/AUS system for HMI is made up of a signal collection module and a composite sensor armband. The former arranges two different types of sensors at the same muscle location, while the latter makes it possible to acquire sEMG and AUS data simultaneously. The system can generate signals with good SNR and time-frequency characteristics, as shown by the equipment assessment experiment. Because the hybrid sEMG/AUS feature greatly enhances identification accuracy when compared to the findings of ultrasonographic features and sEMG features, respectively, the hand gesture recognition experiment also shows the complementary nature of sEMG-based and AUS-based HMI (Xia et al. 2019).

USING A MULTILAYER BINARY DECISION TREE, RECOGNIZE EMOTIONS:

An essential component of computational research on communication patterns is automated emotion state tracking. Both to improve analytical skills to assist decision and help to build human–machine interfaces that promote effective connection, it is imperative to implement robust and reliable emotion identification systems. A hierarchical computational system was created by Lee et al. to identify emotions. This structure uses several stages of binary categorization to translate a voice utterance into various emotion classes. The fundamental concept was that the layers of the tree are built to handle the simplest categorization jobs first, enabling them to reduce error propagation. They tested the classification framework utilizing acoustic characteristics on the AIBO dataset and the USC IEMOCAP database, two separate emotional databases. Using this hierarchical structure, they are able to recall each of the many emotion types fairly in the case of the AIBO database. On the consideration dataset, the average unweighted recall performance metric maximizes by 3.37% absolute (8.82% relative) beyond a baseline SVM model. They achieve an efficiency level of 7.44% (14.58%) across a benchmark SVM modeling. The outcomes show that the proposed hierarchical technique was successful in categorizing emotional utterances across various database environments (Lee et al. 2011).

The advanced HMI, with special application to develop connection, is used in health care and rehabilitation. This method depends on the development of an entirely new compact system for the discrete monitoring of a biological signal utilizing customizable or embedded sensors incorporated into advanced HMI. The BCI application's primary objective is to improve the opportunities this technology presents for rehabilitation and medical care while also enabling communication for persons who are severely impaired ("Frontiers | Brain–Computer Interface (BCI)" 2022). Its integration into a more intricate ambient intelligence system that enables the control of core parts at residence or through the diversification of discrete set platforms encouraging other services should make this achievable. Individuals who have significant communication disabilities may find a solution in a brain–computer interface, where messages and instructions can be communicated through electrophysiological events linked to the user's intention. Modern BCI systems offer a sluggish and inaccurate communication route; however, research on algorithms and electrical devices is aimed at resolving this sort of issue. While a BCI can't replace the typical pathways of the human body or drive a car in the near future, it can improve the standard of living for those whose interface with the outside world is severely limited (Alkawadri 2019).

7.8 ADVANCED NANO-FUNCTIONAL MATERIALS FOR HMI DESIGN

With the rapid innovation of the world in the present century, particularly in the area of medical engineering and technology, innovative brilliant systems with the neural interfaces are developing rapidly. In the past, the simultaneous functional controls of a gadget in real time have been improved largely with the recent HMI interface systems, with the help of various technological applications real-time control of functions of machines is possible with great accuracy (Yonan 2018; Tian et al. 2019). In the living cells of the human body, various biochemical reactions take place, which produce biochemical signals that are used by HMI to control gadgets accurately (He et al. 2022).

Advanced machines have various well-functional parts then the traditional ones. Nanotechnology is a current new technique for making the fabrication of the machine. Nanomaterials are composed of structures at the nanoscale, usually achieved via specifically designed self-assembly processes. They acquire unique properties such as magnetic, catalytic, optical, mechanic, and electrical, which can be achievable by their nano-architecture ("Fabrication of 3D Micro-Architected/ Nano-Architected Materials" 2022). These advanced nanomaterials give chance to unfold new occasions for tuning their properties widely. It is a rapidly developing area of today's research with a broad range of applications from nano-electronics as well as power collecting to biology including nano-medicine. Modern polymers and nanocomposites, multifunctional materials and materials of battery, medicine supply, tissue engineering, and hybrid nanomaterials are some examples of research areas in which advanced nanomaterials gain an essential role (Valdevit and Bauer 2016).

With rapid advancement of information technology and the pressing needs of the IoT, HMI systems have garnered a lot of interest (Valdevit and Bauer 2016; Singh and Kumar 2021). To realize the function of HMI gadgets effectively, it should contain: of object's signal monitoring multiblock sensors, microcontroller (MCU) modules for the reception of signals, analyzing and further processing, and terminal material to accept the processed signals and give response according to it (Yonan 2018). Moreover, the energy supply is unavoidable for giving power to the whole system for constant operation (Yonan 2018; Tian et al. 2019). Various technologies and machines, which are developed these HMI devices, are greatly helpful for the human use. There are several major categories of HMI that have been developed as a result of technological advancements such as reduction in the size of sensors and the creation of new technologies. A video camera that was once long enough to match the size of foot may now fit into a smaller object like your hand. Each has different human–machine interactions, which will be discussed as follows.

7.8.1 OPTIC-BASED TECHNOLOGY

In optic-based technology, the principal used is generally a camera HMI. In a slightly different way LEDs and lasers also fit in the optic heading. For the interaction of optic devices gestures and simple hand, motions are used instead of touching anything that also improves public interest greatly (He et al. 2022).

The modern optical implementation of nanomaterials also increases the interest in optical HMI. This is because of the fact that their interior electronic construction, whose optical qualities depend on them, can be modified to meet the needs of various optical applications by changing their size and dimension. However, firstly one needs to recognize the fundamental of optics and the quantum confinement outcome of nanomaterials, which are accountable for the extraordinary optical properties ("Optical Applications of Nanomaterials" 2022).

In general, "material optics deals with the interchange of light with matter which results in manipulation of the flow of light involving reflection, refraction, absorbance, fluorescence, dispersion, frequency modifications, and focusing or splitting of an optical beam." Strict characteristic related to light wavelength used is the attribute of optical objects. Since the nanoscale, which is occasionally smaller than the wavelength of light, is where nanomaterials' dimensions are defined, the knowledge of light–matter interaction take place the main step under the optical applications of nanomaterials scope. Thus, the characteristics of these materials may be classified into two main sets, firstly linear and secondly nonlinear properties. Including graphene, fullerene, nanotubes, different biosensors and more of them functional materials are used now are graphene, ultrathin films, super lattices, discs, and quantum days largely (Peng et al. 2020).

Different vision-based implementations are handy, but the technology that concentrates primarily on HMI is trustworthy for movement and recognition statistics. Self-parking automobiles (Yonan 2018), head gesture-controlled wheelchairs (Tian et al. 2019), eye movement control interface ("Optical Applications of Nanomaterials" 2022), the ability to control music software with gestures (He et al. 2022), tabletop interaction tracing of an eye for computer input control ("How Nanotechnology Enables Wearable Electronics" 2022), and multitouch interfaces are a few examples of such applications.

7.8.2 Acoustic-Based Technology

Acoustic-based technology, with primarily emphasis on voice recognition, might be utilized to handle or control a device, convert spoken words to text, or interact with the sound-based machine. With the different sounds of everyone, accurate recognition of all persons' sounds with minimal error is the goal of sound recognition. It has significant potential, and phones have software for voice recognition.

Since science-fiction authors and film producers made voice recognition interesting, it has always been widely used. In order for speech recognition to function, users must first speak into a microphone, where the frequency of actual voice waves is turned into an electrical impulse. Then, a microcontroller or computer may utilize that straightforward component. Analyzing the voice is the hard part. For recognition of your voice computer extracts some necessary features from the received voice by pattern recognition techniques. It may sound easy but in actual difficult to do because for improving the signal many processing techniques are required (He et al. 2022).

Throughout the many years different types of materials have been getting useful in microphones as well as in loudspeakers. Nowadays permanent magnets of loudspeakers are replaced with nanoparticles and same performance is given by thin film of carbon nanotubes. The use of graphene is also tried to reproduce sound for the microphones.

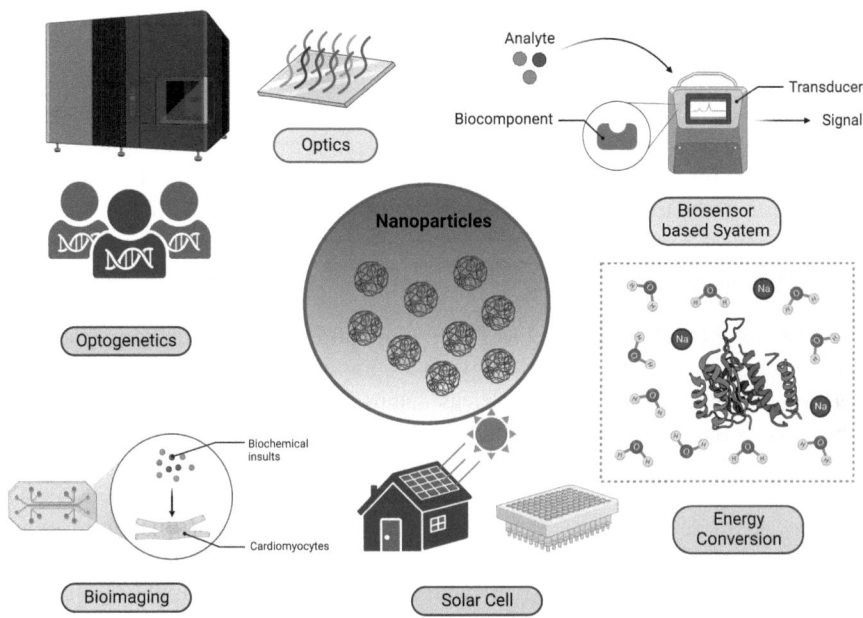

FIGURE 7.3   Different optical applications of nanomaterials. (Created with Biorender.Com)

For generating pliable loudspeakers or microphones research was done using UNIST technology by the researchers as they utilized silver nanowires to create a nano-membrane. This can be attached to the surface of the skin to playing a final movement or signal of the violin concerto, namely "*La Campanella* by Niccolo Paganini."

Using polymer-based nanomembrane, Korean researchers inserted a silver nanowire network which is based on polymer hybridization of nanowires being comparatively easy than another one. Less than the nanoscale thickness of 100 nanometers there are substances like conducive, and transparent hybrid nanomembrane are the biggest advancement of the research of the Ko. The demonstration of a skin-attachable indiscernible loudspeaker and microphone is made possible by the exceptional optical, electrical, and mechanical capabilities of nanomembranes ("Nanotechnology in Healthcare" 2022).

In the acoustics field there is the first time that graphene is applied and the result suggested and created a multilayered graphene sound source device. The output sound performance is measured between 1 and 50 kHz. The frequency band between 20 and 50 kHz is found to have a flat sound spectrum. Comparative analysis is done on the effectiveness of multilayered graphene sound source gadgets with various thicknesses. The system programming language (SPL) value of graphene was discovered to be greater in thinner varieties.

By using laser scribing, low-cost graphene earphones are made at the water level. Compared to conventional earphones, graphene earbuds have a wider and flatter audio spectrum output. Additionally, animal behavior can be controlled via graphene earphones. Using laser scribed graphene (LSG), a wearable false throat that is produced in a single step has been put into practice. On account of its superior thermoacoustic and piezoresistive qualities, the LSG device can emit and detect sound simultaneously (Kim, Cho, and Yu 2018). Due to the sound sources' resonance-free oscillations, the artificial throat in the LSG has a rather wide frequency spectrum. In addition, the LSG artificial throat's sound detector can precisely record the mechanical vibration of the throat cords ("Nanotechnology in Healthcare" 2022).

7.8.3 BIONIC TECHNOLOGY

To perform any function HMI devices have to monitor the biological feature and for that technologies like robotics, biology, and computer science in combination are used. We can have bionic arms, feet, legs, hands, eyes, and ears. Electrodes collect the necessary data from bodies for utilizing biological signals. Electrodes for all of the above HMI techniques collect the electrical data. By altering the level of amplification and frequency we can use similar electrodes for different bionics aspects. All strategies are more or less got the pioneer from EEG. For checking that muscles and nervous system are correctly working, EOG and EMG are used and for heart ECG is used. Bionics can be divided into two major categories: (1) EEG-based BCI (Janapati et al. 2020) and (2) EMG myoelectric control. However, there are minimal areas in electrooculography (EOG) and electrocardiography (ECG). Diagnostic approaches are getting improved with the help of various carbon nanomaterials like graphene, fullerene, nanotubes, and carbon quantum dots, and attention toward nanomaterials made from carbon is increased in past few years (He et al. 2022).

With a variety of advancements such as nanoscale therapies, drug delivery systems, imaging technologies, biosensors, and implantable devices, drug nanotechnology is increasingly serving as a major catalyst for medical and healthcare innovation (Chavda et al. 2022). Degree programs for nanomedicines are also started by various universities (Janapati et al. 2020).

Certain properties of graphene such as ultra-flexibility and high conductivity make it a crucial part of wearable electronics such as:

- Nowadays in hospitals, patients have to wear a tag on their arm, which is of Radiofrequency identification printed graphene. These are wireless patient care devices that give readings of heartbeat and temperature with the help of some other 2D materials.
- In the diagnosis of certain diseases, nanomaterials are also used, like in cancer cell diagnosis carbon nanotube chip is investigated, which captures the data and analyzes the blood tumor cell. It is an advanced technique for detecting circulating tumor cells rather than using magnetic and microfluidic methods ("Graphene-Based Wearable Sensors for a Human-Machine Interface" 2022).

Use of nanosensors that can formulate as biocompatible hydrogel and implanted this material in subcutaneous to detect the concentration of analyte in its locality.

This is continuous monitoring of *in vivo* biomarkers made possible by fluorescent nanosensors ("Graphene-Based Wearable Sensors for a Human-Machine Interface" 2022).

7.8.4 TACTILE (TOUCH) TECHNOLOGY

Until now the chapter discussed technology that monitors thing in an untouched manner; now the tactile technology allows the object to touch the gadgets. An example of keyboard is easiest to learn tactile technology. The keyboard is now having a place on every device ranging from computers to mobile phones, home entertainment gadgets, and big industrial machines (He et al. 2022).

Many wearable technologies use nanomaterials: (1) Thermally managed facemask made of nanofibers; (2) apparel with adjustable heating and cooling; (3) a glove with heat; (4) thermal management fabric. fabric exoskeleton; and (5) an extensible heater. Stretch-activated medication delivery ("How Nanotechnology Enables Wearable Electronics" 2022).

Various materials like power supply, actuators, electronics, and sensors make wearable electronics. In the first generation, detachable devices are used, which get renovated in the second generation by the textile-embedded things like actuators and sensors.

The need to attain and integrate several qualities, such as adaptability, convenience, and the possibility for the gadget to be downsized and trendy, is one of the main obstacles. Researchers are using a variety of materials, including carbon nanotubes, graphene, polymers, dielectric elastomers, and composites, to achieve this. These are designed for certain purposes based on their many distinctive actions in response to various inputs ("How Nanotechnology Enables Wearable Electronics" 2022).

7.9 CHALLENGES

Challenges in human–machine interface:

7.9.1 A NEW MINDSET

Various developed machines are nowadays available with their sophisticated features. New machines have complex features, for sufficient use of machine both humans, and machines have to work collaboratively. For accepting a new feature of the machine and working with it like a team, a new mindset is required. A machine and human have to work together by using their cognitive resources for the continuous human–machine interaction. This lowers risk of consequences ("ICHMS 2020" 2022).

7.9.2 DETERMINE THE HUMAN STATE

Machines must be able to adjust their behavior in accordance with algorithms that analyze human behavior. To gain an accurate assessment of the human's condition, these technologies will need to incorporate a variety of sources of data and recognize minute clues in gaze movements, body position, and control inputs. Additionally, because no two people are alike, it is difficult to construct a generic algorithm. Some of the challenges occurring in this field will likely be solved by computer science and other types of AI (Alvarez-Cortes et al. 2009). Various human–machine interface devices like emotion detectors are invented nowadays so that machines have to adapt to human behavior by monitoring their body posture and human input. Furthermore, no two humans are 100% identical, so a general algorithm composite is needed. Machines have to collect data and need to analyze the speech of the subject, the facial gesture of the subject, and appropriate intervening variables.

7.9.3 Communicate Clearly

When work assignments are flexible and dynamic, effective and secure collaboration depends on mutual understanding between humans and machines. There could be catastrophic repercussions if a task is transferred from a machine to a human without the human being aware of it. An interface that is well designed is balanced, giving the user the proper amount of information without overwhelming them. Intelligent interfaces should modify the amount and structure of information provided based on the user's situation in order to prevent alarm floods in the future ("5 Common HMI Failures: Automation & Control: Rowse" 2022).

7.9.4 Troubles Powering Up

For the machine to get started we need the proper power supply for it; then and then only we start the machine constantly. If the machine takes some trials or needs frequent power cycles or gets difficulties in initiating gadgets it is indicated that internal energy supplies fail in a short period. To treat, these appropriate devices are required in machines (Tsarouchi, Makris, and Chryssolouris 2016). It is also necessary to reduce the power level for the working of the machine, which will automatically reduce the overall cost of the machine (Rocha-Jácome et al. 2021).

7.9.5 Damage in Touchscreen

In tactile devices, it is observed that if keys have more usage in operating machines, then the chances of degrading them also increase; for example, the keyboard enter key will be degraded firstly if it is in high usage while others are working properly. In touchscreen devices screen failure or specific areas of the screen sometimes get damaged because of being highly pressed sometimes. Those areas get irresponsive because of damage that occurs in that part. It is surely a failure of the HMI (Tsarouchi, Makris, and Chryssolouris 2016).

7.9.6 Problem in Screen

A dull or flickering screen is one of the most annoying problems out of others. It takes almost months for this problem to arise, which provides enough time to replace it with the new one. Lines on the screen either horizontal or vertical also sometimes occur, which need work to solve it. It gives a single for the failure of the backlight of the machine (Tsarouchi, Makris, and Chryssolouris 2016).

7.9.7 A Problem in Voice Recognition

In voice recognition devices it is important to analyze the language of humans, knowing the subject, sentence analysis, and then convert it to physical command (Hentout et al. 2019). On the work floor recognition and analysis of voice is difficult because of so much noise surrounding the machines. The complex system also gets failed sometimes with voice recognition (Tsarouchi, Makris, and Chryssolouris 2016).

7.10 CONCLUSION AND WAY FORWARD

Nowadays machines are everywhere from home appliances to industries to hospital care units. We are all connected to different kinds of machines differently. Interaction between humans and machines is simply known as human–machine interface. In the traditional machines now, new materials are also included like graphics, joysticks, single-handed keyboards, and many new functional devices range are included. Examples we consider as a range of joypads and sticks.

Self-healable, as well as biocompatible, stretchable, flexible, imperceptible devices of sensors, power suppliers, logic circuits, and data storage systems are common in all HMI devices. Industry 4.0 and 5.0 are giving good knowledge about HMI like knowing the various part and concepts of operations in the industry and making as much as possible use of on-the-spot data to improve results, procedures, and overall growth. Industry 4.0 is helpful for any owner of the business (Bobra et al. 2014).

To increase the machine's outcome and manage it with the environment, conventional machines must be transformed into self-aware, self-learning machines. The goal of industry 4.0 is to build an open, intelligent manufacturing platform for a networked industrial information application. The essential requirements of industry 4.0 are real-time data monitoring, tracking a product's status and locations, as well as holding the instructions to regulate production operations.

In our society, industry 5.0 also plays an important role. It is a human-oriented program that turns the industry 4.0 perspective to focus on mankind. The interaction of humans and machines is the main focal point of industry 5.0. HMI in the pharmaceutical industry is give a great opportunity in the pharmaceutical field for the humans. HMI works in various ways of human activity, such as gesture control and the body language of humans, and it is also linked to the emotions of a human. Other activities of human-like gestures, tabletop interaction, and eye movement monitor help for input control in computers, and tactile HMI multitouch action is also utilized for designing devices' work (Bogaerts 2014).

Voice recognition like difficult activity now days successfully used for HMI design. In the therapeutic and diagnostic part, there is a wide range of HMI that is very useful in various techniques like in analyzing the nervous system, which covers the brain that is the first and most valuable part, neurons, and also the spinal cord from varying earlier time. Myoelectric (EMG) HMI electrocardiogram ECG and EKG Electrooculography (EOG) are some examples of abundant HMI.

The use of advanced nanophotonic readout circuits gives various new facilities through HMI; some of them are wearable THMI, integration of diamond-based quantum emitters, OCTANE: optical coherence tomography, direct thermal infrared vision, nanophotonic circuits: diamond nanophotonic circuits, A very good interaction InP Nanophotonic circuits for vision sensing occasions reducing light-based technology through Nanophotonic.

Utilizing SDO/HMI vector magnetic field data with a machine learning system to predict solar flares, a clever vision-based method for identifying helmets to ensure worker safety, an individual who creates adaptive human–machine interfaces for portable devices ("The Helioseismic and Magnetic Imager (HMI) Vector Magnetic Field Pipeline: SHARPs—Space-Weather HMI Active Region Patches" 2022). An effective real-time 3D gesture recognition system for vehicle HMI, avionics for manned and unmanned aircraft: human–machine interfaces and interactions ADAS System Interaction Framework for Personalized HMI.

A fresh interest in developing intuitive and immersive interaction is sparked by recent developments in HMI. To build a glove-based multidimensional HMI, we created a simple-structured, high-resolution bending angle triboelectric sensor called the turning site triboelectric nanogenerator (BA-TENG). The modern, HMI, developed complementary metal-oxide semiconductor (CMOS) ("Complementary Metal-Oxide-Semiconductor: An Overview" 2022) or microelectromechanical system (MEMS) solution is based on sophisticated sensing tools, such as the camera, microphone, and inertial measurement unit (IMU) (Mitternacht, Hermann, and Carqueville 2022).

Nowadays there are various highly sophisticated parts in the advanced machines than traditional machines for HMI. Nanotechnology is a current new technique for making the fabrication of the machine. Nanoscale structures are achievable because of the self-assembly procedure, which is specifically made for it. Nanomaterials are made from these various nanoscale structures. Nanostructures are well-developed materials having optical, catalytic, magnetic, electronic, and mechanical features that help provide unparalleled opportunities for modulating their function in a broad range.

In optic-based technology, there are nanomaterials like nanocapsules, fullerenes like zero-dimensional, quantum wires, nanorods like dimensional, and graphene-like two-dimensional nanomaterials are used. Nanomaterials are useful in also acoustic technological devices example we take graphene and carbon nanotubes which are used in that. Bionic electrodes do important work of gathering data and performing according to that data. In diagnostic approaches nanomaterials like carbon quantum dots, fullerenes, and graphene are used for designing devices (Korah and Mathew 2022).

With these great advantages, some challenges also arise in human–machine interfaces like a new mindset for humans because of human and machines have to work in collaboration in fulfilling the responsibility of doing teamwork. For better work, we have to understand machines and work according. Also, to determine the state of humans we have to gather various data like human body gesture patterns and many more things.

Designing the interaction of humans and machines is also an important task for human–machine interface designers. For making a good interface, it is necessary to give only a sufficient amount of guidance and data to resist overload on the subject or ease of understanding. Safety plays an important role in designing and using any machine. Noise recognition in a crowded environment is also challenging for humans. Difficulty in the supply of power, improper touch sensor or screen, which are dull or continuous flaring. Lining on screen like the one annoying. Challenges are also sometimes seen in HMI that designers have to focus and work up.

ABBREVIATION

ADAS:	Advanced driver-assisted systems
AIN:	Aluminium nitride
AUS:	A-mode ultrasound
BA-TENG:	Bending angle triboelectric nanogenerator
BOLD:	Blood-oxygen-level-dependent
CMOS:	Complementary metal-oxide-semiconductor
CNT:	Carbon nanotube materials
CVD:	Chemical vapor deposition
DPN:	Dip-pen nanolithography
EEG:	Electroencephalogram
EMG:	Electromyographic
EOG:	Electrooculography
fMRI:	Functional magnetic resonance imaging
fNIRS:	Functional near-infrared spectrography
HAL:	Human assisted limb
HCI:	Human–computer interaction
HMI:	Human–machine interface
H-SVM:	Hierarchical support vector machine
IMU:	Inertial measurement unit
IoT:	Internet of Things
L-DPN:	Lipid- dip-pen nanolithography
MCU:	Microcontroller
MEMS:	Microelectromechanical system
MMHF:	Model-driven mobile HMI framework
OCTANE:	Optical coherence tomography advanced nanophotonic engine
PCB:	Print circuit board
PDMS- CB:	Polydimethylsiloxane-carbon black
QSAR:	Quantitative structural activity relationship
RBSD:	Real-time brain signal decoding
SCADA:	Supervisory control and data acquisition
SD-OCT:	Spectral domain optical coherence tomography
sEMG:	Surface electromyography

SNR:	Signal-to-noise ratio
SPL:	System programming language
SVM:	Support vector machine
THMIs:	Triboelectric–human–machine interface
ToF:	Time of flight
UML:	Unified modeling language
YOLO:	You only look once

REFERENCES

"5 Common HMI Failures: Automation & Control: Rowse." 2022. Accessed November 27. https://www.rowse.co.uk/blog/post/5-common-hmi-failures.

Adir, Omer, Maria Poley, Gal Chen, Sahar Froim, Nitzan Krinsky, Jeny Shklover, Janna Shainsky-Roitman, Twan Lammers, and Avi Schroeder. 2020. "Integrating Artificial Intelligence and Nanotechnology for Precision Cancer Medicine." *Advanced Materials* 32 (13): e1901989. doi:10.1002/adma.201901989.

Alkawadri, Rafeed. 2019. "Brain–Computer Interface (BCI) Applications in Mapping of Epileptic Brain Networks Based on Intracranial-EEG: An Update." *Frontiers in Neuroscience* 13 (March): 191. doi:10.3389/fnins.2019.00191.

Alvarez-Cortes, Victor, Victor H. Zarate, Jorge A. Ramirez Uresti, Benjamin E. Zayas, Victor Alvarez-Cortes, Victor H. Zarate, Jorge A. Ramirez Uresti, and Benjamin E. Zayas. 2009. *Current Challenges and Applications for Adaptive User Interfaces. Human-Computer Interaction.* IntechOpen. doi:10.5772/7745.

Arsalan, Muhammad, and Avik Santra. 2019. "Character Recognition in Air-Writing Based on Network of Radars for Human-Machine Interface." *IEEE Sensors Journal* 19 (19): 8855–64. doi:10.1109/JSEN.2019.2922395.

Bach-y-Rita, Paul, and Stephen W Kercel. 2003. "Sensory Substitution and the Human-Machine Interface." *Trends in Cognitive Sciences* 7 (12): 541–46. doi:10.1016/j.tics.2003.10.013.

Belkhiria, Chama, and Vsevolod Peysakhovich. 2020. "Electro-Encephalography and Electro-Oculography in Aeronautics: A Review Over the Last Decade (2010–2020)." *Frontiers in Neuroergonomics* 1 (December): 606719. doi:10.3389/fnrgo.2020.606719.

Bezbaruah, Rajashri, Vivek P. Chavda, Lawandashisha Nongrang, Shahnaz Alom, Kangkan Deka, Tutumoni Kalita, Farak Ali, Bedanta Bhattacharjee, and Lalitkumar Vora. 2022. "Nanoparticle-based delivery systems for vaccines." *Vaccines* 10 (11): 1946. doi:10.3390/vaccines10111946.

Bissoli, Alexandre, Daniel Lavino-Junior, Mariana Sime, Lucas Encarnação, and Teodiano Bastos-Filho. 2019. "A Human–Machine Interface Based on Eye Tracking for Controlling and Monitoring a Smart Home Using the Internet of Things." *Sensors* 19 (4). Multidisciplinary Digital Publishing Institute: 859. doi:10.3390/s19040859.

Bobra, M. G., X. Sun, J. T. Hoeksema, M. Turmon, Y. Liu, K. Hayashi, G. Barnes, and K. D. Leka. 2014. "The Helioseismic and Magnetic Imager (HMI) Vector Magnetic Field Pipeline: SHARPs: Space-Weather HMI Active Region Patches." *Solar Physics* 289 (9): 3549–78. doi:10.1007/s11207-014-0529-3.

Bogaerts, J. 2014. "Complementary Metal-Oxide-Semiconductor (CMOS) Image Sensors for Use in Space." In *High Performance Silicon Imaging*, 250–80. Elsevier. doi:10.1533/9 780857097521.2.250.

"Character Recognition in Air-Writing Based on Network of Radars for Human-Machine Interface | IEEE Journals & Magazine | IEEE Xplore." 2022. Accessed November 27. https://ieeexplore.ieee.org/document/8735896.

Chaudhary, Vrantika, Sumit Jangra, and Neelam R. Yadav. 2018. "Nanotechnology Based Approaches for Detection and Delivery of MicroRNA in Healthcare and Crop Protection." *Journal of Nanobiotechnology* 16 (1): 40. doi:10.1186/s12951-018-0368-8.

Chavda, Vivek P. 2019. "Chapter 4: Nanobased Nano Drug Delivery: A Comprehensive Review." In *Applications of Targeted Nano Drugs and Delivery Systems*, edited by Shyam S. Mohapatra, Shivendu Ranjan, Nandita Dasgupta, Raghvendra Kumar Mishra, and Sabu Thomas, 69–92. Micro and Nano Technologies. Elsevier. doi:10.1016/B978-0-12-814029-1.00004-1.

Chavda, Vivek P., Yavuz Nuri Ertas, Vinayak Walhekar, DhartiModh, Avani Doshi, Nirav Shah, Krishna Anand, and Mahesh Chhabria. 2021a. "Advanced Computational Methodologies Used in the Discovery of New Natural Anticancer Compounds." *Frontiers in Pharmacology* 12 (August): 702611. doi:10.3389/fphar.2021.702611.

Chavda, Vivek P., Zeel Patel, Yashti Parmar, and Disha Chavda. 2021b. "In Silico Protein Design and Virtual Screening." In *Computation in Bioinformatics*, 85–99. John Wiley & Sons, Ltd. doi:10.1002/9781119654803.ch5.

Chavda, Vivek P., Amit Sorathiya, Disha Valu, and Swati Marwadi. 2021c. "Role of Data Mining in Bioinformatics." In *Computation in Bioinformatics*, 69–84. John Wiley & Sons, Ltd. doi:10.1002/9781119654803.ch4.

Chavda, Vivek P, Gargi Jogi, Ana Cláudia Paiva-Santos, and Ajeet Kaushik. 2022a. "Biodegradable and Removable Implants for Controlled Drug Delivery and Release Application." *Expert Opinion on Drug Delivery* 0 (0). Taylor & Francis: 1–5. doi:10.10 80/17425247.2022.2110065.

Chavda, Vivek P., Gargi Jogi, Nirav Shah, Mansi N. Athalye, Nirav Bamaniya, Lalitkumar K Vora, and Ana Cláudia Paiva-Santos. 2022b. "Advanced Particulate Carrier-Mediated Technologies for Nasal Drug Delivery." *Journal of Drug Delivery Science and Technology* 74 (August): 103569. doi:10.1016/j.jddst.2022.103569.

"Complementary Metal-Oxide-Semiconductor: An Overview | ScienceDirect Topics." 2022. Accessed November 28. https://www.sciencedirect.com/topics/engineering/complementary-metal-oxide-semiconductor.

Damiati, Safa A. 2020. "Digital Pharmaceutical Sciences." *AAPS PharmSciTech* 21 (6): 206. doi:10.1208/s12249-020-01747-4.

Dong, Bowei, Yanqin Yang, Qiongfeng Shi, Siyu Xu, Zhongda Sun, Shiyang Zhu, Zixuan Zhang, et al. 2020. "Wearable Triboelectric–Human–Machine Interface (THMI) Using Robust Nanophotonic Readout." *ACS Nano* 14 (7): 8915–30. doi:10.1021/acsnano.0c03728.

Elbestawi, Mo, Dan Centea, Ishwar Singh, and Tom Wanyama. 2018. "SEPT Learning Factory for Industry 4.0 Education and Applied Research." *Procedia Manufacturing* "Advanced Engineering Education & Training for Manufacturing Innovation"8th CIRP Sponsored Conference on Learning Factories (CLF 2018) 23 (January): 249–54. doi:10.1016/j.promfg.2018.04.025.

Esposito, Daniele, Jessica Centracchio, Emilio Andreozzi, Gaetano D. Gargiulo, Ganesh R. Naik, and Paolo Bifulco. 2021. "Biosignal-Based Human–Machine Interfaces for Assistance and Rehabilitation: A Survey." *Sensors* 21 (20). Multidisciplinary Digital Publishing Institute: 6863. doi:10.3390/s21206863.

"Fabrication of 3D Micro-Architected/Nano-Architected Materials: ScienceDirect." 2022. Accessed November 28. https://www.sciencedirect.com/science/article/pii/B978032 3353212000182.

Ferreira, Andre, Wanderley C. Celeste, Fernando A. Cheein, Teodiano F. Bastos-Filho, Mario Sarcinelli-Filho, and Ricardo Carelli. 2008. "Human-Machine Interfaces Based on EMG and EEG Applied to Robotic Systems." *Journal of Neuroengineering and Rehabilitation* 5 (March): 10. doi:10.1186/1743-0003-5-10.

Feyissa, Anteneh M., and William O. Tatum. 2019. "Adult EEG." In *Handbook of Clinical Neurology*, 160: 103–24. Elsevier. doi:10.1016/B978-0-444-64032-1.00007-2.

"Frontiers | Brain–Computer Interface (BCI) Applications in Mapping of Epileptic Brain Networks Based on Intracranial-EEG: An Update." 2022. Accessed November 28. https://www.frontiersin.org/articles/10.3389/fnins.2019.00191/full.

Gao, Jianye, Jinfeng Wang, and Jing Xie. 2022. "Application of PLC and HMI in the CO_2 Transcritical Refrigeration Experimental Platform." *Scientific Reports* 12 (1): 15199. doi:10.1038/s41598-022-19602-w.

Grabowska, Sandra, Sebastian Saniuk, and Bożena Gajdzik. 2022. "Industry 5.0: Improving Humanization and Sustainability of Industry 4.0." *Scientometrics* 127 (6): 3117–44. doi:10.1007/s11192-022-04370-1.

"Graphene-Based Wearable Sensors for a Human-Machine Interface: (A)... | Download Scientific Diagram." 2022. Accessed November 27. https://www.researchgate.net/figure/Graphene-based-wearable-sensors-for-a-human-machine-interface-a-schematic-illustration_fig2_326279447.

Greentree, Andrew D., Barbara A. Fairchild, Faruque M. Hossain, and Steven Prawer. 2008. "Diamond Integrated Quantum Photonics." *Materials Today* 11 (9): 22–31. doi:10.1016/S1369-7021(08)70176-7.

Handelzalts, Shirley, Giulia Ballardini, Chen Avraham, Mattia Pagano, Maura Casadio, and Ilana Nisky. 2021. "Integrating Tactile Feedback Technologies into Home-Based Telerehabilitation: Opportunities and Challenges in Light of COVID-19 Pandemic." *Frontiers in Neurorobotics* 15 (February): 617636. doi:10.3389/fnbot.2021.617636.

He, Yang, Chenyan Hu, Zhijia Li, Chuan Wu, Yuanyuan Zeng, and Cheng Peng. 2022. "Multifunctional Carbon Nanomaterials for Diagnostic Applications in Infectious Diseases and Tumors." *Materials Today. Bio* 14 (March): 100231. doi:10.1016/j.mtbio.2022.100231.

Hentout, Abdelfetah, Mustapha Aouache, Abderraouf Maoudj, and Isma Akli. 2019. "Human–Robot Interaction in Industrial Collaborative Robotics: A Literature Review of the Decade 2008–2017." *Advanced Robotics* 33 (15–16). Taylor & Francis: 764–99. doi:10.1080/01691864.2019.1636714.

Houssein, Essam H., Asmaa Hammad, and Abdelmgeid A. Ali. 2022. "Human Emotion Recognition from EEG-Based Brain–Computer Interface Using Machine Learning: A Comprehensive Review." *Neural Computing and Applications* 34 (15): 12527–57. doi:10.1007/s00521-022-07292-4.

"How Nanotechnology Enables Wearable Electronics." 2022. Accessed November 27. https://www.nanowerk.com/spotlight/spotid=52432.php.

"ICHMS 2020–1st IEEE International Conference on Human-Machine Systems." 2022. Accessed November 27. http://www.sensyscal.it/ichms2020/index.html.

Jalali, Bahram, and Sasan Fathpour. 2006. "Silicon Photonics." *Journal of Lightwave Technology* 24 (12): 4600–4615. doi:10.1109/JLT.2006.885782.

Janapati, Ravichander, Vishwas Dalal, and Rakesh Sengupta. 2021. "Advances in Modern EEG-BCI Signal Processing: A Review." *Materials Today: Proceedings*, July. doi:10.1016/j.matpr.2021.06.409.

Janapati, Ravichander, Vishwas Dalal, Rakesh Sengupta, Usha Desai, P. V. Raja Shekar, and Sreedhar Kollem. 2022. "Towards a More Theory-Driven BCI Using Source Reconstructed Dynamics of EEG Time-Series." *Nano LIFE* 12 (02): 2250005. doi:10.1142/S1793984422500052.

Janapati, Ravichander, Viswas Dalal, N. Govardhan, and Rakesh Sen Gupta. 2020. "Review on EEG-BCI Classification Techniques Advancements." *IOP Conference Series: Materials Science and Engineering* 981 (3): 032019. doi:10.1088/1757-899X/981/3/032019.

Janapati, Ravichander, Viswas Dalal, G. Mahesh Kumar, P. Anuradha, and P. V. Raja Shekar. 2022. "Web Interface Applications Controllers Used by Autonomous EEG-BCI Technologies." In *030038*. Warangal, India. doi:10.1063/5.0081780. https://pubs.aip.org/aip/acp/article/2418/1/030038/2822128/Web-interface-applications-controllers-used-by

Javaid, Mohd, Abid Haleem, Ravi Pratap Singh, and Rajiv Suman. 2022. "Enhancing Smart Farming through the Applications of Agriculture 4.0 Technologies." *International Journal of Intelligent Networks* 3: 150–64. doi:10.1016/j.ijin.2022.09.004.

Johannsen, Gunnar, John E. Rijnsdorp, and Andrew P. Sage. 1983. "Human System Interface Concerns in Support System Design." *Automatica* 19 (6): 595–603. doi:10.1016/0005-1098(83)90023-7.

Kala, Rahul. 2016. "Advanced Driver Assistance Systems." In *On-Road Intelligent Vehicles*, 59–82. Elsevier. doi:10.1016/B978-0-12-803729-4.00004-0.

Kim, Taemin, Myeongki Cho, and Yu Ki. 2018. "Flexible and Stretchable Bio-Integrated Electronics Based on Carbon Nanotube and Graphene." *Materials* 11 (7): 1163. doi:10.3390/ma11071163.

Korah, Binila K., and Beena Mathew. 2022. "Sustainable Carbon Quantum Dots from Vitex Negundo Leaves as a Synergistic Nanoplatform for Triple Object Sensing and Anticounterfeiting Applications." *Materials Today Sustainability*, November, 100273. doi:10.1016/j.mtsust.2022.100273.

Kwon, Young-Tae, Yun-Soung Kim, Shinjae Kwon, Musa Mahmood, Hyo-Ryoung Lim, Si-Woo Park, Sung-Oong Kang, et al. 2020. "All-Printed Nanomembrane Wireless Bioelectronics Using a Biocompatible Solderable Graphene for Multimodal Human-Machine Interfaces." *Nature Communications* 11 (1): 3450. doi:10.1038/s41467-020-17288-0.

Lee, Chi-Chun, Emily Mower, Carlos Busso, Sungbok Lee, and Shrikanth Narayanan. 2011. "Emotion Recognition Using a Hierarchical Binary Decision Tree Approach." *Speech Communication, Sensing Emotion and Affect: Facing Realism in Speech Processing*, 53 (9): 1162–71. doi:10.1016/j.specom.2011.06.004.

Legrand, Mathilde, Manelle Merad, Etienne de Montalivet, Agnès Roby-Brami, and Nathanaël Jarrassé. 2018. "Movement-based control for upper-limb prosthetics: Is the regression technique the key to a robust and accurate control?" *Frontiers in Neurorobotics* 12 (July): 41. doi:10.3389/fnbot.2018.00041.

Lim, Yixiang, Alessandro Gardi, Roberto Sabatini, Subramanian Ramasamy, Trevor Kistan, Neta Ezer, Julian Vince, and Robert Bolia. 2018. "Avionics Human-Machine Interfaces and Interactions for Manned and Unmanned Aircraft." *Progress in Aerospace Sciences* 102 (October): 1–46. doi:10.1016/j.paerosci.2018.05.002.

Liu, Yuhao, James J. S. Norton, Raza Qazi, Zhanan Zou, Kaitlyn R. Ammann, Hank Liu, Lingqing Yan, et al. 2016. "Epidermal Mechano-Acoustic Sensing Electronics for Cardiovascular Diagnostics and Human-Machine Interfaces." *Science Advances* 2 (11): e1601185. doi:10.1126/sciadv.1601185.

Malerba, F. 1999a. "'History-Friendly' Models of Industry Evolution: The Computer Industry." *Industrial and Corporate Change* 8 (1): 3–40. doi:10.1093/icc/8.1.3.

———. 1999b. "'History-Friendly' Models of Industry Evolution: The Computer Industry." *Industrial and Corporate Change* 8 (1): 3–40. doi:10.1093/icc/8.1.3.

Mendoza, E., J. Andramuño, J. Núñez, and L. Córdova. 2021. "Human Machine Interface (HMI) Based on a Multi-Agent System in a Water Purification Plant." *Journal of Physics: Conference Series* 2090 (1). IOP Publishing: 012122. doi:10.1088/1742-6596/2090/1/012122.

Mitternacht, Jürgen, Aljoscha Hermann, and Patrick Carqueville. 2022. "Acquisition of Lower-Limb Motion Characteristics with a Single Inertial Measurement Unit: Validation for Use in Physiotherapy." *Diagnostics* 12 (7): 1640. doi:10.3390/diagnostics12071640.

"Nanotechnology in Healthcare." 2022. Accessed November 27. https://www.nanowerk.com/spotlight/spotid=47031.php.

Nasrollahzadeh, Mahmoud, Zahra Issaabadi, Mohaddeseh Sajjadi, S. Mohammad Sajadi, and Monireh Atarod. 2019. "Types of Nanostructures." In *Interface Science and Technology*, 28:29–80. Elsevier. doi:10.1016/B978-0-12-813586-0.00002-X.

Olmo, Manuel del, and Rosario Domingo. 2020. "EMG Characterization and Processing in Production Engineering." *Materials* 13 (24): 5815. doi:10.3390/ma13245815.

"Optical Applications of Nanomaterials." 2022. *Springerprofessional.De*. Accessed November 27. https://www.springerprofessional.de/en/optical-applications-of-nanomaterials/17077134.

Peng, Zhiwei, Qinglin Lin, Yu-An Angela Tai, and YuHuang Wang. 2020. "Applications of Cellulose Nanomaterials in Stimuli-Responsive Optics." *Journal of Agricultural and Food Chemistry* 68 (46): 12940–55. doi:10.1021/acs.jafc.0c04742.

Ribeiro, José, Francisco Mota, Tarique Cavalcante, Ingrid Nogueira, Victor Gondim, Victor Albuquerque, and Auzuir Alexandria. 2019. "Analysis of Man-Machine Interfaces in Upper-Limb Prosthesis: A Review." *Robotics* 8 (1): 16. doi:10.3390/robotics8010016.

Rocha-Jácome, Cristian, Ramón González Carvajal, Fernando Muñoz Chavero, Esteban Guevara-Cabezas, and Eduardo Hidalgo Fort. 2021. "Industry 4.0: A Proposal of Paradigm Organization Schemes from a Systematic Literature Review." *Sensors* 22 (1): 66. doi:10.3390/s22010066.

Roh, Eun, Byeong-Ung Hwang, Doil Kim, Bo-Yeong Kim, and Nae-Eung Lee. 2015. "Stretchable, Transparent, Ultrasensitive, and Patchable Strain Sensor for Human-Machine Interfaces Comprising a Nanohybrid of Carbon Nanotubes and Conductive Elastomers." *ACS Nano* 9 (6): 6252–61. doi:10.1021/acsnano.5b01613.

Sang, Jun, Zhongyuan Wu, Pei Guo, Haibo Hu, Hong Xiang, Qian Zhang, and Bin Cai. 2018. "An Improved YOLOv2 for Vehicle Detection." *Sensors* 18 (12): 4272. doi:10.3390/s18124272.

Saniuk, Sebastian, Sandra Grabowska, and Martin Straka. 2022. "Identification of Social and Economic Expectations: Contextual Reasons for the Transformation Process of Industry 4.0 into the Industry 5.0 Concept." *Sustainability* 14 (3). Multidisciplinary Digital Publishing Institute: 1391. doi:10.3390/su14031391.

Sankai, Yoshiyuki. 2010. "HAL: Hybrid Assistive Limb Based on Cybernics." In *Robotics Research*, edited by Makoto Kaneko and Yoshihiko Nakamura 66:25–34. Springer Tracts in Advanced Robotics. Berlin, Heidelberg: Springer. doi:10.1007/978-3-642-14743-2_3.

Schrinner, Philip P. J., Jan Olthaus, Doris E. Reiter, and Carsten Schuck. 2020. "Integration of Diamond-Based Quantum Emitters with Nanophotonic Circuits." *Nano Letters* 20 (11): 8170–77. doi:10.1021/acs.nanolett.0c03262.

Sen, Abir, Tapas Kumar Mishra, and Ratnakar Dash. 2022. "Design of Human Machine Interface through Vision-Based Low-Cost Hand Gesture Recognition System Based on Deep CNN." *arXiv*. doi:10.48550/arXiv.2207.03112.

Singh, Harpreet Pal, and Parlad Kumar. 2021. "Developments in the Human Machine Interface Technologies and Their Applications: A Review." *Journal of Medical Engineering & Technology* 45 (7): 552–73. doi:10.1080/03091902.2021.1936237.

Sukhavasi, Susrutha Babu, Suparshya Babu Sukhavasi, Khaled Elleithy, Ahmed El-Sayed, and Abdelrahman Elleithy. 2022. "Deep Neural Network Approach for Pose, Illumination, and Occlusion Invariant Driver Emotion Detection." *International Journal of Environmental Research and Public Health* 19 (4): 2352. doi:10.3390/ijerph19042352.

Sun, Zhongda, Minglu Zhu, and Chengkuo Lee. 2021. "Progress in the Triboelectric Human–Machine Interfaces (HMIs)-Moving from Smart Gloves to AI/Haptic Enabled HMI in the 5G/IoT Era." *Nanoenergy Advances* 1 (September): 81–120. doi:10.3390/nanoenergyadv1010005.

Swathi, M., K. C. Sreedhar, Meeravali Shaik, and V. Prabhakar. 2022. "A Novel Approach to Improve the Training of Deep Convolutional Networks." In *Proceedings of the 2nd International Conference on Recent Trends in Machine Learning, IoT, Smart Cities and Applications*, edited by Vinit Kumar Gunjan and Jacek M. Zurada, 199–209. Lecture Notes in Networks and Systems. Singapore: Springer Nature. doi:10.1007/978-981-16-6407-6_19.

"The Helioseismic and Magnetic Imager (HMI) Vector Magnetic Field Pipeline: SHARPs – Space-Weather HMI Active Region Patches | SpringerLink." 2022. Accessed November 28. https://link.springer.com/article/10.1007/s11207-014-0529-3.

Tian, He, Guang-Yang Gou, Fan Wu, Lu-Qi Tao, Yi Yang, Tian-Ling Ren, He Tian, et al. 2019. *Graphene Acoustic Devices. Graphene and Its Derivatives: Synthesis and Applications*. IntechOpen. doi:10.5772/intechopen.81603.

Tsarouchi, Panagiota, Sotiris Makris, and George Chryssolouris. 2016. "Human–Robot Interaction Review and Challenges on Task Planning and Programming." *International Journal of Computer Integrated Manufacturing* 29 (8). Taylor & Francis: 916–31. doi:10.1080/0951192X.2015.1130251.

Vaidya, Saurabh, Prashant Ambad, and Santosh Bhosle. 2018. "Industry 4.0: A Glimpse." Procedia Manufacturing, 2nd International Conference on Materials, Manufacturing and Design Engineering (iCMMD2017), 11-12 December 2017, MIT, Aurangabad, Maharashtra, India, 20 (January): 233–38. doi:10.1016/j.promfg.2018.02.034.

Valdevit, Lorenzo, and Jens Bauer. 2016. "Fabrication of 3D Micro-Architected/Nano-Architected Materials." In *Three-Dimensional Microfabrication Using Two-Photon Polymerization*, 345–73. Elsevier. doi:10.1016/B978-0-323-35321-2.00018-2.

Vinitha, K., R. A. Prabhu, R. Bhaskar, and R. Hariharan. 2020. "Review on Industrial Mathematics and Materials at Industry 1.0 to Industry 4.0." *Materials Today: Proceedings, International Conference on Nanotechnology: Ideas, Innovation and Industries* 33 (January): 3956–60. doi:10.1016/j.matpr.2020.06.331.

Wallhoff, F., M. Ablassmeier, A. Bannat, S. Buchta, A. Rauschert, G. Rigoll, and M. Wiesbeck. 2007. "Adaptive Human-Machine Interfaces in Cognitive Production Environments." In 2246–49. doi:10.1109/ICME.2007.4285133.

Wang, Hong, Xiaohua Ma, and Yue Hao. 2017. "Electronic Devices for Human-Machine Interfaces." *Advanced Materials Interfaces* 4 (4): 1600709. doi:10.1002/admi.201600709.

Wang, Hongman, Hui Qiao, Jingyu Lin, Rihui Wu, Yebin Liu, and Qionghai Dai. 2021. "Model Study of Transient Imaging With Multi-Frequency Time-of-Flight Sensors." *IEEE Transactions on Pattern Analysis and Machine Intelligence* 43 (10): 3523–39. doi:10.1109/TPAMI.2020.2981574.

Wang, Nianfeng, Kunyi Lao, and Xianmin Zhang. 2017. "Design and Myoelectric Control of an Anthropomorphic Prosthetic Hand." *Journal of Bionic Engineering* 14 (1): 47–59. doi:10.1016/S1672-6529(16)60377-3.

Xia, Wei, Yu Zhou, Xingchen Yang, Keshi He, and Honghai Liu. 2019. "Toward Portable Hybrid Surface Electromyography/A-Mode Ultrasound Sensing for Human–Machine Interface." *IEEE Sensors Journal* 19 (13): 5219–28. doi:10.1109/JSEN.2019.2903532.

Yonan, Matthew. 2018. "Flexible Loudspeaker Made of Nanowires Will Stick to Your Skin and Play Music." *DexMat*. August 4. https://dexmat.com/application/flexible-loud-speaker-made-of-nanowires-will-stick-to-your-skin-and-play-music/.

Zambon, Ilaria, Massimo Cecchini, Gianluca Egidi, Maria Grazia Saporito, and Andrea Colantoni. 2019. "Revolution 4.0: Industry vs. Agriculture in a Future Development for SMEs." *Processes* 7 (1): 36. doi:10.3390/pr7010036.

Zander, Thorsten O., and Christian Kothe. 2011. "Towards Passive Brain-Computer Interfaces: Applying Brain-Computer Interface Technology to Human-Machine Systems in General." *Journal of Neural Engineering* 8 (2): 025005. doi:10.1088/1741-2560/8/2/025005.

8 An Engineering Approach of Noninvasive Detection and Analysis of Human Heart Sound for Alarm Generation

Iti Saha Misra

CONTENTS

DOI: 10.1201/9781003326830-8

8.1   INTRODUCTION

Cardiovascular diseases (CVDs) have escalated 34% in India, resulting 209 deaths per one lakh people in the country. Another statistic states that out of 30 million people with CVDs in India, 14 million cases are from urban areas and 16 million from rural areas [1]. The number of death per year due to heart diseases is shown in Figure 8.1 [2] among different age groups. It shows that coronary heart disease and the stroke take a vital role in mortality rate.

The population growth over the world creates three major challenges:

1. Demographic peak of baby boomers;
2. Increase of life expectancy leading to aging population; and
3. Rise in health care costs.

The average life expectancy is in increasing order as reported by the World Health Organization (WHO). This growth rate is higher in eastern side of the world as compared to the western side. Among the Asian countries, India and China have this growth rate extremely high. According the latest WHO publication in 2020, the life expectancy of male is 69.5, female: 70.2, giving total life expectancy in India 70.8 years with world ranking of 117 [3]. Living longer may always not better. Lack of timely diagnosis may not save many lives. Prevention is better than cure—this old proverb may guide us for timely diagnosis to save lives.

Figure 8.2 shows the number of deaths per year throughout the world for different CVDs. The medical facility for the rural economically poor people is constrained because of lack of sufficient number of doctors, hospitals, and many other factors related to the man power and infrastructure. Further, according to the 2020 Human Development report, India has 8.6 doctors, 5 hospital beds for every 10,000 people and ranks 155th in the developmental index in this contest. From Figure 8.3 we can see the situation of the ratio of physicians and people in different countries. It shows

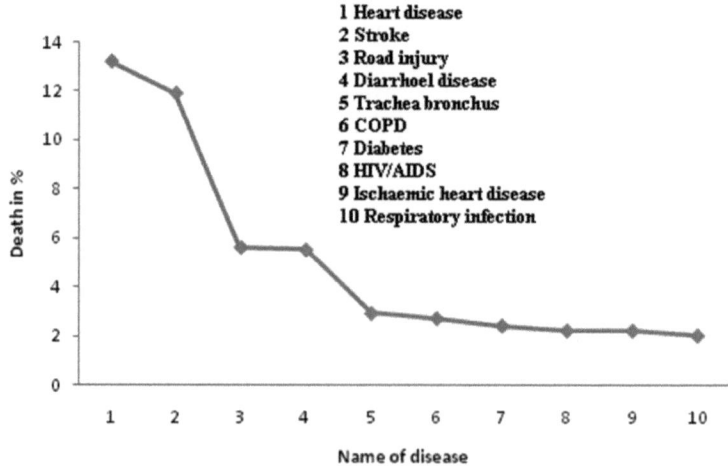

FIGURE 8.1   Percentage death of 10 diseases in the world.

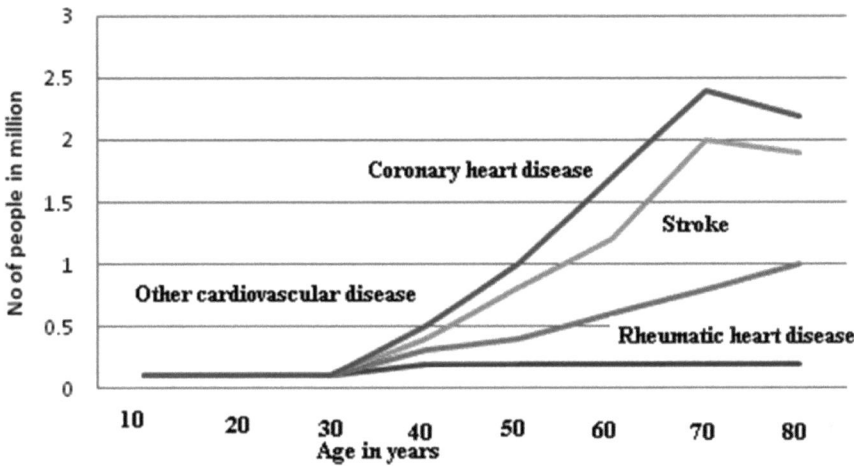

FIGURE 8.2   The number of death per year throughout the world for different cardiovascular diseases.

FIGURE 8.3   Density of physicians per 1,000 people in different countries.

that India is lagging in terms of the ratio of physician to the people [4]. According to the WHO, India should increase doctor–patient ratio from 1:1700 to 1:1000.

The death toll for CVDs is highest due to insufficient timely support from hospitals and other medical establishments. Here is the significance of this sort of research where low-cost and user-friendly health checkup system for early detection of possible life-threatening CVDs to save millions of common lives. Studies on heart diseases and the treatment of heart diseases including the diagnosis cost reveal that the cost of heart-related treatment is very high. Even for a well-to-do family bearing the cost of treatment is difficult and it is much challenging to the people of weaker economic background as they are sometimes rejected for a loan or mostly do not have health insurance. The alternative way is the early detection of CVD and lifestyle maintenance. Therefore, awareness generation of heart diseases through cost-effective heart diagnosis method becomes the need of the hour for the common mass.

From early days, heart sound signal (HSS) has been taken as an important parameter for the diagnosis of CVDs. In 1816, Hyacinth Laennec invented acoustic stethoscope to listen to the heart sound. Since then to till now, the direct auscultation (understanding of heart sound) is becoming the traditional way of heart disease diagnosis. But the understanding of heart sound is not an easy job. Only specialist doctors and trained medical practitioners can understand the heart sound by the auscultation method. Therefore for the diagnosis process, recording or detection of heart sound is required through analysis, which helps to get the proper diagnosis of heart disease. For the enhancement of clinical diagnosis process, the audio recording and thereafter graphical representation of heart sound were achieved. The first recording of heart sound was made by Hurthle in 1895 [5]. Later on, the graphical representation of heart sound, which is known as phonocardiogram (PCG), has been developed for visual inspection of HSS. To detect valvular cardiac dysfunctions, cheap and efficient computer-based advanced auscultation techniques are being used recently throughout the world [6]. Recent developments in CVDs show that there are two stages involved in the process of heart disease diagnosis by auscultation technique: acquisition and analysis of HSS [7]. The acquisition means taking of heart sound data from the patient while analysis means finding of different parameters of the heart sound, which are taken as the pathological parameters for the diagnosis purpose.

Electronic stethoscope converts the received heart sound signal to an electrical signal [8], which can be electronically amplified for better listening. This electrical signal is an analog signal and it can also be converted to digital signal where the entire signal processing task can be done. PCG that requires very affordable equipment and skills, common to all physicians, is another alternative way of heart disease diagnosis [6]. The disadvantage of PCG is that the stethoscope does not store any heart sound data for analysis, so the results are best on the experience and abilities of the physicians. Analyzing PCG signal with modern signal processing techniques is gaining importance for heart anomaly detection.

PCG signal carries information about the time duration, frequency, and other important parameters of heart sound to determine the current conditions and functionality of the heart. By hearing heart sounds through a stethoscope only needs expert skill for diagnoses. A variety of electronic stethoscopes are available, which help to design some automatic heart diagnosis system [9, 10]. In [11], it shows that the pathological sound can be analyzed in both time and frequency domains of heart sound. In [12], an artificial neural network-based heart sound analysis has been described but the detection of heart sound and purification of the signal have been ignored. Wavelet-based denoising [13] techniques are used by some researchers on heart sound and ECG but it is not an instantaneous result. In very recent time heart sound signal detection and analysis have been done using various artificial and machine learning techniques. In [14], a deep learning method to denoise, compress, segment, and classify PCG signals effectively has been done with accuracy level of 95%. Mel-scaled power spectrogram (MSPS) and Mel-frequency cepstral coefficients (MFCCs) for feature extraction and five-layer feed-forward deep neural network (DNN) for classification are applied in this paper. In [15], an automatic detection and classification of systolic and diastolic profiles of PCG signal corrupted

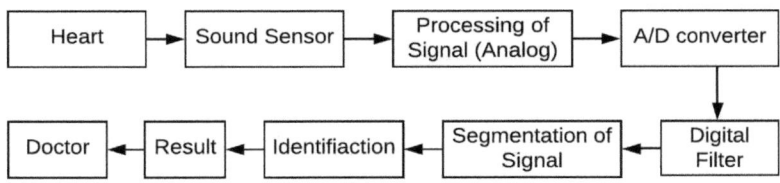

FIGURE 8.4   General heart sound detection process.

with noise during the capture of electronic stethoscope (because the subjects may suffer from respiratory dysfunction, cough, mental illness, etc.) recording has been shown. In this chapter, time duration features obtained from segmentation are used for the classification of systolic and diastolic profiles of PCG, and the automated classification is done using hidden semi Markov model (HSMM), multilayer perceptron (MLP), support vector machine (SVM), and k-nearest neighbor (KNN) classifiers. The performance comparison metrics of the classifiers are also given. The classification accuracy for systolic and diastolic profiles of PCG is shown to be 96.74%. Though these methods show accuracy level of more than 95% for classification but are computationally heavy and complex. With a little less accuracy PCG signal detection and analysis is possible.

The objective of this work is to provide a simple but complete solution from heart sound acquisition to detection with reasonable accuracy (>90%) to be handled by persons having only basic education and knowledge of system handling. It follows low cost acquisition procedure with reliability. Component identification algorithm is less complex with simple mathematical basis. Heart sound acquisition is done using developed electronic circuits in the laboratory and signal processing by DSP Kit, which is easy to handle, less installation complex, and is also portable. HSS denoising is performed using traditional wavelet analysis and with a developed power law algorithm. For component identification, a simple SEPD algorithm is used. Finally, HSS analysis is done based on the features of heart rate, duration of S1, duration of S2, duration between S1 and S2, duration between S2 and S1, and power spectral density (PSD) of each component to be more surer for component identification. Figure 8.4 is the general block diagram of HSS detection process.

8.2   STRUCTURE OF HUMAN HEART AND DIFFERENT HEART SOUNDS

The human heart is roughly the size of a fist having a mass of 250–350 g for an adult heart. Figure 8.5 is the structure of the human heart [16]. To understand heart sound signal or PCG with clarity, some basic ideas about the following points are outlined.

- Location of the heart
- Physical structure of the heart
- Function of the heart
- Heart sound mechanism

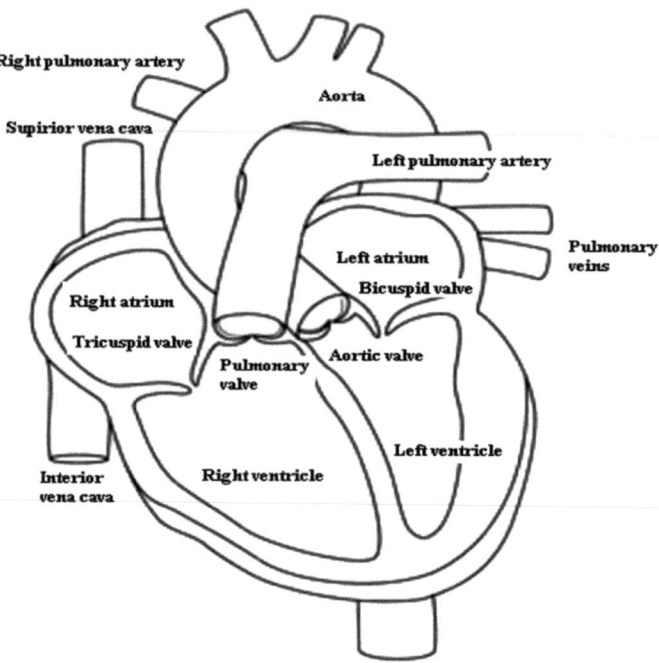

FIGURE 8.5   Structure of the human heart.

The heart is situated anterior to the vertebral column and posterior to the sternum [16]. It is enclosed with a double-walled sac, which is known as the pericardium, and the main part of this sac is known as fibrous pericardium [17]. The main role of this sac is to give the proper structure and to protect the heart from overfilling with blood. There are three layers in the outer wall of the human heart. These layers are known as epicardium, myocardium, and endocardium.

Figure 8.5 shows the structure of the human heart consisting of four different chambers; two are recognized as superior atria and two are inferior ventricles. The atria is working as the blood receiving chambers, and the ventricle chambers are working as discharging chambers. The pulmonary circuit and the systemic loop make the pathway of blood in the heart, which assists to circulate blood to and from the body [18]. Four valves take the vital role to control the blood flow within these four chambers. These are: tricuspid valve, mitral valve, aortic valve, and pulmonary valve. Due to the expansion and compression of the heart muscles, blood enters into the two chambers and leaves from other chambers through the valves [18]. When the blood enters the heart, two valves remain opened and two remain closed. Now due to the closure of the heart valves, a sound is produced, which is known as heart sound (HS). Now if there are no problems in the valves or in the whole heart, a normal HS is produced and if there is any problem in the valves or in another part of the heart an abnormal HS is the result [17]. Blood comes from the body to the receiving chambers (superior atria) through pulmonary veins. When the receiving chambers are filled up with blood, the valves *start to open* to enter the blood to the discharging chambers (inferior ventricles). When these two discharging chambers are filled up with blood the two valves known as *the tricuspid valve* and the *mitral valve* start to be closed and after a certain time, they are closed completely. Due to the closure of these two valves, a sound is produced and it is known as first heart sound S1.

In the next happenings, the other two valves known as the *pulmonary valves* and the *aortic valves* start to be opened and allow the blood to pass to the body. After a certain time, these two valves are closed. Due to the closure of these two valves, another heart sound is produced and it is known as the second heart sound S2 [19]. Some extra heart sounds may be produced due to other factors like the improper closure of the heart valves, leakages, some problem in the walls, etc. These extra sounds are known as murmurs.

Some important parameters related to heart are:

- Cardiac cycle
- Heartbeat rate
- Heart sound and heart sound components
- Normal and abnormal heart sounds

8.2.1 CARDIAC CYCLE

We know that the human heart produces an average of 72 heartbeats in each minute. The term 'cardiac cycle' refers to the time taken by each heartbeat. In other way, it intimates the events occur due to the flow of blood inside the heart from beginning of one heartbeat to the beginning of the next heartbeat. The *frequency of the cardiac cycle is the heart rate.*

A separate pathway to the lungs is created through the four chambers of atrium and ventricles [20]. The blood circulation pathways are known as *pulmonary circulation* (lead to the lungs) and *systemic circulation* (lead to the body) [18]. A single cardiac cycle consists of two basic phases: diastole and systole. When the ventricles are relaxed (without contraction), the time period is the diastole. Systole represents the time duration for which the left and right ventricles contract and eject blood into the aorta and pulmonary artery, respectively. At the end of the diastolic period, both atria contract injecting an excess amount of blood into the ventricles [16].

8.2.2 HEART SOUNDS

From the above discussions, it is understood that the main cause of the heart sound is the vibration of the flaps of the heart valves due to the closure of the heart valves. Heart sound consists of few heart sound components and murmurs, which play a significant role to diagnose heart diseases. For healthy adults, two normal heart sounds, a "lub" and a "dub," occur in sequence with each heartbeat [21].

8.2.3 HEART SOUND COMPONENTS

The two main heart sound components are the first heart sound component S1 and the second heart sound component S2. Other heart sounds components S3 and S4 are also called as heart murmurs.

The first heart sound S1: Though there are several high-frequency components in S1, only the first two are normally audible related to the closure of the mitral and tricuspid valves. When these two valves close, vibration is caused by the flaps of the valves and circulation of the blood inside the ventricles producing the first heart sound S1. The contraction of the ventricles coincides with the first heart sound, identifying the onset of ventricular systole and the end of mechanical diastole. The first heart sound S1 has finite duration and is composed of components M1 (mitral valve closure) and T1 (tricuspid valve closure).

The second heart sound S2: S2 occurs mainly due to the closer of aortic and pulmonary valves. Two audible components of S2 are the aortic closure sound (A2) and the pulmonic closure sound (P2). A2 and P2 are normally split on inspiration and merge on expiration [22]. The A2 sound is usually louder than the P2 due to higher pressures in the left side of the heart. S2 appears at the end of ventricular contraction and also recognizes the start of ventricular diastole and the end of systole [22]. The S2 is of shorter duration and higher frequency than the S1. To distinguish two audible sounds A2 and P2, they must be separated by more than 20 msec. Figure 8.6 shows the S1 and S2 components with their splitting parts.

8.2.4 SIGNIFICANCE OF HEART SOUNDS

The clinical significance of heart sounds is very important for analyzing to arrive at a decision. S1 and S2 are clinically evaluated by their intensity and the degree of splitting. In general S1 has a lower pitch and longer duration than S2. The maximum intensity of S1 occurs when the PQ interval lies between 80 and 120 msec. If PQ interval is greater than 200 msec, this implies the first-degree heart block [22]. A loud S1 is a definite mark of hemodynamically significant mitral stenosis (narrowing of the mitral valve). Mitral valve controls the flow of blood from the heart's left atrium to the left ventricle. For young people, S1 sounds are louder, which is also common in patients with thin chest walls. S1 splits into two audible components in normal cardiac auscultation, M1 and T1, separated by 20–30 msec. First-degree heart block is the most common cause of a soft S1.

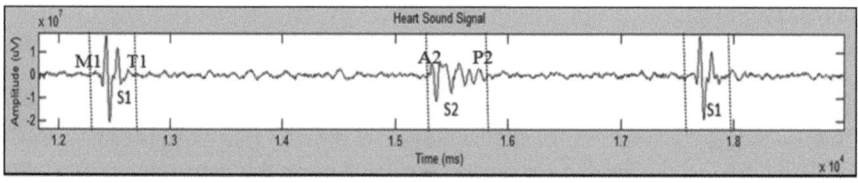

FIGURE 8.6 Heart sound signal (PCG) showing two primary components S1 and S2 with their splits.

8.2.5 EXTRA HEART SOUNDS

Besides principal components S1 and S2, sometimes few extra heart sound components occur; those are generally known as murmurs in the name of S3 and S4. These extra heart sound components have significance of having various heart diseases but difficult to detect. S3 is known as the ventricular gallop or proto-diastolic gallop [23], occurs just after S2 when the mitral valve opens (early diastole), and is lower in pitch than S1 or S2. S3 allows passive filling of the left ventricle (LV) and occurs when a large amount of blood striking a very compliant LV. S3 may be listened in case of young people, child, or sometimes in women during pregnancy. If it exists for a long time or become permanent then it becomes a symptom of any heart disease. Normally it indicates the increased volume of blood in the ventricles and could be a sign of systolic congestive heart failure.

The extra heart sound S4 is called as presystolic gallop or atrial gallop [22]. In general, S4 is caused by the sound of blood flowing forcefully into a stiff ventricle. The S4 sound occurs just after atrial contraction at the end of diastole and immediately before S1. Any abnormal condition that generates a noncompliant LV will produce an S4, while any condition that creates an overly compliant LV will produce an S3. Presence of S4 indicates abnormality in heart.

8.3 HEART SOUND SIGNAL ACQUISITION

For HSS acquisition a digital stethoscope is used. A chest piece receiving the heart sound of a subject is fed to a microphone to convert the sound signal into weak electrical signal (generally 1–10 mV, maximum up to 50 mV), which needs to be passed through an amplifier circuit to boost up the signal (200 mV). A preamplifier is an electronic amplifier that enhances the power level of the weak electrical signal into an output signal strong enough for processing. The amplified signal is fed to an analog low-pass filter (LPF) to remove high frequencies present in the heart sound. Generally, HSS components including murmurs have the frequency spectrum within a range of 10–400 Hz [22]. Beyond this range, the signal is considered as noise or unwanted signal. A LPF with cut-off frequency of 350 Hz was designed. A DSP Kit (TMS320C6713) is employed at the output of the LPF for signal processing using the inbuilt CODEC of DSP processor to convert the analog HSS to digital signal, sampled at the rate of 8 kHz [24]. The advantage of the DSP Kit is to use the digital filters for further processing to get the PCG signal in favorable graphical representation to be stored in personal computer in Excel file format. After the filtering stage, an overvoltage protection circuit was designed. The overvoltage protection circuit is used to ensure that the input voltage to the DSP kit should always be less than 1.5 volts (RMS). The block diagram of the HSS acquisition system is shown in Figure 8.7. The collected heart data signals of a subject having no heart problem of age 23 (normal) and from a subject with known heart problem (abnormal) of age 62 have been shown in Figure 8.8(a) and (b), respectively.

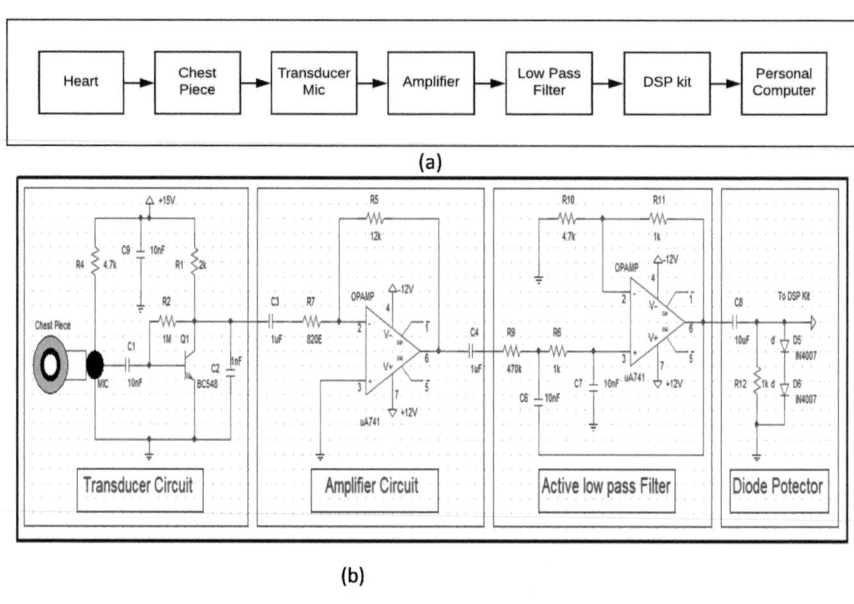

(a)

(b)

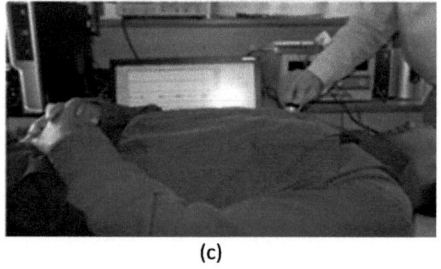

(c)

FIGURE 8.7 (a) Block diagram, (b) actual circuit diagram, and (c) DSP kit-based heart sound acquisition system.

8.4   HEART SOUND DETECTION TECHNIQUES

HSS processing steps include denoising, segmentation, feature extraction, and classification. During the recording of PCG, signals are subjected to noise and artifacts, which come from external influence. These unwanted noises that are mixed with fundamental heart sounds create problems for the analysis and diagnosis of heart condition. So denoising is the very first step of HSS detection to be further processed. The primary purpose of heart sound segmentation is to decompose HSS into a meaningful cardiac cycle. Further segmentation of HSS is to find out starting and ending points of the HS components. Segmentation can be performed on raw PCG signals or denoised signals. The first heart sound is S1, systolic period, and the second heart sound is S2, the diastolic period. Segmentation provides the location of first and second heart sounds (S1 and S2), which in turn helps to locate and analyze murmurs.

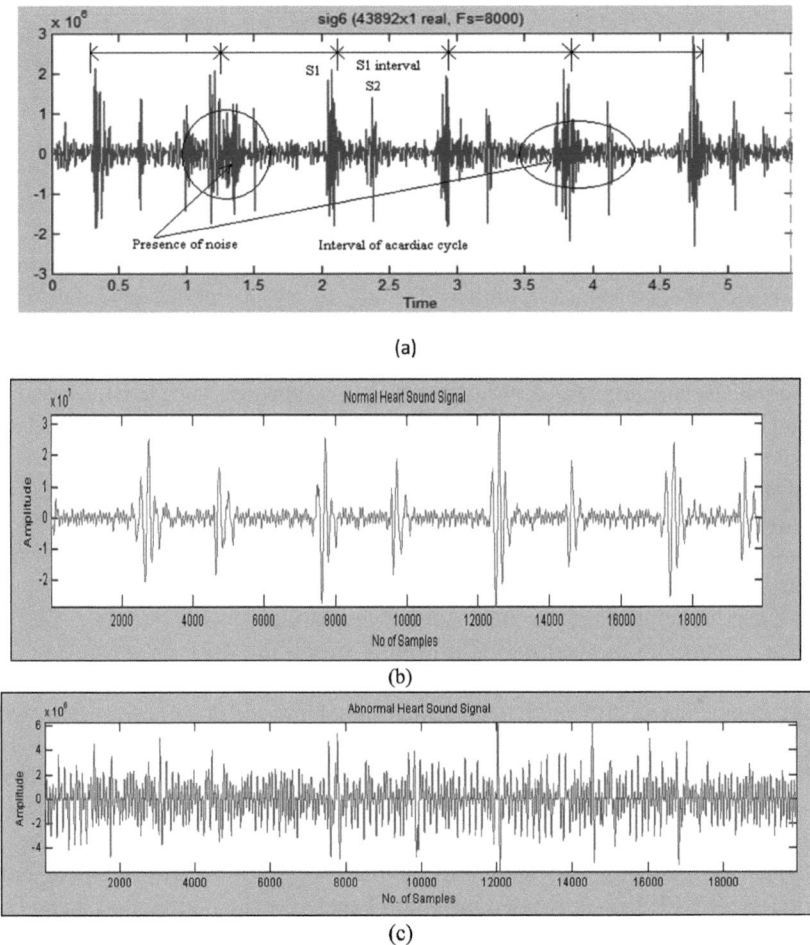

FIGURE 8.8 Heart sound signal acquired with designed heart acquisition system for (a) noisy signal from DSP kit, (b)normal heart sound, and (c) abnormal heart sound after filtering by DSP kit.

Feature extraction on the other hand is a process of transforming raw data into numerical features with reduced dimension that would be manageable and processed while preserving the information in the original datasets. Finally, the classification means separating the PCG signals into normal or abnormal groups. Classification can be performed in raw PCG signals, PCG signals after denoising, PCG signals after segmentation, or PCG signals after feature extraction. The goal of classification is to present the qualitative results of the detection, dividing the heart sound signals into normal and abnormal groups. By analyzing extracted features from abnormal sets, further detection of heart diseases is possible for which various machine learning tools are being applied. Figure 8.9 summarizes the various steps of detection method.

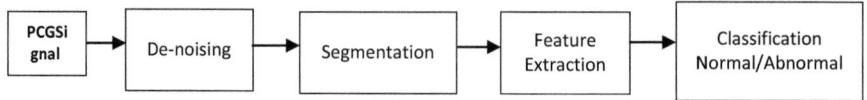

FIGURE 8.9   Basic block diagram for PCG signal detection steps.

8.4.1   Denoising of Heart Sound Signal

Heart sound is generally affected by low- and high-frequency noises from several internal and external noise sources. External noise may be any environmental sounds such as speech, the noise of the device, etc. On the other hand, internal noise includes muscular contraction, respiration, coughing, swallowing, and others. Denoising of HSS is a crucial step for reliable and accurate identification of HSS components such as S1, S2, and others, as HSS gets easily corrupted with random internal and external noise sources.

PCG is a nonstationary signal for which application of traditional Fourier transform or the short-time Fourier transform is difficult for denoising. Thus wavelet transform is known for best effort result to analyze nonstationary signal like HSS denoising [25]. Wavelet-based hard and soft thresholding filtering are the most common for denoising of HSS, for which selection of correct wavelet, level of decomposition, and selection of best threshold are important necessitating domain expertise. But low-frequency murmurs in abnormal HSS cannot be eliminated fully using transform domain analysis as the murmur frequency overlaps with HSS component frequency. Thus components identification in abnormal signal finds difficulty even after denoising. In this section, at first, the useful denoising techniques generally used for HSS such as wavelet transform techniques and their mathematical basis are discussed. Results have been generated and discussions are made with the pros and cons of those methods. Finally, a very simple and efficient power law algorithm (PLA) [26] to perform denoising of HSS in the time domain is given. The low-amplitude noise/murmurs using PLA is suppressed in comparison to the high-amplitude signal components both for normal and abnormal HSSs. Comparative study for qualitative and quantitative analysis on real-time HSSs clearly shows PLA-based method improves quality components identification than wavelet transform. Quantitative analysis is based on two parameters: signal-to-noise ratio and fitness coefficient while qualitative analysis seeks visual inspection.

The wavelet transform is a mathematical tool that fits up data or function into different frequency components. Each component is studied with a resolution matched to its scale [27]. Wavelet transformation is applied in both discrete and continuous domains, called discrete wavelet transform (DWT) and the continuous wavelet transform (CWT). DWT operates over scales and positions and is widely used in data compression and feature extraction [28], as it is efficient and sufficient for exact reconstruction.

8.4.1.1   Signal Decomposition

The very first step in the wavelet denoising is the decomposition of the signal. For that, at first the mother wavelet is selected from a list of Daubechies wavelets (db1–db10) [29]. On the selected mother wavelet scaling ($s = s_0{}^m$) and sifting ($p = kp_0$) operations are applied to determine the frequency spectrum range of the signal and decompose it accordingly. The wavelet function can be represented as:

$$\varpi_{m,k}(n) = s_0^{-\frac{m}{2}}\,\varpi\!\left(\frac{n - kp_0}{s_0^m}\right) \tag{8.1}$$

Where m and k are integers, $\varpi_{m,k}(n)$ is the wavelet function, with scale and sifting factor s_0 and p_0.

Decomposition system constitutes of low-pass filter (LPF) and a high-pass filter (HPF) (Figure 8.10) forming filter bank tree [29]. The entire frequency spectrum of the signal is divided into two parts: (1) detail coefficients "d_i," which are small and contain high-frequency components, while (2) approximation coefficients "a_i," contain a low-frequency component of the original signal.

The center frequency is the frequency of the detected HSS. Wavelet transform is applied on heart signal *x(n)* and decomposed at different levels by repetition through two sets of numbers a1(n) and d1(n) to get actual heart signal frequency band. The original heart signal when passed through one of the two complementary filters, i.e., a LPF, performs a convolution operation of the noisy heart signal and a half-band LPF function (*g*).The signal will now have half the number of points. The decomposed signals are expressed as:

$$Y_{\text{low}}[n] = \sum_{k=-\infty}^{\infty} s[k]\,g[2n - k]$$

$$Y_{\text{high}}[n] = \sum_{k=-\infty}^{\infty} s[k]\,h[2n - k] \tag{8.2}$$

Where $Y_{\text{low}}[n]$ and $Y_{\text{high}}[n]$ are output of HPF and LPF after subsampling by 2 and h(n) and g(n) are the impulse functions of the LPF and the LPF, respectively.

This process has been iterated several times to achieve the exact heart signal frequency band, which can be noticed at the approximation level 5 or 6. After repeating convolution operation and low-pass filtration, the heart signal loses the part of the

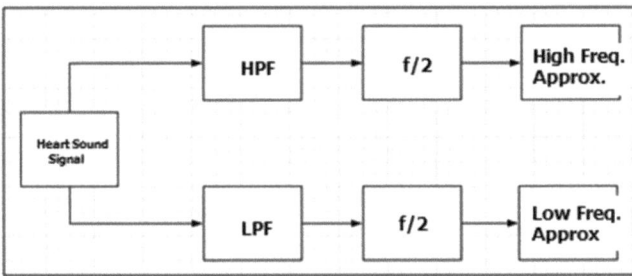

FIGURE 8.10   Wavelet decomposition method by filter bank.

noise of high frequency. Now the result of HPF (h) and LPF (g) are summed up for reconstruction. The reconstructed signal can be represented as:

$$C[n] = \sum_{k=-\infty}^{\infty} s[k]g[2n-k] + \sum_{k=-\infty}^{\infty} s[k]h[2n-k] \qquad (8.3)$$

Reconstruction has been performed by the inverse wavelet transform. Decomposition levels are shown in Figure 8.11 as an illustration.

Figure 8.12 shows the decomposition level of HSS up to levels 5 and 6 using db10 wavelet for normal and db2 for abnormal-type signals. The visual inspection of Figures 8.13(a) and 8.13(b) for a5 and a6 explains the quality difference at level 5 and level 6, respectively. Increasing one level from 5 to 6 enhances the quality of approximate coefficients a6 greatly, especially for abnormal signal. Depending on the type of the HSS this decomposition level though may vary.

Thresholding in wavelet transform may be hard and soft. The soft thresholding function can be defined as:

$$D_s\left(d_{jk}|t\right) = \begin{cases} 0, & for\ |d_{jk}| \le t \\ d_{jk} - t, & for\ |d_{jk}| > t \\ d_{jk} + t, & for\ |d_{jk}| < -t \end{cases} \qquad (8.4)$$

Where d_{jk} represents the kth wavelet coefficient at jth decomposition level with "t" as the threshold level [30].

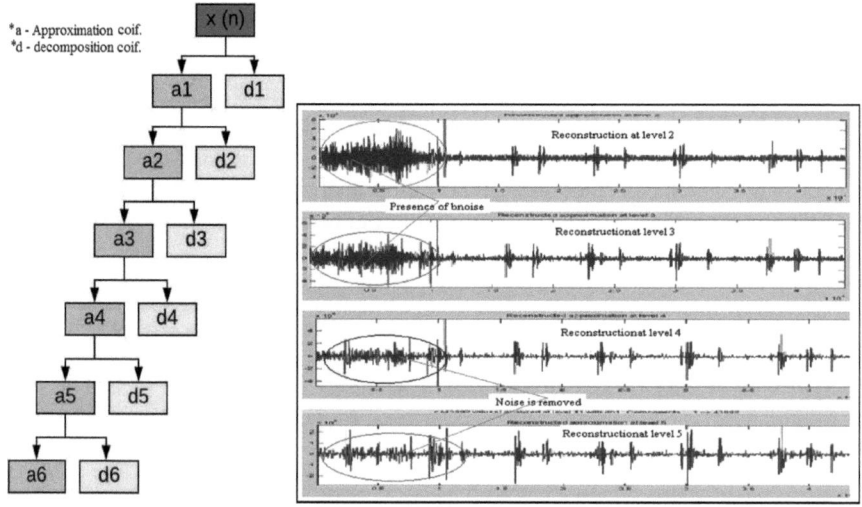

FIGURE 8.11 Wavelet decomposition using Daubechies wavelet of order 10 (db10) and reconstruction to level 5.

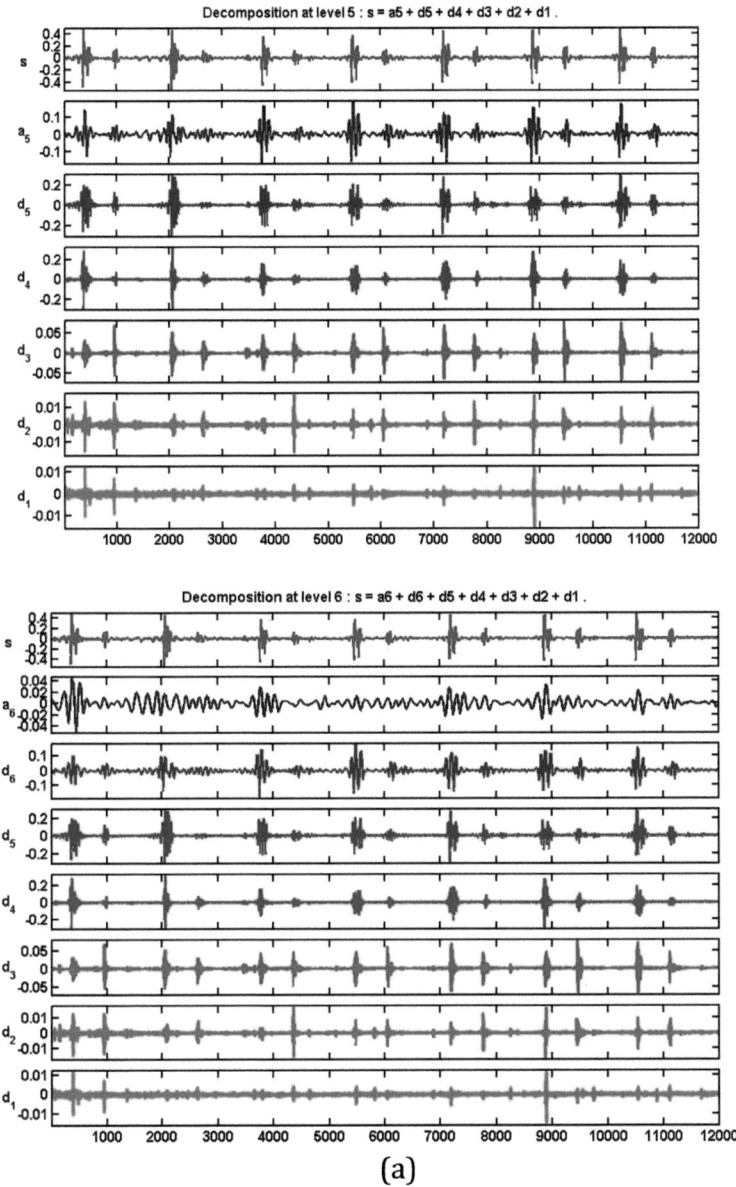

(a)

FIGURE 8.12   Wavelet decomposition (a) normal type and (b) abnormal type HSS with levels 5 and 6.

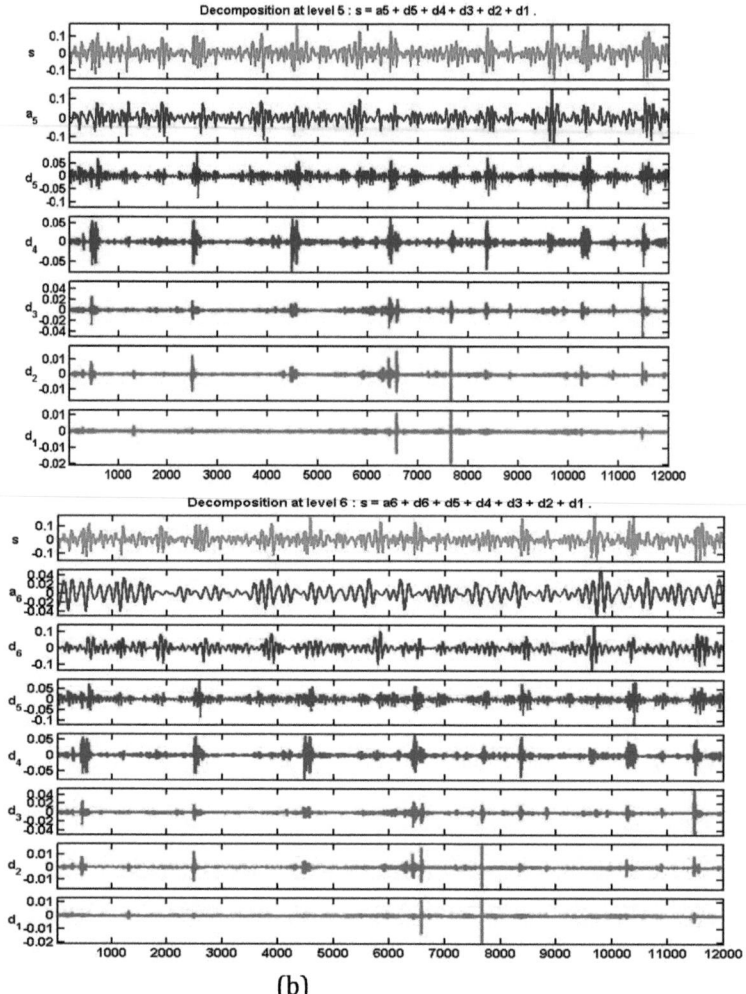

(b)

FIGURE 8.12 (CONTINUED)

The hard thresholding is the binary decision and the condition is defined as [31]:

$$D_H\left(d_{jk}|\mathrm{t}\right) = \begin{cases} 0, & for\left|d_{jk}\right| \le \mathrm{t} \\ d_{jk}, & for\left|d_{jk}\right| > \mathrm{t} \end{cases} \tag{8.5}$$

Hard thresholding results in a suppression of noise to a great extent.

Figures 8.13 and 8.14 show the DWT denoising using hard thresholding method for the normal type and abnormal type, respectively. From these results, it can be observed that for normal HSS, DWT can provide an excellent outcome with clearly visible HSS components. The presence of components is localized at the lighter

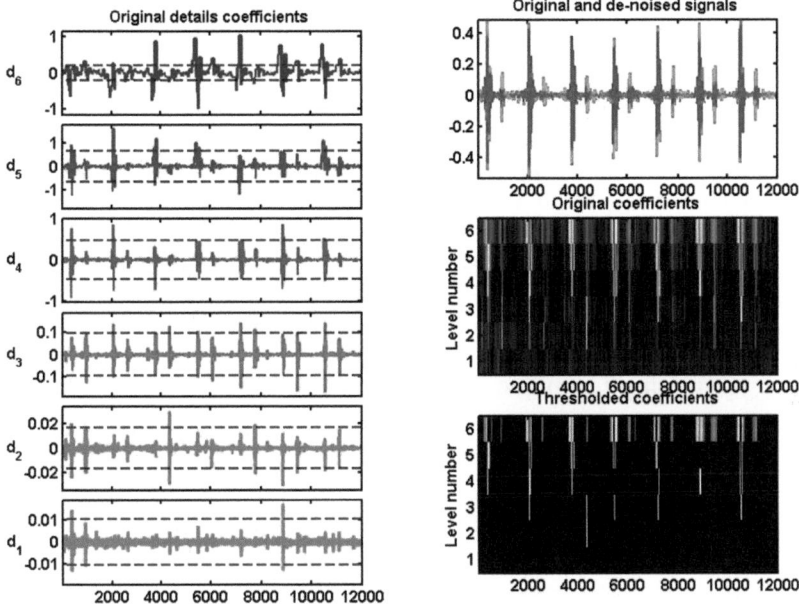

FIGURE 8.13   Denoised signal using hard thresholding method for normal-type HSS signal.

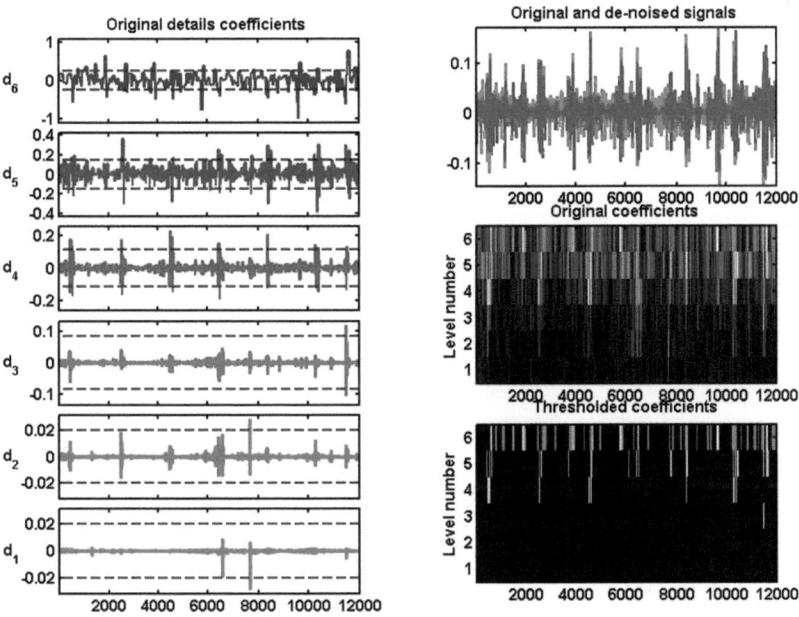

FIGURE 8.14   Denoised signal using hard thresholding method for abnormal type HSS signal (the blue lines in the figures denote threshold level).

colored region in denoised part of the signal. But in case of abnormal HSS, challenge is there to extract the components of the signal because of the strong presence of murmurs. Murmurs frequency spectrum overlaps with HSS component's spectrum and both reside in the same decomposition level [32]. A high-amplitude threshold for murmurs can affect the signal component by losing vital information of HSS. Thus trade-off is always required while picking threshold for HSS denoising.

On the other hand, due to the nonideal wavelet filter, redundant frequency components in the signal will be present [33]. For proper functioning in real system, selection of best threshold for wavelet denoising needs expertise. To overcome these problems, PLA is proposed [26] for clear identification of HSS components both for normal and abnormal cases. PLA is simple but effective one.

8.4.1.2   Denoising Technique Using Power Law Algorithm

PLA algorithm is described in Figure 8.15. Before discussing the working of PLA, some basic terms are explained on which the algorithm is based. Figure 8.17 is the

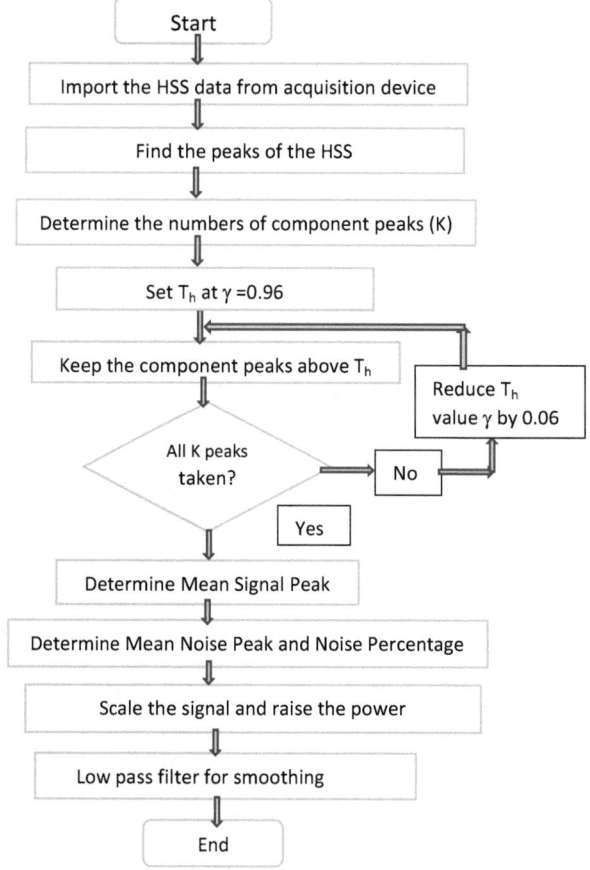

FIGURE 8.15   Flow diagram for PLA.

important illustration to understand the fundamental entities of the PLA. Normally the main heart sound components S1 and S2 hold a greater amplitude level in the signal in comparison to S3, S4, and noise [22]. Therefore, component amplitude can be distinguished by locating the peaks of the heart sound signal within a complete heart cycle. S1 and S2 are known as principal components of HSS.

The approximate numbers of principal component peaks can be determined from the duration of the signal acquisition (t_{acq}), heartbeat rate (HBR), and the number of the peaks in a single component (i). If "C_p" is denoted as the possible principle component peaks of HSS, then:

$$Cp = 2*i*\frac{BPM}{60}*t_{acq} \qquad (8.6)$$

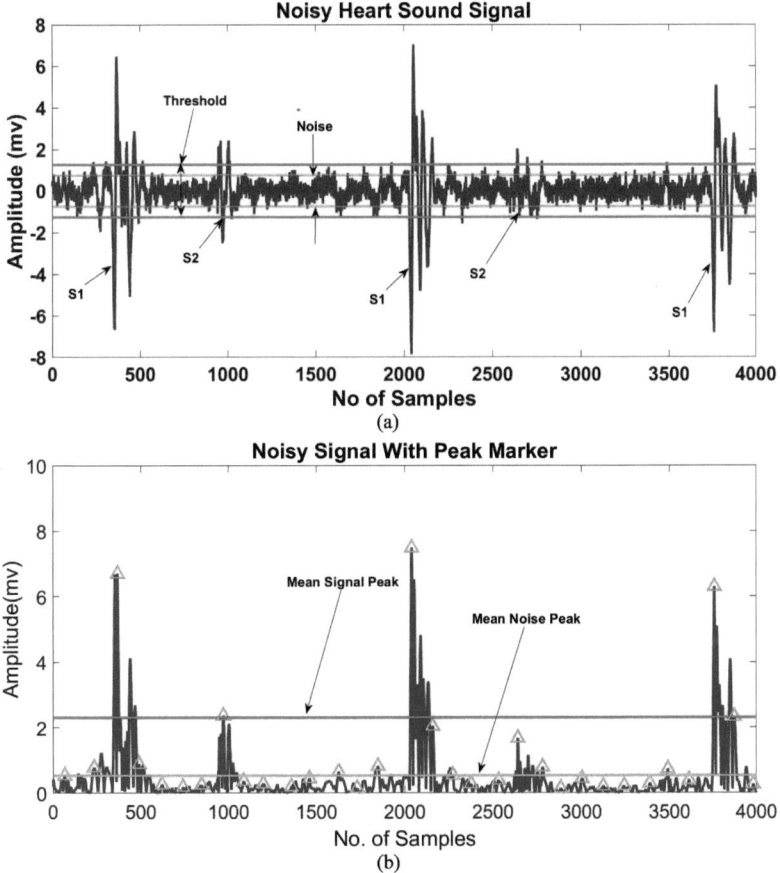

FIGURE 8.16   (a) Mean peak signal and (b) noise amplitude of a standard noisy heart sound signal.

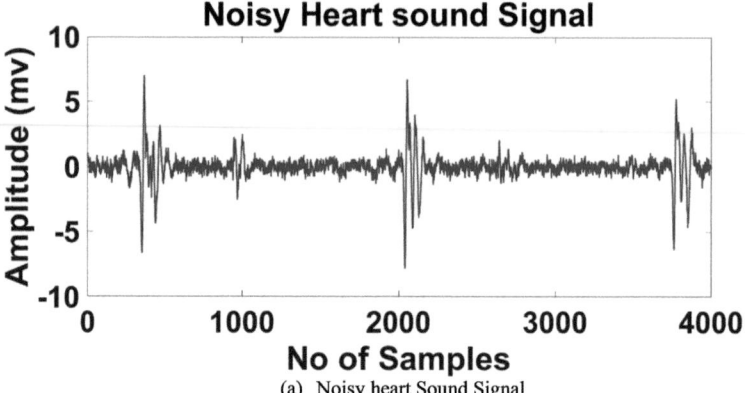

(a) Noisy heart Sound Signal

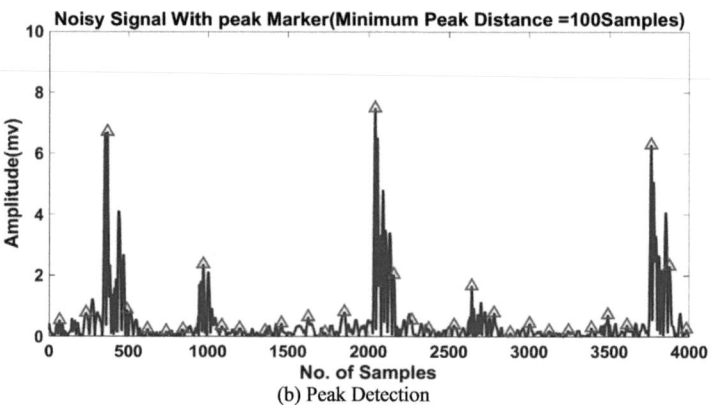

(b) Peak Detection

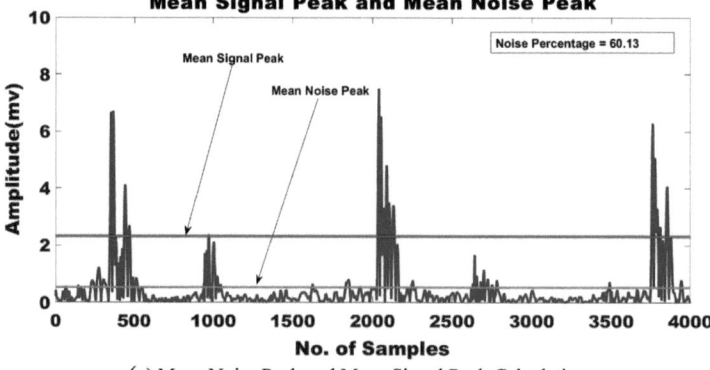

(c) Mean Noise Peak and Mean Signal Peak Calculation

FIGURE 8.17 (a) Noisy heart sound signal, (b) detected peaks of the noisy signal, (c) determined mean signal peak and mean noise peak, (d) noise compressed signal with Power raised to 5, and (e) output of Savitzky–Golay filter.

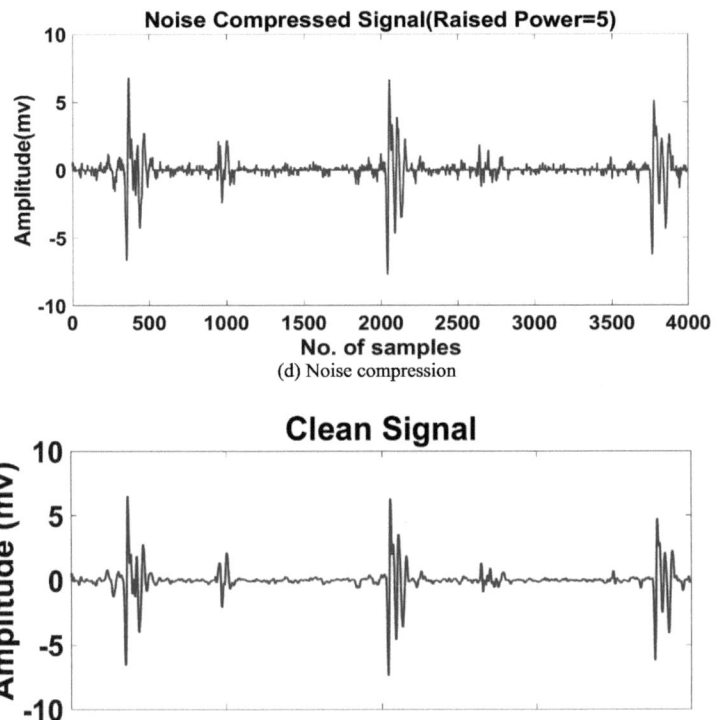

(d) Noise compression

(e) Smoothing using Savitzky–Golay filter of order 20

FIGURE 8.17 (CONTINUED)

BPM = heartbeat rate, i = number of peaks or local maxima within signal acquisition time t_{acq}, the factor 2 comes for two principal components S1 and S2.

The principal components peak or local maxima has higher amplitude than the noise peak as shown in Figure 8.16. Thus using "findpeaks" function in MATLAB, it finds the tallest peaks within a minimum peak distance avoiding undesired local maxima. By applying proper threshold, the mean peak amplitude of the signal components is evaluated segregating noise peaks. The absolute value of signal sample peaks having a higher amplitude than the threshold value is estimated as the mean signal peak. If x(n) is the time series sequence of the HSS, then the threshold T_h is denoted as:

$$T_h = {}^*\max\big(x(n)\big) \tag{8.7}$$

Where γ is the percentage limit considered for the threshold determination, i.e., "T_h" is the "$\gamma\%$" of the highest peak of the HSS. At first PLA sets $\gamma = 96\%$ and counts the number of peaks. If all "C_p" principle component peaks reside within this limit, PLA proceeds to the next stage, else it will decrease the limit by 3% or 6% and computes again until all C_p peaks are covered. The mean peak amplitude is the mean value of all "C_p" peaks. Mathematically:

$$MeanSignalPeak\left(P_{sig}\right) = \frac{1}{Cp}\sum_{n=1}^{Cp} p_h\left(n\right) \tag{8.8}$$

Where $p_h(n)$ is the amplitude of the nth peak above the threshold.

8.4.1.2.1 Finding Noise Percentage

Selection of noise% is important for HSS de-noising for normal and abnormal HSSs. Noise amplitude level varies randomly during HSS acquisition depending on the active external and internal noise sources. So consideration of noise amplitude within a range of 0%–75% of the mean signal peak is justified as noise amplitude is lower than principle amplitude of S1 and S2. The noise% will be low within 5% for noise-free HSS and as high as 75% for signals badly affected with noise and is an unrecoverable signal. Noise percentage determines the percentage of mean noise peak amplitude (P_{noi}) of the signal to the mean signal peak amplitude (P_{sig}) of the signal, i.e.,

$$Noise\ percentage\left(N_p\right) = \frac{P_{noi}}{P_{sig}} * 100 \tag{8.9}$$

Now, for mean noise peak (P_{noi}) calculation, all the peaks below ($T_s = P_s*80\%$) are accounted.

Mean is the average value of peak amplitudes below the threshold, and the standard deviation is the variation of the peaks from its mean value. So mean noise peak has been defined as:

$$Mean\ Noise\ Peak\left(P_{noi}\right) = \left(\frac{1}{N}\sum_{n=1}^{N} p_k\left(i\right)\right) \tag{8.10}$$

Here, $p_k(i)$is the amplitude of the nth peaks with amplitude less than "T_h" and "N" is no. of peaks below the threshold.

8.4.1.2.2 Amplitude Scaling of the Signal

Enhancing amplitude level of a signal by multiplying a constant is the process of amplitude scaling which is required to suppress the added noise without altering heart sound components (S1, S2) in the time domain of a signal. Amplitude scaling plays a vital role to keep noise amplitude below a certain value (<1), while keeping other portion of the signal above that. Mean noise peak amplitude can be derived referring to Figure 8.17 $\left(P_{sig} * N_p\%\right)$, multiplying the noisy HSS with a factor, $\left(\dfrac{1}{P_{sig} * N_p\%}\right)$, HSS sample having amplitude $A \le (P_{sig}*N_p\%)$ will scale to ≤ 1 and sample having amplitude A'> ($P_{sig}*N_p\%$) will scale to greater than 1. So, amplitude scaled signal Z(n) can mathematically be represented as:

$$Z(n) = \frac{x(n)}{(P_{sig}N_p / 100)} \tag{8.11}$$

Where x(n) is the original heart sound signal sequence, N_p is the noise percentage, and P_{sig} is the mean signal peak amplitude of the signal.

8.4.1.2.3 Power Law for Noise Minimization

The basic arithmetic operation, "raised power," has been performed to suppress the noise with the logic that a fraction (<1) "raised to any power greater than 1" will give a lower value than the original. This is the principal idea of the PLA. Amplitude scaling brings down the noise signal amplitude below 1. Now power has been raised for each sample having an amplitude less than resulting the suppression of noise to a great extent without affecting the actual signal components. Mathematically:

$$R(n) = Z(n)^t \text{ For } Z(n) < 1$$

$$R(n) = Z(n) \text{ For } Z(n) > 1 \tag{8.12}$$

Where t = Raised power, t > 1.

8.4.1.3 Savitzky–Golay Filtering for Smoothing Operation

After application of raised power to the HSS, the signal amplitudes near to the zero of the principal component (S1 and S2) of HSS get distorted, termed as "base/cross-over distortion." For recovering the original signal trends from the cross-over distortion, signal smoothing is needed by applying Savitzky–Golay filter [34]. Signal smoothing reserves the data precision without distorting the signal frequency. Savitzky–Golay filter works on the time domain by fitting successive subsets of adjacent data points with a low-degree polynomial by the method of linear least mean squares. Considering the order of the polynomial (N), the coefficient of a polynomial can be represented as $p(n) = \sum_{k=0}^{N} a_k n^k$.

Considering (2M + 1) input samples centered at n = 0, M samples will be at the left of n = 0 and M samples will be at the right of n = 0; the mean approximation error is represented as:

$$\xi_N = \sum_{n=-M}^{M} \left(\sum_{k=0}^{N} a_k n^k - x[n] \right)^2 \tag{8.13}$$

Here, x[n] is the samples within the frame length (2M + 1). By evaluating p(n) at the central point n = 0, components smoothing is obtained. Mathematically, $y[0] = p(0) = a_0$.

8.4.1.4 PLA Algorithm: Working Principal

PLA algorithm works as described in the flowchart in Figure 8.16. After finding local peak amplitudes of the signal, PLA determines the number of approximate principle component peaks present in the signal with I = 2, HBR = 72 (this value can change depending on age and different health parameters of the patients). The threshold limit is set with $\gamma = 0.96$ or 96% and checks if all of the component peaks (C_p) lie above that threshold. If all peaks lie above that limit it proceeds to the next stage, else decreases 'γ' by 0.06 or 6% and checks again. Once all C_p peaks are included, mean signal peak (P_{sig}) is computed. As a next step PLA searches for all peaks having amplitude below the threshold ($T_h = P_s * 90\%$) and calculates the mean of the amplitude values to calculate mean noise peak (P_{noi}) and then noise percentage of the signal using P_{noi} and P_{sig}. Original HSS is scaled using the noise percentage calculated. Then power raising method is applied according to Equation 8.12 and the noise affection level of the signal. For example, t = 2 for the noise percentage below 20%, t = 3 for Np < 40%, t = 5 for Np < 70% are taken as default values. Once the noise effect is minimized using raised power, the signal is passed through the Savitzky–Golay filter of order 7 and frame length of 51. Figure 8.17 describes the stepwise PLA outcome. Figures 8.18 and 8.19 show the denoising of normal and abnormal HSS by PLA method, respectively.

8.4.1.4.1   Quantitative Analysis of PLA

Quantitative analysis has been performed based on the signal-to-noise ratio (SNR) and the FIT coefficient. SNR is calculated using the expression as given by Equation 8.14 [35, 36].

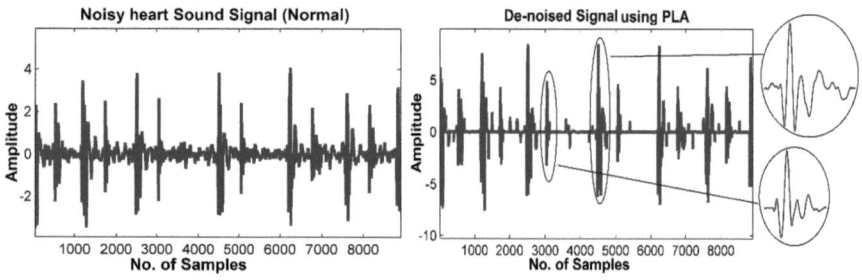

FIGURE 8.18   Denoising of normal HSS using PLA.

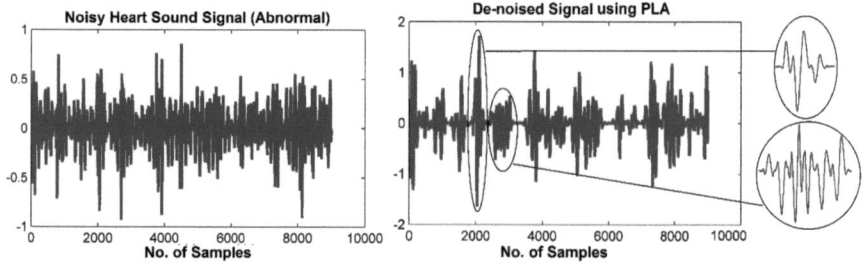

FIGURE 8.19   Denoising of abnormal HSS using PLA.

$$SNR = 10log_{10} \frac{\frac{1}{N}\sum_{n=1}^{N}\left[C(n)\right]^2}{\frac{1}{N}\sum_{n=1}^{N}\left[C(n)-D(n)\right]^2} \qquad (8.14)$$

Here, D(n) is the denoised signal and C(n) is the clean signal.

The FIT coefficient defines how much patient's health information is preserved after the application of cleaning algorithms and processing, which is expressed in percentage: 0% = all information gets lost, 100% = all information of the signal is preserved [36].

$$FIT = 100 * \left(1 - \frac{\sum_{n=1}^{N}\left[C(n)-D(n)\right]^2}{\sum_{n=1}^{N}\left[C(n)-\frac{1}{N}\sum_{n=1}^{N}C(n)\right]^2}\right) \qquad (8.15)$$

Random White Gaussian noise of specified level +5 to −10 dB is added to clean HSSs for normal and abnormal types considering low, medium, and high noise power levels. These noisy signals are then denoised using PLA and the wavelet transform separately for the comparative evaluation. After both quantitative and qualitative evaluations, it is observed that PLA performs best under noisy environment (low SNR) of the heart sound. To preserve important health information for normal HSS, N_p should be less than 72% and for abnormal HSS, N_p should be less than 50%. PLA outperforms wavelet de-noising in all cases of SNR for abnormal HSS than DWT, which is a significant observation. Table 8.1 shows the comparative quantitative analysis of denoising using PLA and wavelet methods.

8.5 FEATURE EXTRACTION, ANALYSIS, AND COMPONENTS IDENTIFICATION FOR HEART SOUND SIGNAL USING THE PROPOSED SEPD ALGORITHM

The main objective of the segmentation is to detect the boundaries of heart sound components with their locations of the fundamental heart sounds (FHs), (S1) and second (S2), for the components identification. An efficient segmentation provides valuable diagnostic information of subjects from whom HSSs are collected. Heart sound segmentation method may combine the time and/or frequency domain analysis. Apart from the heart cycle identification, the systolic and diastolic regions of the heart sound depend on the accurate localization of the FHs. Thus, the characteristics of different pathological situations in the region of one heart cycle are used to classify different heart sound categories [37–40].

Determining the envelop of the heart sound signal is the direct segmentation of the HS, where envelop of the HS has been calculated using the Shannon energy formula [41] as:

$$E_s = 1/N \sum_{k=1}^{N} x^2(i)logx^2(i) \qquad (8.16)$$

TABLE 8.1

Quantitative Performance Analysis of PLA and Wavelet Transform for HSS Denoising

Type of HSS	SNR of added noise in dB	Raised power for PLA	SNR of denoised signal by PLA in dB	FIT coefficient of denoised PLA	SNR of denoised signal by DWT in dB	FIT coefficient of denoised DWT	Remarks
Normal	5	4	13.853	98.77	14.334	99.31	PLA provides a better FIT coefficient and improved SNR than DWT in low SNR desirable in the denoising process keeping SNR at significant level with slow variation. FIT coefficient value greater than 92% indicates very low information loss for PLA.
	0	4	13.705	96. 345	10.62	91.31	
	-5	5	12.825	94.88	6.75	78.85	
	-10	5	10.903	92.99	3.4155	54.45	
Abnormal	5	4	12.707	94.623	10.78	92.71	For abnormal HSSPLA denoising outperforms DWT in all SNR conditions, keeping both SNR and FIT coefficient almost same in PLA, significant for denoising.
	0	4	12.105	93.61	9.245	88.06	
	-5	5	11.885	93. 12	5.227	69.90	
	-10	5	10.780	92.73	1.845	34.41	

A simple solution of heart sound components identification and analysis with reasonable accuracy (>90%) is provided to be handled by engineers and persons with science education. A very simple algorithm namely SEPD (start and endpoint detection of heart sound) [24] is proposed, which has low mathematical complexity and can be done in a procedural way. The abnormality identification can be done from "parametric value" and the "pictorial view" of heart sound components. SEPD algorithm considers both the segmentation and feature detection in combination as it finds the location of components and then analyses. On the denoised signal some significant HSS features have been extracted through principal components S1 and S2, their duration, and finally the cardiac cycle from which HBR is calculated. The opening and closing valves of the heart muscle create the natural sounds, namely S1 and S2, respectively. In between each sound, the presence of a silence period is called systolic and diastolic phase. So, the systolic interval is confirmed by the S1 sound and the systolic phase; the diastolic interval by the S2 sound and diastolic phase. In case of any heart abnormality other occurrence of sounds in between S1 and S2 will be there between systolic or diastolic phases. This whole compound of fundamental components forms the heart sound signal over periods called cardiac cycles and is the basic unit for heart sound analysis. The analysis is performed on the basis of heart rate, duration of S1, duration of S2, duration between S1 and S2, duration between S2 and S1, and PSD of each component.

Detection of the presence of S3 and S4 components is a more tricky and difficult task. Presence of murmurs and S3 and S4 parameters is more likely to occur in case of abnormality of the heart. Duration of heart components, i.e., longer or shorter, energy or strength of S1 and S2, cardiac cycle, HBR, etc., are the few main things to check the preliminary condition of the heart from the collected PCG signals. For a more accurate diagnosis, detection and position of S3 and S4 are important using various machine learning methods.

The cardiac cycle and HBR are generated for several HSSs collected from different subjects in which some with known heart problems. Comparative results are given both for PLA and wavelet transform denoising methods. For superior detection of S1 and S2 parameters from the denoised HSSs and separation of components with SEPD algorithm, the frequency domain power spectral density of the S1 and S2 is also given to observe the frequency components present in the detected signal as we know S1 component lies within 0–150 HZ and S2 from 0 to 200 Hz. The higher frequency in the cycle indicates the presence of murmurs. We observed that the very simple proposed SEPD algorithm can detect the heart sound components for diagnosis purpose with good accuracy. This process will be extremely helpful in a place with the low medical establishment to take mass-scale computation and interpretation to be finally sent to the doctor for consultation and treatment.

8.5.1 Description of SEPD Algorithm

The recorded denoised (using PLA and wavelet) digital heart sound signal is imported in MATLAB workspace. Then to design SEPD algorithm, at first peak value of the signal x(n) is determined by using findpeak. Then thresholds values for different heart sound components are set. The first heart sound S1 is the strongest (η). To find out the maximum peak of S1, the maximum value of the buffer is chosen. But peak values of S1 are not equal for all signals because the loudness of S1 is not equal but lies within a certain range. Within a chosen η% (say 90% for normal HS, different value for abnormal HS), maximum of x(n) is set to determine threshold for S1, $Th_{S1} = *max\{x(n)\}$ as shown in Figure 8.20. S1 detection steps are given below:

> **Step 1: S1 detection**: Identify all peaks above 90% using "findpeaks" -> select highest peaks within a neighborhood -> find an envelope of that neighborhood -> find start point in the left using the slope of envelope -> find endpoint in the right using the slope of the envelope -> extract S1
> Step 2: Set S1 positions to zero as these positions are no longer required

The positions of peaks of all S1$_i$ are found. Once the peaks and their corresponding locations are found, the SEPD algorithm identifies the start and end positions of each S1 from the envelope. Let the start and end positions of S1$_i$ are η_{1Si} and η_{1Ei}, respectively. Once η_{1Si} and η_{1Ei} are found, S1 can be represented as:

$$S1(n) = x(n) \text{ if} \Delta_{1Si} < n < \Delta_{1Ei}$$

$$= 0 \text{ otherwise} \tag{8.17}$$

The duration of S1 can be calculated as:

$$S1_i = \left\{(T_{1i} + \Delta_{1Ei}) - (T_{1i} - \Delta_{1Si})\right\} * \frac{1}{f_s} \tag{8.18}$$

Where f_s = sampling frequency, T_{1i} = position of maxima of S1$_i$, Δ_{1Si} = starting position of S1$_i$, Δ_{1Ei} = end position of S1$_i$, and I = 1, 2, 3 . . . (ith S1 component present in the signal).

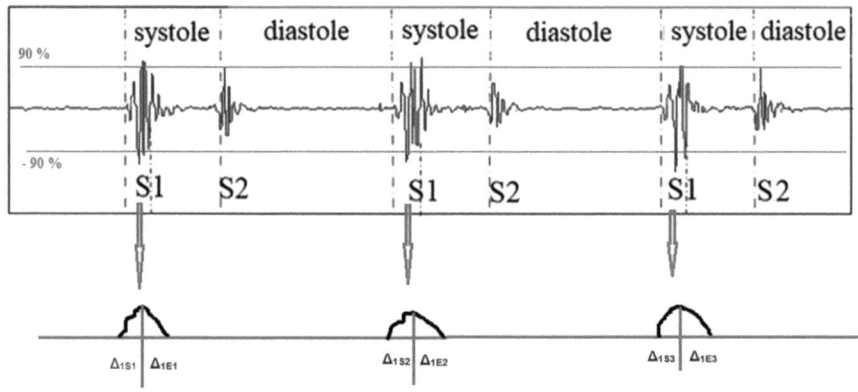

FIGURE 8.20   Extraction of S1 components.

Similarly, the second principal component S2 detection process is described by steps 3 and 4 and in Figure 8.21. Finally, step 5 describes the murmur detection.

Step 3: S2 detection: Identify all peaks above 70% using findpeaks -> select highest peaks within a neighborhood -> find an envelope of that neighborhood -> find start point in the left using the slope of envelope -> find endpoint in the right using the slope of the envelope -> extract S2

Step 4: Set S2 positions to zero as these positions are no longer required.

For S2, SEPD sets different threshold, $Th_{S2} = *max\{x(n)\}$, where η may be 70% for normal heart sound and may vary for abnormal signals of the maximum of x(n). Positions of all the S2 are determined from the corresponding positions of the samples with the starting (Δ_{2Si}) and ending positions (Δ_{2Ei}) of S2 envelop.

$$S2(n) = x(n) \text{ if} \Delta_{2Si} < n < \Delta_{2Ei} \& n \neq \text{Positions range of S1}$$

$$= 0 \text{ otherwise}$$

(8.19)

The duration of S2 will be:

$$S2_i = \left\{(T_{2i} + \Delta_{2Ei}) - (T_{2i} - \Delta_{2Si})\right\} * \frac{1}{f_s}$$

(8.20)

Where f_s = sampling frequency, T_{2i} = position of maxima of $S2_i$, Δ_{2Si} = starting position of $S2_i$, Δ_{2Ei} = end position of $S2_i$, and i = 1, 2, 3 . . . (ith S2 component present in the signal).

Heart murmurs are produced when aberrant blood flows across the heart valve due to any kind of heart disorder. An average threshold value $\eta\%$ (60% for normal heart sound) is considered to detect the murmurs. Murmurs signal can be expressed as:

$$Mur(n) = Sm = x(n) \text{ if } x(n) \geq *Max\{x(n)\} \& n \neq \text{Positions of S1 and S2}$$

$$= 0 \text{ otherwise}$$

(8.21)

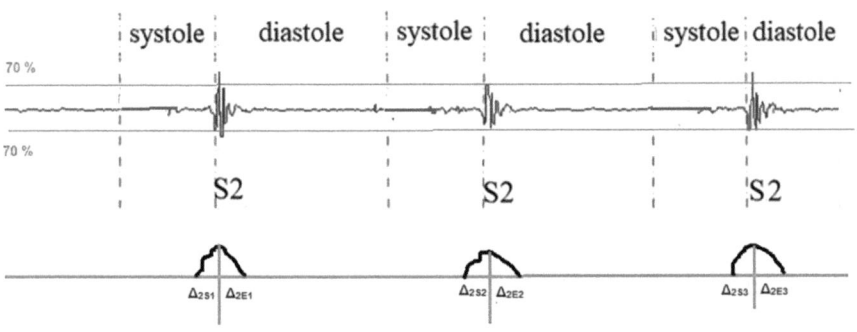

FIGURE 8.21 Extraction of S2 components

Step 5: **murmur detection**: Identify all peaks above 60% (approx.: need to change based on the signal complexity) using findpeaks -> select highest peaks within a neighborhood -> find an envelope of that neighborhood -> find start point in the left using the slope of envelope -> find endpoint in the right using the slope of the envelope -> extract Sm.

For component separation LPF may be used for simplicity. But Shannon energy equation will perform better.

8.5.1.1   Heartbeat Calculation

Heart rate is interpreted as the number of contractions of the heart per minute (bpm). At resting situation, heart rate of an adult human is 60–100 bpm [42]. Heartbeat rate can change according to the body's physical needs (absorb oxygen and excrete carbon dioxide) or as an effect of physical exercise, sleep, anxiety, stress, illness, and ingestion of drugs. Irregular heart rate can lead to many serious diseases like hypertensive heart disease, paroxysmal supraventricular tachycardia (PSVT), etc. [42]. From SEPD algorithm, the heartbeat rate is determined using the distance between two consecutive peaks of S1. Mathematically it can be expressed as:

$$HBR = 60 * f_s / (\text{Location of } S1_{i+1} - \text{Location of } S1_i) \text{ beats/min} \qquad (8.22)$$

The SEPD algorithm is applied to HSSs collected from an online database physiobank [43]. All the data has been denoised by proposed PLA method for feature extraction for different heart sound components, heartbeat rates, etc. Table 8.2 is given by PLA followed by SEPD for known datasets. The abnormal observations are marked as a bold letter. The extracted components of HSS by both PLA and DWT are given as an example for better visual inspection for the HSS components in Figure 8.22 along with the power spectral density plots, which show better denoising effect for PLA.

Finally, the frequency domain analysis for power spectral density of the HSS in normal and abnormal signals is given for surer analysis. The PSD of complete HSS within a cardiac cycle and PSD of S1 and S2 components after applying SEPD algorithm for components identification have been analyzed. PSD of HSS elucidates spectral ranges of the heart sound components and the presence of the murmurs. In general, the frequency spectrum of the S1 and S2 lies between 0–150 Hz and 0–200 Hz, respectively. Further, the presence of S3 (25–75 Hz), S4 (15–75 Hz), and low-frequency murmurs (100–400 Hz) indicates abnormality in HSS [44]. This information has clearly reflected in the PSD plot from PLA-based denoised signal. It is observed from Figure 8.23 that for normal signal (1) PSD takes sharp decrement for the slope within the frequency range 0–150 Hz than the abnormal signal, (2) which varies slowly because of the presence of murmurs in other frequencies. In normal HSS, the PSD spectrum within cardiac cycle shows a clear presence of S1 within 100 Hz and S2 between 100 and200 Hz, but in abnormal case presence of murmurs is visible. The amplitude of the power spectrum for abnormal cases is more than normal because of the noise/murmur presence, and both the S1 and S2 components spectra spread over usual frequency ranges. The PSD evaluation for S1 and S2 shows that SEPD is a good method of detection even for an abnormal heart condition.

TABLE 8.2
Extracted Feature Value Using SEPD from PLA-Based Denoised Signal (Known Samples)

Normal heart sound signal

HSS sample	Duration		S1–S2 in sec Systole	S2–S1 in sec Diastole	Heart beat rate beats/mins
	S1 in sec	S2 in sec			
Normal range	0.100–0.180	0.080–0.140			60–90
1	0.1550	0.1096	0.2949	0.5767	69
2	0.1290	0.1323	0.3143	0.6550	62
3	0.1437	0.1248	0.2962	0.5500	71
4	0.1832	0.1092	0.3209	0.6120	64
5	0.1399	0.1059	0.2987	0.5142	74

Abnormal heart sound signal

HSS sample	Duration		S1–S2 in sec Systole	S2–S1 in sec Diastole	Heart beat rate beats/mins
	S1 in sec	S2 in sec			
	0.100–0.180	0.080–0.140			60–90
1	**0.2467**	0.1021	0.2424	0.4636	85
2	0.1489	**0.2254**	0.2765	**0.8719**	**52**
3	0.1928	0.1172	0.4008	0.8355	**48**
4	**0.2041**	**0.6578**	0.3705	0.8431	**49**
5	**0.3970**	0.0870	0.4915	0.3403	72

*Bold sign in the result column indicates some anomaly of the heart to generate alarm.

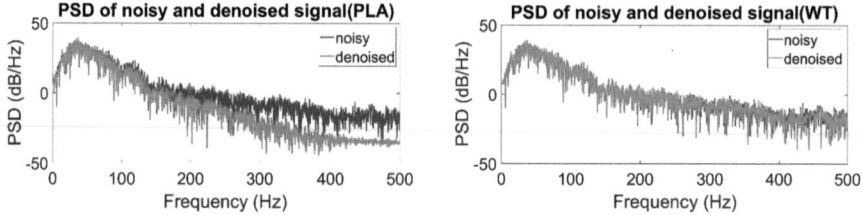

FIGURE 8.22   Component separation for S1 and S2 using SEPD for power spectral density (PSD) of the denoised signal for PLA and DWT.

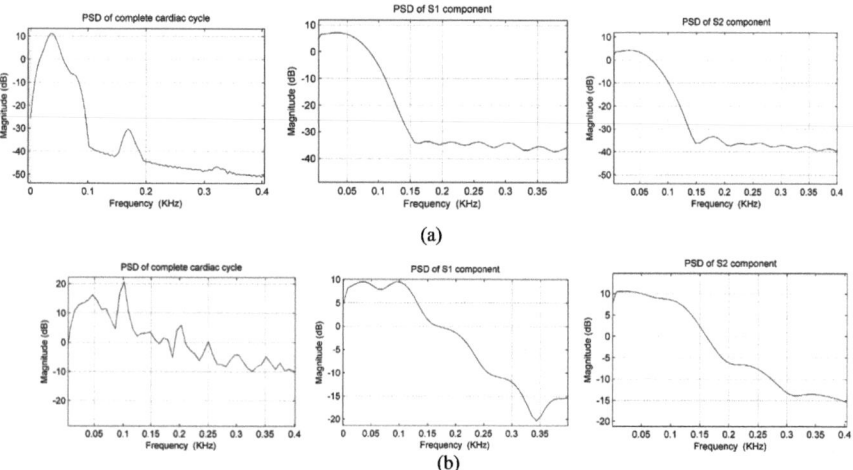

FIGURE 8.23   Power spectral density for complete cardiac cycle, S1 and S2 components for (a) normal and (b) abnormal heart sound Signal after PLA denoising.

By observing the PSD of the complete heart cycle, or the PSD of individual HS component may generate alarm for subject's heart condition. PSD of one complete heart cycle, S1, and S2 components are given separately with PLA and wavelet-based denoising for frequency domain study. From Figures 8.24 and 8.25 for PSD of normal and abnormal heart sound using PLA- and DWT-based denoised signal followed by SEPD, it is clearly visible that PLA spectrum components have sharp falling in frequency range, which helps to detect the heart sound components clearly whether for DWT wide variation of spectrum is observed in which murmurs may be overlapped in the same frequency ranges. As the S1 intensity is higher than S2, the spectrum magnitude is also higher for S1. Moreover, close observation exhibits that murmurs in abnormal heart sound signals can be identified more prominently in the PLA-based method than the DWT-based method in the low-frequency range.

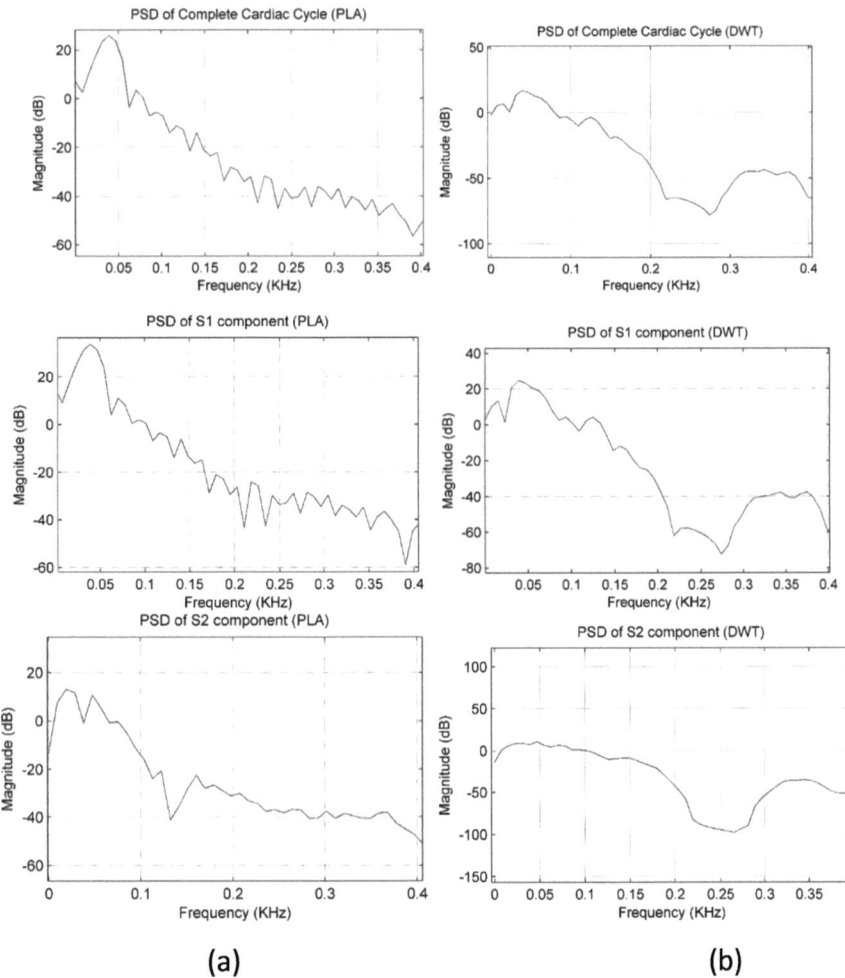

FIGURE 8.24 PSD for normal heart sound signal components: (a) PLA, (b) DWT.

8.6 CONCLUSION

This chapter describes the complete procedure of noninvasive way of human heart sound acquisition to detect for normal and abnormal hearts. For this a cost-effective and simple heart sound acquisition system is designed with simple electronic circuitry and DSP processor TMS320C67XX that takes the advantage of high speed and accuracy. Further advantage of the DSP processor is taken to apply proper FIR filter design of desired order for normal and abnormal heart sounds. Denoising of heart sound signal is an important method for components identification through which the condition of heart can be detected. Wavelet-based denoising though common but is critical for abnormal HSSs for overlapping

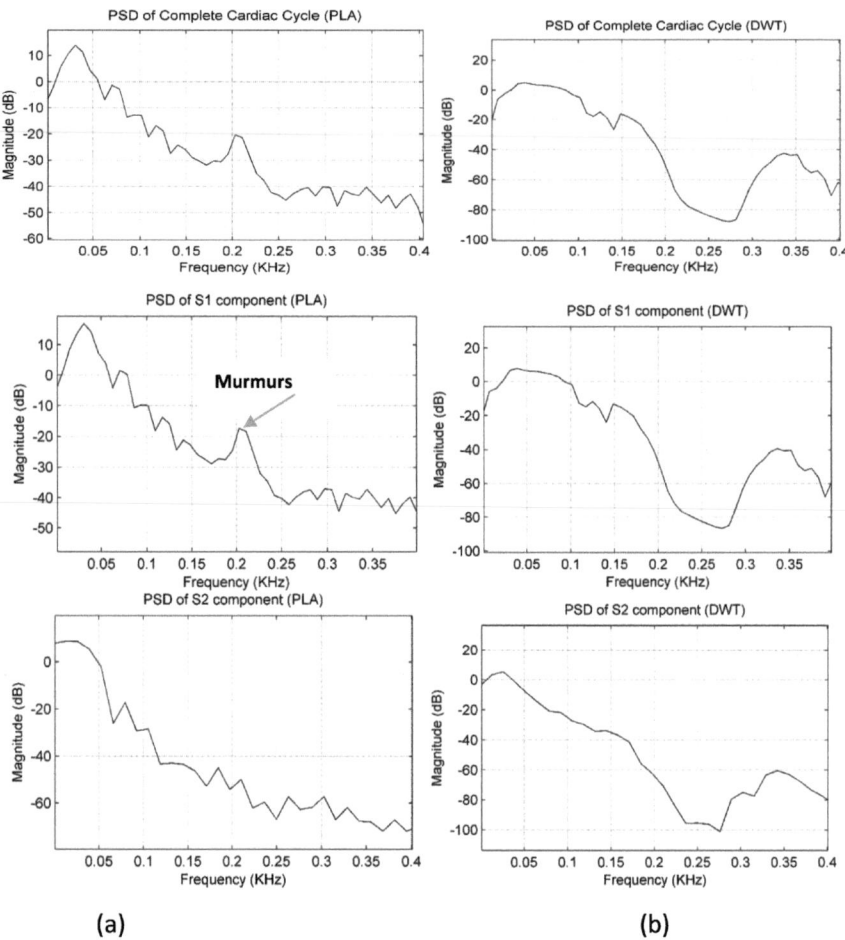

FIGURE 8.25 PSD for abnormal heart sound signal components: (a) PLA, (b) DWT.

murmur frequency. The proposed PLA for HSS denoising is employed and studied with wavelet transform in time and frequency domain. Though qualitative visual observation reveals the clear components identification, PLA denoising method is evaluated quantitively by finding SNR and FIT coefficients both for normal and abnormal HSSs. It is found that the application of PLA improvs SNR and FIT coefficients for low SNR signals. The feature extraction becomes easy and more reliable as PLA identifies each principal component precisely. The PSD evaluation for S1 and S2 shows that SEPD is a good method of detection even for an abnormal HSS. Separating each heart sound components by applying SEPD algorithm with good accuracy provides correct PSD information of each component. As the frequency content of the heart sound signal provides some indication of abnormality, PSD is considered an abnormality detection parameter. PSD provides a more desirable frequency spectrum of the principal components and a clear indication of murmur presence in case of abnormal signals considering this technique be efficient in accurate and reliable feature extraction and interpretation of HSS for the early detection. It is found that the presence of murmurs is more prominent in abnormal heart signals than the normal, one which is an indication for awareness generation and it is more prominent in PLA-based denoising signifying the importance of PLA over wavelet.

ACKNOWLEDGMENT

I am thankful to my students Mr. Arnab Maity and Dr. Dulal Mandal and Mr. Sukant Behera for their support in generating some figures in this work.

REFERENCES

1. https://timesofindia.indiatimes.com/india/heart-disease-deaths-rise-in-india-by-34-in-15-years/articleshow/64924601.cms
2. https://www.who.int/news-room/fact-sheets/detail/the-top-10-causes-of-death
3. https://www.worldlifeexpectancy.com/india-life-expectancy
4. https://economictimes.indiatimes.com/industry/healthcare/biotech/healthcare/less-than-one-doctor-for-1000-population-in-india-government-to-lok-sabha/articleshow/59697608.cms
5. G. A. Gibson, *Diseases of the Heart and Aorta* (Vol. 6), Young J. Pentland, Edinburgh and London, 1898.
6. T. R. Reed, N. E. Reed, & P. Fritzson, Heart sound analysis for symptom detection and computer-aided diagnosis. *Simulation Modelling Practice and Theory*, 12(2), 129–146, 2004.
7. S. Sun, Z. Jiang, H. Wang, & Y. Fang, Automatic moment segmentation and peak detection analysis of heart sound pattern via short-time modified Hilbert transform. *Computer Methods and Programs in Biomedicine*, 114(3), 219–230, 2014.
8. C. Ahlström, *Processing of the Phonocardiography Signal: Methods for the Intelligent Stethoscope* (Doctoral dissertation, Institutionenförmedicinskteknik), 2006.
9. S. Gupta, S. Pandey, & F. K. Jiavana, Low noise electronic stethoscope. *Advances in Natural and Applied Sciences*, 10(14), 52–58, 2006.
10. J. P. Tourtier, N. Libert, P. Clapson, K. Tazarourte, M. Borne, L. Grasser, & Y. Auroy, Auscultation in flight: Comparison of conventional and electronic stethoscopes. *Air Medical Journal*, 30(3), 158–160, 2011.
11. W. C. Lang, & K. Forinash, Time-frequency analysis with the continuous wavelet transform. *American Journal of Physics*, 66(9), 794–797, 1998.
12. C. G. DeGroff, S. Bhatikar, J. Hertzberg, R. Shandas, L. Valdes-Cruz, & R. L. Mahajan, Artificial neural network–based method of screening heart murmurs in children. *Circulation*, 103(22), 2711–2716, 2001.
13. P. Varady, Wavelet-based adaptive denoising of phonocardiographic records. In 2001 Conference Proceedings of the 23rd Annual International Conference of the IEEE Engineering in Medicine and Biology Society (Vol. 2, pp. 1846–1849), Oct 2001.
14. T. H. Chowdhury, K. N. Poudel, & Y. Hu, Time-frequency analysis, denoising, compression, segmentation, and classification of PCG signals. *IEEE Access*, 8, 160882–160890, 2020, doi: 10.1109/ACCESS.2020.3020806.

15. A. B. Kambhampati and B. Ramkumar, Automatic detection and classification of systolic and diastolic profiles of PCG corrupted due to limitations of electronic stethoscope recording. *IEEE Sensors Journal*, 21(4), 5292–5302, 2021, doi: 10.1109/JSEN.2020.3028373.

16. *Cardiovascular Physiology Concepts*, 2nd edition, a textbook published by Lippincott Williams & Wilkins, 2012.

17. A. G. Fredriksson, *Blood Flow Specific Assessment of Ventricular Function* (Vol. 1598). Linköping University Electronic Press, 2017.

18. H. Frost, *The Circulatory System*. Capstone, 2000.

19. A. Leatham, Splitting of the first and second heart sounds. *The Lancet*, 264(6839), 607–614, 1954.

20. D. Srivastava, Building a heart: Implications for congenital heart disease. *Journal of Nuclear Cardiology*, 10(1), 63–70, 2003.

21. P. S. Reddy, R. Salerni, & J. A. Shaver, Normal and abnormal heart sounds in cardiac diagnosis: Part II. Diastolic sounds. *Current Problems in Cardiology*, 10(4), 1–55, 1985.

22. *Clinical Methods: The History, Physical, and Laboratory Examinations*, 3rd edition, Editors: H. K. Walker, W. D. Hall, & J. W. Hurst, Butterworths, Boston, 1990.

23. P. S. Mehta, N. J., & Khan, I. A., Third heart sound: Genesis and clinical importance. *International Journal of Cardiology*, 97(2), 183–186, 2004.

24. D. Mandal, A. Maity, & I. S. Misra, Low cost portable solution for real-time complete detection and analysis of heart sound components. *Wireless Personal Communications*, Springer, 2019.

25. D. L. Donoho, De-noising by soft-thresholding. *IEEE Transactions on Information Theory*, 41(3), 613–627, 1995.

26. A. Maity, D. Mandal and I. S. Misra, A simple proposition for heart sound signal de-noising for effective components identification in normal and abnormal cases. *Biomedical Signal Processing and Control*, 71 (2022) 103264, Elsevier.

27. I. Daubechies, Ten lectures on wavelets: Society for Industrial and Applied Mathematics (SIAM), Rutgers University and AT & T Bell Laboratories, 1992.

28. P. Du, W. A. Kibbe, & S. M. Lin, Improved peak detection in mass spectrum by incorporating continuous wavelet transform-based pattern matching. *Bioinformatics*, 22(17), 2059–2065., 2006.

29. P. S. Addison, *The Illustrated Wavelet Transform Handbook: Introductory Theory and Applications in Science, Engineering, Medicine, and Finance*. CRC Press, 2017.

30. R. D. da Silva, R. Minetto, W. R. Schwartz, & H. Pedrini, Adaptive edge-preserving image denoising using wavelet transforms. *Pattern Analysis and Applications* 16 (4) (2013) 567–580.

31. L. Kaur, S. Gupta, & R. C. Chauhan, Image denoising using wavelet thresholding. In *ICVGIP* (Vol. 2, pp. 16–18), 2002, December.

32. G. R. Johnson, R. J. Adolph, & D. J. Campbell, Estimation of the severity of aortic valve stenosis by frequency analysis of the murmur. *Journal of the American College of Cardiology*, 1(5), 1315–1323, 1983.

33. E. Koutsiana, L. Hadjileontiadis, A. H. Khandoker, & I. Chouvarda, A comparative phonocardiography study: Two wavelet based methods for fatal heart sound detection. In 2018 Computing in Cardiology Conference (CinC) (Vol. 45, pp. 1–4), 2018, September.

34. R. W. Schafer, What is a Savitzky-Golay filter. *IEEE Signal Processing Magazine*, 28(4), 111–117, 2011.

35. A. H. Salman, N. Ahmadi, R. Mengko, A. Z. Langi, & T. L. Mengko, Performance comparison of denoising methods for heart sound signal. In 2015 IEEE International Symposium on Intelligent Signal Processing and Communication Systems (ISPACS) (pp. 435–440), Nov 2015.
36. D. Gradolewski, & G. Redlarski, Wavelet-based denoising method for real phonocardiography signal recorded by mobile devices in noisy environment. *Computers in Biology and Medicine*, 52, 119–129, 2014.
37. P. S. Molcer, I. Kecskes, V. Delić, E. Domijan, & M. Domijan, Examination of formant frequencies for further classification of heart murmurs. In IEEE 8th International Symposium on Intelligent Systems and Informatics (pp. 575–578). 2010, September.
38. A. Djebbari, & F. B. Reguig, Short-time Fourier transform analysis of the phonocardiogram signal. In ICECS 2000, 7th IEEE International Conference on Electronics, Circuits and Systems (Cat.No. 00EX445) (Vol. 2, pp. 844–847). 2000, December.
39. G. Livanos, N. Ranganathan, & J. Jiang, Heart sound analysis using the S transform. In *Computers in Cardiology 2000. Vol. 27 (Cat. 00CH37163)* (pp. 587–590), 2000, September.
40. W. C. Kao, C. C. Wei, J. J. Liu, & D. Hsiao, Automatic heart sound analysis with short-time Fourier transform and support vector machines. In 2009 52nd IEEE International Midwest Symposium on Circuits and Systems (pp. 188–191), 2009, August.
41. M. Nath, M. Srivastava, N. Kulshrestha, & D. Singh, Detection and localization of S 1 and S 2 heart sounds by 3rd order normalized average Shannon energy envelope algorithm. *Proceedings of the Institution of Mechanical Engineers*, 235(6), 615–624, 2021 Jun. doi: 10.1177/0954411921998108, Epub 2021 Mar 30.
42. *All about Heart Rate (Pulse)*, American Heart Association, 22 Aug 2017, Retrieved 25 Jan 2018.
43. https://physionet.org/physiobank/database/challenge/2016/training-a
44. R. L. Donnerstein, Continuous spectral analysis of heart murmurs for evaluating stenotic cardiac lesions. *American Journal of Cardiology*, 64(10), 625–630, 1989.

9 Image Denoising
An Overview of Important Methods in Spatial and Transform Domain

B. N. Aravind, K. V. Suresh, H. D. Nataraj Urs,
N. Yashwanth, and Usha Desai

CONTENTS

9.1 INTRODUCTION

Multimedia has gained lot of importance now a days. It includes areas such as audio, video, image, etc. There is an increased demand for good-quality images. Many a times, due to several circumstances, the captured image will get degraded. Noise is one of the major causes of degradation. It occurs during acquisition of image and/ or transmission. During acquisition, imaging sensors are affected by several factors like atmospheric conditions, improper lighting, camera sensors, etc., which will contribute to the amount of noise. The statistical behavior of noise is characterized

DOI: 10.1201/9781003326830-9

by its probability distribution function. Literature records several probability density functions (PDFs) such as Gaussian, Rayleigh, exponential, uniform, impulse, etc. Further, noise can also be modeled as either additive or multiplicative. The performance of applications like object detection, recognition, segmentation, tracking, and many other industrial as well as general applications are affected due to the presence of noise. Hence, denoising is a prerequisite in most of the applications. In this chapter, an effort has been done in investigating the well-known methods of spatial and transform domains. In transform domain, specifically the wavelet transform methods are highlighted.

9.2 NOISE IN IMAGES

Noise can be defined as an unwanted disturbance that interferes with desired signal. It is introduced into image at different levels of image formation. Figure 9.1 illustrates some examples of images affected by noise. Figure 9.1(a) is a synthetic image with added noise; whereas Figure 9.1(b) is a real image of traffic signal captured at low light condition. In a capturing device like camera, noise arises due to electronic circuit, sensors, poor illumination, and/or high temperature. Thermal noise in cameras increases with increase in duration of exposure. Analog-to-digital conversion introduces quantization noise, but it can be reduced by using sufficient number of bits per pixel. Shot noise is introduced due to the amount of photons falling on sensors. This is due to the fact that the photons received by sensors from a uniform scene is not uniform [1]. Coherent imaging applications like synthetic aperture radar (SAR), LASER, ultrasound scanning, etc., are affected by dark and bright spots. It is due to the interference of echoes of a transmitted waveform [2]. The superposition of acoustical echoes in random phase and amplitude produces a granular-like

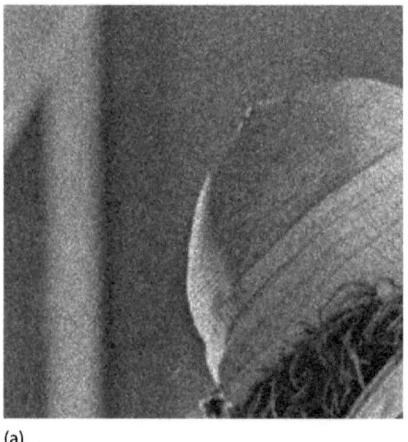 

(a)   (b)

FIGURE 9.1 Noisy images. (a) A part of synthetic image. (b) A traffic signal image captured at low light condition.

interference patterns. Noise in transmission channels occurs due to modification in bit pattern and/or interference due to light sources. Depending on the medium of transmission, shot noise or granular noise may appear.

9.3 NOISE PDFS

Noise, being random in nature, can be characterized by a suitable PDF for modeling the statistical behavior of intensity values. Some of the most commonly used noise PDFs in image processing are listed in Table 9.1 [3]. The PDFs are useful tools for modeling noise that occurs at various situations. Noise due to electronic circuits and poor illumination is modeled as Gaussian. Speckle noise is common in applications that involve coherent radiation such as SAR, ultrasound, and high-intensity LASERs. The Poisson noise is inherent in functional MRI, fluorescence microscopy, low-intensity LASERs, X-ray film, and infrared photometers [4, 5]. In [5], Rician noise for MRI and Gaussian noise for computerized tomography (CT) are used. Joao et al. [4] approximated Rice distribution as Gaussian distribution for low-intensity regions of MRI and as Rayleigh distribution for high-intensity regions.

Literature records exhaustive work on image denoising. In most of these methods, noise is either assumed to be additive [6] or multiplicative [6, 7]. If x designates the

TABLE 9.1

Various Types of Noise and Its PDF [3]

Noise	PDF	Mean	Variance
Gaussian	$p(z) = \dfrac{1}{\sqrt{2\pi}\sigma} e^{-(z-m)^2/2\sigma^2}$	m	σ^2
Rayleigh	$p(z) = \begin{cases} \dfrac{2}{b}(z-a)e^{-\frac{(z-a)^2}{b}} & \text{for } z \ge a \\ 0 & \text{for } z < a \end{cases}$	$m = a + \sqrt{\pi b/4}$	$\sigma^2 = \dfrac{b(4-\pi)}{4}$
Erlang (gamma)	$p(z) = \begin{cases} \dfrac{a^b z^{b-1}}{(b-1)!}e^{-az} & \text{for } z \ge 0 \\ 0 & \text{for } z < 0 \end{cases}$	$m = \dfrac{b}{a}$	$\sigma^2 = \dfrac{b}{a^2}$
Exponential	$p(z) = \begin{cases} ae^{-az} & \text{for } z \ge 0 \\ 0 & \text{for } z < 0 \end{cases}$	$m = \dfrac{1}{a}$	$\sigma^2 = \dfrac{1}{a^2}$
Uniform	$p(z) = \begin{cases} \dfrac{1}{b-a} & \text{if } a \le z \le b \\ 0 & \text{otherwise} \end{cases}$	$m = \dfrac{a+b}{2}$	$\sigma^2 = \dfrac{(b-a)^2}{12}$
Impulse (salt and pepper)	$p(z) = \begin{cases} P_a & \text{for } z = a \\ P_b & \text{for } z = b \\ 0 & \text{otherwise} \end{cases}$	-	-

noise-free original image and η indicates multiplicative noise, then, the degraded observation is represented as:

$$y = x \cdot \eta \tag{9.1}$$

where η is the noise. Speckle noise is commonly categorized under multiplicative noise [6, 7]. Additive noise model is described by:

$$y = x + \eta \tag{9.2}$$

where η is the noise [6]. In this chapter, the assumption is of an image degraded/corrupted by additive white Gaussian noise (AWGN).

9.4 IMAGE DENOISING

The denoising methods can be broadly classified into two categories:

Spatial domain methods: These methods involve the manipulation of pixel values directly. In a given location (i, j), the pixel value is decided by a mask or a window function. The simplest method is the weighted average of neighboring pixels or it can make use of a mathematical model.

Transform domain methods: Transform domain representation clusters an image according to its frequency distribution and is concentrated to few high-valued coefficients. Transform domain methods involve the manipulation of coefficients. Thresholding is a simple and efficient technique, where coefficient values are manipulated by comparing against a threshold value.

9.4.1 SPATIAL DOMAIN DENOISING METHODS

In these methods, to achieve denoising, the pixel values are manipulated directly. Denoising is achieved by considering local neighboring pixels. Wallis [8] defined local mean and variance as:

$$m_{i,j} = \frac{1}{(2r+1)^2} \sum_{k=i-n}^{n+i} \sum_{l=j-m}^{j+m} y_{k,l} \tag{9.3}$$

$$v_{i,j} = \frac{1}{(2r+1)^2} \sum_{k=i-n}^{n+i} \sum_{l=j-m}^{j+m} \left(y_{k,l} - m_{i,j} \right)^2 \tag{9.4}$$

where $m_{i,j}$ is mean and $v_{i,j}$ is variance w.r.t. (i, j).

This method is applied for filtering scan line noise and good results are obtained.

Lee [9] extended the above idea by considering a priori mean and variance. A priori mean $m_{i,j}$ remains same as Equation 9.3, whereas the variance is represented by:

$$\hat{v}_{i,j} = \frac{1}{(2r+1)^2} \sum_{k=i-n}^{n+i} \sum_{l=j-m}^{j+m} (y_{k,l} - \hat{m}_{i,j})^2 - \sigma_n^2 \qquad (9.5)$$

Now, the final estimated image $\hat{x}_{i,j}$ is obtained by:

$$\hat{x}_{i,j} = \hat{m}_{i,j} + k_{i,j}(y_{i,j} - \hat{m}_{i,j}) \qquad (9.6)$$

where $k_{i,j}$ is gain and is given by:

$$k_{i,j} = \frac{\hat{v}_{i,j}}{\hat{v}_{i,j} + \sigma_1^2} \qquad (9.7)$$

σ_1^2 is a constant value and is assumed as 300 in [8].

If $\hat{v}_{i,j}$ is small compared to σ_1^2, $k_{i,j} \cong 0$, then a low signal-to-noise ratio is obtained and the estimated $\hat{x}_{i,j}$ becomes same as $\hat{m}_{i,j}$. If $\hat{v}_{i,j}$ is much larger than σ_1^2, $k_{i,j} \cong 1$ and $\hat{x}_{i,j} \cong y_{i,j}$, then higher signal-to-noise ratio is obtained.

Lee filter found its application in video denoising. Later, Jin et al. [10] identified the presence of annoying noise around the images edges because of this filter. It makes an assumption that the gray values inside the local window are from the same band. This assumption becomes invalid for sharp edges that fall inside the window.

As a solution, a weighted version of Equation 9.4 is introduced [11] and is given by:

$$v_{i,j} = \sum_{k=i-n}^{i+n} \sum_{l=j-m}^{j+m} w(i,j,k,l)[y_{k,l} - m_{i,j}]^2 \qquad (9.8)$$

An adaptive approach is proposed by Jin et al. [10] to select $w(\cdot)$.

$$w(i,j,k,l) = \frac{K(i,j)}{1 + a\left\{\max\left[\epsilon^2, (y(i,j) - y(k,l))^2\right]\right\}} \qquad (9.9)$$

where $K(i,j)$ is gain and is represented as:

$$K(i,j) = \left\{ \sum_{p,q} \frac{1}{1 + a\left\{\max\left[\epsilon^2, (y(i,j) - y(k,l))^2\right]\right\}} \right\}^{-1} \qquad (9.10)$$

An updated definition for local mean and variance was provided in [10] by introducing a weight function in Equation 9.8 and is given as:

$$\hat{m}_{i,j} = \sum_{k=i-n}^{i+n} \sum_{l=j-m}^{j+m} w(i,j,k,l) y(k,l) \tag{9.11}$$

$$\hat{v}_{i,j} = \sum_{k=i-n}^{i+n} \sum_{l=j-m}^{j+m} w(i,j,k,l) [y(k,l) - \hat{m}(i,j)]^2 \tag{9.12}$$

Bilateral filter is a nonlinear filtering method introduced by Tomasi and Manduchi [12]. It is a noniterative procedure and uses a local approach to smooth the image by preserving edges. It utilizes the idea that two pixels are similar not only if they are close to each other spatially but also if they possess similar intensity values. Hence, the intensity value at every location is obtained by taking into account of both photometric similarity and geometric mean between neighboring pixels within the spatial window. Bilateral filter makes use of weighted average of pixels on the neighborhood to recover original image from its degraded version [13].

$$\hat{x}(i,j) = \frac{\sum_{k,l=-N}^{N} w(i,j,k,l) y(k,l)}{\sum_{k,l=-N}^{N} w(i,j,k,l)} \tag{9.13}$$

Equation 9.13 is the normalized weighted average of neighborhood samples around $(i,j)^{th}$ pixel. The weight $w(\cdot)$ is computed by using two factors: temporal (w_s) and radiometric weights (w_R).

$$w_S(i,j,k,l) = \exp\left\{ -\frac{(i-k)^2 + (i-k)^2}{2\sigma_S^2} \right\} \tag{9.14}$$

$$w_R(i,j,k,l) = \exp\left\{ -\frac{\| y(i,j) - y(k,l) \|^2}{2\sigma_R^2} \right\} \tag{9.15}$$

The final weight function $w(\cdot)$ is realized by:

$$w(\cdot) = w_S(\cdot) \cdot w_R(\cdot) \tag{9.16}$$

Temporal weight measures geometric distance between the center pixel (i,j) and the neighboring pixel (k,l). Radiometric distance refers to the intensity value difference between the center and the neighboring pixel. Hence, neighborhood has influence on the final result. Bilateral filter depends on three parameters: size of the neighborhood N, σ_S, and σ_R. Higher the value of N gives more smoothness. σ_S and σ_R control decay of weights. Higher the value leads to uniform nonadaptive filtering; lower values reduce smoothing effect [13].

Barash [14] revealed that a wide spatial window can only make the bilateral filtering noniterative. Bigger the spatial window may lead to oversmoothing of edges.

Consequently, in bilateral filtering, maintaining a balance between number of iterations and size spatial window is important. Trilateral filter [15] adds local structural similarity to the bilateral filter to smoothen images in a narrow spatial window by preserving edges. In homogeneous regions, low-pass filter is applied. Whereas for edges, smoothing is achieved by taking consideration of photometric, geometric, and local structural similarities among neighboring pixels.

For continuous signals, Rudin and Osher [16] proposed total variation (TV) minimization technique. Later, Chen et al. [17] introduced its discrete version. In order to identify the edges, edge derivatives and graphs are used. Digital TV filter consists of mainly two parameters:

- Regularization parameter a—small positive value.
- Filtering parameter λ—a positive number.

TV filtering at any location α involves the following steps:

1. Computation of local variation $|\nabla u|_\alpha$ at α and its neighborhoods β
2. Computation of weights $w_{\alpha,\beta}$ using

$$w_{\alpha,\beta}(u) = \frac{1}{|\nabla_\alpha u|_\alpha} + \frac{1}{|\nabla_\beta u|_\alpha} \qquad (9.17)$$

3. Computation of filter coefficients $h_{\alpha\alpha}$, $h_{\alpha\beta}$:

$$h_{\alpha\beta} = \frac{w_{\alpha\beta}(u)}{\lambda + \displaystyle\sum_{\gamma\sim\alpha} w_{\alpha\gamma}(u)} \qquad (9.18)$$

$$h_{\alpha\alpha} = \frac{\lambda}{\lambda + \displaystyle\sum_{\gamma\sim\alpha} w_{\alpha\gamma}(u)} \qquad (9.19)$$

4. Filtering $F_\alpha(u) = \displaystyle\sum_{\gamma\sim\alpha} h_{\alpha\beta} u_\beta + h_{\alpha\alpha} u_\alpha^0$

TV involves iterative procedure. The term λ is responsible for controlling fidelity and regularity terms. Smaller the value of λ will increase the weight of the regularity term. Also, this leads to the oversmoothing of details and texture [18].

Nonlocal means (NLM) is a well-known spatial domain denoising method introduced by Buades et al. [19–21]. The NLM uses the ideology that an image includes several similar repeated structures, and averaging across those similar structures leads to the reduction of noise. At any location i in an image, the estimated value $NL\big[y(i)\big]$ is given by the weighted average of all pixels in the image:

$$NL[y](i) = \sum_{j\in I} w(i,j) y(j) \qquad (9.20)$$

where the weight $\{w(i,j)\}_j$ depends on the similarity between i and j and satisfies the conditions $0 \le w(i,j) \le 1, \sum_j w(i,j) = 1.$

The similarity between the pixels i and j is dependent on the similarity of the gray-level vectors $v(N_i)$ and $v(N_j)$, where N_k is the square neighborhood centered at k. The weight function is represented as:

$$w(i,j) = \frac{1}{z(i)} e^{-\frac{\|v(N_i)-v(N_j)\|_{2,a}^2}{h^2}} \tag{9.21}$$

where $z(i)$ indicates normalizing constant and is defined as:

$$z(i) = \sum_j e^{-\frac{\|v(N_i)-v(N_j)\|_{2,a}^2}{h^2}} \tag{9.22}$$

and the parameter h indicates degree of filtering.

NLM is computationally very intensive and also it is not practical [22]. For an image of M pixels, it is required to calculate M weights for each pixel. If the complete image is considered, the process will become inefficient and intensified with calculation.

The K-SVD [23]-based denoising algorithm basically makes use of nonlocal similarity concept. It makes use of the dictionary encoding, which is optimized by using nonlocal similarities present in the image. It involves three steps:

1. Sparse coding: Initial dictionary is used to create sparse approximations of fixed-size patches of image.
2. Updating of dictionary: Sparse approximation quality is increased by this.
3. Reconstruction: From the collection of denoised patches, denoised image is recovered.

Before obtaining the reconstructed image, the steps (1) and (2) takes K iterations.

Another spatial domain approach for image denoising is by modeling the image as Markov random field (MRF), which preserves edges. The next few paragraphs will cover a brief introduction about MRF and is further continued with its application towards denoising.

MRF is a two-dimensional (2D) random process that is defined over a discrete lattice. Lattice is considered as a regular grid on the 2D plane. The MRF uses local conditional probability distribution in order to obtain global representation. Because of this, MRF is applicable for several applications of image processing [24].

A random field $F = \{F_{ij}\}$ designed over lattice S is a MRF with respect to the neighborhood system $\mathcal{N}$, if and only if:

$$P(F = f) > 0, \forall f \in \mathbb{F}(positivity)$$

$$P\left[F_{ij} = f_{ij} | F_{kl} = f_{kl} \forall (k,l) \neq (i,j)\right] = P\left[F_{ij} = f_{ij} | F_{kl} = f_{kl}, (k,l) \in \mathcal{N}_{ij}\right](Markovianity)$$

for all $(i,j) \in S$. Here, f is the realization of random field F and $\mathbb{F}$ is the configuration space (indicates the set of all possible labels for f).

Markovianity is attractive in most of the natural images. It implies that the intensity value of a pixel is dependent on its surrounding neighboring pixel values.

GIBBS RANDOM FIELD (GRF)

Consider $\mathcal{N}$ is the neighborhood system that is defined over S. A random field $F = \{F_{ij}\}$ defined on S is a GRF with respect to $\mathcal{N}$ if and only if its joint distribution is of the form

$$P(F = f) = \frac{1}{Z}\exp\{-U(f)\} \tag{9.23}$$

where $U(f) = \sum_{c \in C} V_c(f)$ and $Z = \sum_{f \in \mathbb{F}} \exp\{-U(f)\}$

Equation 9.23 realizes the Gibbs distribution (GD). Where $U(f)$ represents energy function that is associated with all cliques and Z is the partition function.

MRF and GRF are characterized by local and global property, respectively. The Hammersley and Clifford theorem [25] created an equivalence between GRF and MRF. This increased the utility of MRF in image processing application.

Denoising using MRF is proposed in [26, 27]. It makes use of discontinuity adaptive (DA) function, and the local minima problem is solved by annealing. For optimization, graduated non-convexity (GNC) algorithm, which is a deterministic annealing method, is used. The following is the implementation algorithm:

Initialize: $x^{(0)}$, Iteration count $n = 0, \mu, \lambda$.

Do

Update $x^{(n)}$: $x^{(n+1)} = x^{(n)} - \mu \cdot grad$

Set $n = n+1$

Repeat until $\left(norm\left(x^{(n)} - x^{(n-1)}\right)\right) \leq \epsilon$ and $\lambda^n = k\lambda_{target}$

$$\hat{x} = x^{(n)}$$

where the term μ represents the step size and in every iteration $\lambda^{(n)}$ is modified by a factor of k (less than but very close to unity) and the target modified by a factor k to obtain λ_{target}.

It is an iterative process and continues till final minima is reached.

9.4.2   TRANSFORM DOMAIN DENOISING METHODS

Transform domain denoising is preferred because of the ease of discriminating signal and noise. The basic principle for denoising in transform domain methods

involves shrinkage of coefficients to suppress the effect due to noise. For such thresholding, it is necessary to develop a coefficient mapping method that retains details
in an image.

Fourier representation reveals the spectral details of a signal, but locating particular moment in time is not possible. This makes the analysis of transient signals
inadequate using Fourier transform. In signal and image processing, transients are
considered as essential information. Several bases and transforms are proposed [28,
29] that can analyze transient signals in terms of localizing both in time and frequency. Among them, short-time Fourier transform (STFT) and wavelet transform
are quite popular.

In STFT (also known as Gabor transform/windowed Fourier transform), the
signal is multiplied by a smooth window function (usually Gaussian) and then the
Fourier integral is applied to the windowed signal. Whereas in wavelet transform,
the scale is related as the inverse of frequency. Figure 9.2 presents the tiling representation (in time frequency) for STFT and wavelet transform. It can be identified
that STFT divides the time frequency plane to equal blocks and on other hand the
wavelet transform acts like a microscope [29] by focusing on smaller time phenomenon as the scale decreases. This enables to focus on local characterization of signals.
Additional advantages of the wavelet transform include multiresolution and sparsity.

Since the advent of wavelets, it emerged as a premier tool for several applications in image processing, including restoration. This is due to the statistically useful properties of wavelet coefficients for natural images. Discrete wavelet transform
(DWT) in one dimension (1D) is given by:

$$W_\phi\left(j_0,k\right) = \frac{1}{\sqrt{M}} \sum_{n=0}^{M-1} x\left(n\right)\phi_{j_0,k}\left(n\right) \tag{9.24}$$

$$W_\psi\left(j,k\right) = \frac{1}{\sqrt{M}} \sum_{n=0}^{M-1} x\left(n\right)\psi_{j,k}\left(n\right) \quad j \geq j_0 \tag{9.25}$$

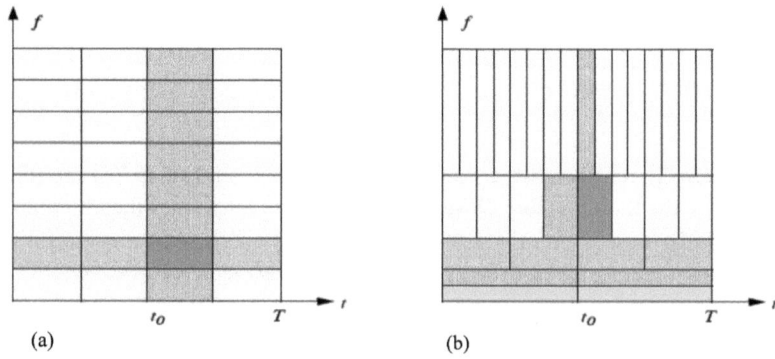

FIGURE 9.2   Time frequency tiling. (a) STFT. (b) Wavelet (courtesy of [29]).

where j is parameter about dilation or visibility in frequency and k is the parameter about position, $x(n)$ is the input signal, and ϕ and ψ are low-pass and high-pass filters, respectively.

DWT is known to provide multiresolution representation of piecewise smooth signals efficiently. Representations presented in [30] project the efficiency of wavelet transform in representing piecewise smooth signal.

9.4.3 WAVELETS FOR DENOISING

Wavelets are available in several flavors. For denoising applications, sparse representation helps to produce visually good quality of reconstructed images. Thus, it is important to make as many wavelet coefficients close to zero, and all the energy being concentrated among few coefficients is important. It is dependent on number of vanishing moments ϑ_N and the supporting size K of analysis wavelet. Higher the vanishing moments ϑ_N, smaller support size is the requirement of the analysis wavelet. The reconstructed signal is associated with error $\in$ and is due to threshold or quantization. The human visual system is tolerable for symmetric errors than asymmetric. Therefore, symmetric wavelets are more preferable. In [31] it is suggested that symlets with vanishing moments ϑ_N provide compact support and are suited for denoising applications. Hence, $sym\vartheta_N$ is the preferred wavelet.

Literature indicates several transforms like Fourier transform (FT) [32], discrete cosine transform (DCT) [33], curvelets [34], contourlets [35, 36], ridgelets [37], surelets [38], wedgelets [39], etc., that are used in image processing. Many papers have been proposed for image denoising using these transforms [40–45].

WAVELET-BASED IMAGE DENOISING

Wavelet is being used as a powerful and efficient tool in several fields of image processing including image denoising. Several denoising techniques were developed using wavelet transform and in an overall view, they can be classified into two categories:

1. Term-by-term denoising
2. Neighborhood consideration

Both uses the following steps:

- Application of wavelet transform
- Thresholding
- Perform inverse wavelet transform

Donoho et al. [46–48] developed a wavelet denoising technique using hard and soft thresholding. Here, each coefficient is modified by comparing it with a threshold value. Soft thresholding (T_{soft}) is shrink or kill process, and hard thresholding (T_{hard}) is keep or kill process. It is given by:

$$T_{soft}(W) = \begin{cases} 0 & if |W| \leq t \\ sign(W)(|W|) - t) & if |W| > t \end{cases} \tag{9.26}$$

$$T_{hard}(W) = \begin{cases} 0 & if |W| \leq t \\ W & if |W| > t \end{cases} \tag{9.27}$$

and $t = \hat{\sigma}\sqrt{2\log(n)}$, where $\hat{\sigma}$ is an estimation of noise standard deviation and n indicates the number of wavelet coefficients in the subband. Wavelet shrinkage is a near-optimal technique and does not require prior knowledge [49].

Gaussian white noise after transformation remains as white noise with same amplitude. Hence, thresholding has an effect of killing the noise while not affecting the signal [50]. It is identified that the thresholding algorithm exhibits artifacts in the neighborhood of edges. To resolve this problem, translation invariant (TI) denoising [51] scheme is proposed. This uses thresholding scheme on the cyclically shifted images. It can be represented as:

$$T(x;(S_h)_{h \in H}) = avg_{h \in H} S_{-h}(T(S_h(x))) \tag{9.28}$$

The artifacts are suppressed by considering all circular shifted signals and averaging over them. The results indicated the TI wavelet denoising as better performer. As an extension to TI denoising, recursive cycle spinning is introduced [52]. For each iteration, the output of previous stage is used as input to the current stage and it is recursive. The results are found to be better than soft thresholding and TI method of denoising. Multiwavelets applied to denoising provide better results compared to single wavelet in non-TI conditions [53]. The idea of multiwavelets is also extended to TI and obtained better results than the single TI method [54].

Hard threshold possesses bigger variance due to discontinuity that happens by the nature of threshold function. Soft threshold also has bigger bias because of shrinkage of bigger coefficients toward zero by t [55]. As a compromise between soft and hard thresholding, firm (semisoft) shrinkage is introduced [56]. It uses two threshold values t_1 and t_2 and is given by:

$$T_{t_1,t_2}(W) = \begin{cases} 0 & if |W| \leq t_1 \\ sgn(W)\dfrac{t_2(|W| - t_1)}{t_2 - t_1} & if t_1 < |W| \leq t_2 \\ W & if |W| > t_2 \end{cases} \tag{9.29}$$

Several variations of thresholding schemes are developed [57–59]:

$$T(W,t,k) = \begin{cases} W+t-\dfrac{t}{2k+1} & \text{if } W < -t \\[2ex] \dfrac{1}{(2k+1)t^{2k}}\,W^{2k+1} & \text{if } |w| \le t \\[2ex] W-t+\dfrac{t}{2k+1} & \text{if } W > t \end{cases} \tag{9.30}$$

$$T(W,t,l) = W + 0.5\left(\sqrt{(W-t)^2 + l} - \sqrt{(W+t)^2 + l}\right) \tag{9.31}$$

$$T(W,t,m,n,k) = \begin{cases} W - 0.5\dfrac{t^m \cdot k}{W^{m-1}} + (k+1)t & \text{if } W > t \\[2ex] 0.5\dfrac{k \cdot |W|^n}{t^{n-1}}\,sign(W) & \text{if } |W| \le t \\[2ex] W + 0.5\dfrac{(-t)^m \cdot k}{W^{m-1}} - (k-1)t & \text{if } W < -t \end{cases} \tag{9.32}$$

$$T(W,t) = \begin{cases} W - 0.5\dfrac{t^2}{W} & \text{if } |W| > t \\[2ex] 0.5\dfrac{W^3}{t^2} & \text{if } |W| \le t \end{cases} \tag{9.33}$$

$$T(W,t,m,k) = \begin{cases} W + (k+1)t - 0.5\dfrac{k \cdot t^m}{W^{m-1}} & \text{if } W > t \\[2ex] 0.5\dfrac{k \cdot |W|^{m+\left[\frac{2-k}{k}\right]}}{t^{m+\left[\frac{2-2k}{k}\right]}}\,sign(W) & \text{if } |W| \le t \\[2ex] W - (k-1)t + 0.5\dfrac{k \cdot (-t)^m}{W^{m-1}} & \text{if } W < -t \end{cases} \tag{9.34}$$

where t, t_1, and t_2 are threshold values, W is wavelet coefficient, $l > 0$ is a user-defined (fixed) parameter, and k and m are shape tuning parameters.

"Multispinning" [60] is a new cyclic shift-based denoising method using wavelet transform. It uses seven versions of an image, namely:

- An image without shift
- Three images having shifted row wise
- Three images having shifted column wise

The idea behind creating multiple images is to capture good amount of edge information when the transform is applied. Soft thresholding is operated on all cyclically shifted images independently. The reconstructed images are then reverse shifted and averaged.

The implementation process involves following steps:

- Apply multispinning to noisy image.

$$y_i = M_{sh}\{x_n\}_i \tag{9.35}$$

- Apply wavelet transform to each image.

$$Y_i = T_m\{y_i\} \tag{9.36}$$

- Denoise transformed images.

$$\hat{Y}_i = D\{Y_i\} \tag{9.37}$$

- Apply inverse transform on all images.

$$\hat{y}_i = T_m^{-1}\{\hat{Y}_i\} \tag{9.38}$$

- Unshift and average them.

$$\hat{x} = \frac{1}{N}\sum_{i=1}^{N} M_{ush}\{\hat{y}_i\} \tag{9.39}$$

All above steps can be realized in single equation as:

$$\hat{x} = \frac{1}{N}\sum_{i=1}^{N} M_{ush}\left[T_m^{-1}\left\{D\left(T_m\left(M_{sh}\left(\{x_n\}_i\right)\right)\right)\right\}\right] \tag{9.40}$$

where $\hat{x}$ is the final denoised image, x_n is noisy representation, M_{sh} is multispin shift operation, T_m is forward wavelet transform, D is denoising process, T_m^{-1} is inverse wavelet transform, M_{ush} is unshift operation, and N indicates total number of images.

Using multiple images in multispin approach has two advantages:

- Reduces Gibb's phenomenon due to averaging.
- Additional edge information captured from shifted version of images adds up to the reconstruction quality.

All wavelet denoising methods discussed till now incorporate term-by-term thresholding. Several methods are introduced that utilize the influence of surrounding wavelet coefficients on the center coefficient. The ideology is that, usually, large-valued wavelet coefficient will be surrounded by large-valued coefficients [61, 62]. By considering neighboring coefficients, Cai and Silverman [62] developed thresholding schemes; Neighblock and Neighcoeff for one-dimensional (1D) signals. First method uses disjoint blocks, whereas second uses overlapping blocks. The experimental results showed an improvement in the result compared to term-by-term wavelet denoising.

The idea of using neighboring wavelet coefficients is also extended to multiwavelets by Chen and Bui [63]. It is tested on standard 1D test signals (Blocks, Bumps, Heavysine, and Doppler), and results are compared with multiwavelet term-by-term thresholding. Results indicate that neighborhood approach is better than the conventional approach. Chen et al. [64–66] proposed neighshrink that incorporates neighboring wavelet coefficients. The experiment is conducted by considering the neighborhood windows of size 3×3 and 5×5. The procedure followed is:

- Calculate $\delta_{j,k}^2$ using neighbor coefficients, $\delta_{j,k}^2 = W_{j-1,k-1}^2 + W_{j-1,k}^2 + W_{j-1,k+1}^2 + \dots$.
- Find $t = \sqrt{2\sigma^2 log n^2}$ (universal thresholding).
- If $S_{j,k}^2 \le t^2$ then set $W_{j,k}$ to zero. Otherwise, shrink according to

$$W_{j,k} = W_{j,k}\left(1 - \frac{t^2}{S_{j,k}^2}\right).$$

It has a disadvantage that it uses universal thresholding, which is dependent on number of coefficients. More number of coefficients yields higher thresholding value leading to oversmooth. Also, it uses identical neighboring size in all wavelet subbands. Zhou and Chang [67] improved this method by using optimal threshold and neighboring coefficients for every subband by using Stein's unbiased risk estimate (SURE). For each subband, threshold values are calculated using various neighborhood sizes (3×3, 5×5, 7×7, etc.). Among all the thresholds, a value will be chosen that has minimum risk for that subband. The estimation is given as:

$$\left(t^s, L^s\right) = arg \min_{t,L}(W_s, tL) \tag{9.41}$$

where t^s and L^s are threshold and neighborhood size in the subband s.

The concept of multispinning [60] and neighshrinksure [67] are combined and multispiinigneighshrink [68] is developed.

Literature indicates that the current coefficient not only depends on neighbors but there is a link with subbands too. Bivariate shrinkage [69, 70] uses parent–child relationship. It considers the dependency between coefficients of successive subbands. The model describes that if noise is present in the parent coefficient, then child coefficient will also have noise. The estimated value depends on parent value. Smaller the parent value leads to higher shrinkage. An improvement to bivariate shrinkage is done by estimating noise variance locally [71, 72]. It proposes the following PDF:

$$p_w(W) = \frac{3}{2\pi\sigma^2}\exp\left(-\frac{3}{\sigma}\sqrt{W_1^2 + W_2^2}\right) \tag{9.42}$$

with this PDF, W_1^2 and W_2^2 are not independent but uncorrelated. σ is the variance, w_2^2 is the parent of w_1^2.

9.5   COMPARISON OF EXPERIMENTAL RESULTS

A detailed discussion about various denoising methods in spatial and wavelet domain is highlighted in previous sections. In this section, a comparative study of denoising methods along with quantitative and qualitative analysis will be presented. At first, the noise estimation method and evaluation criteria (quantitative and qualitative) are introduced.

9.5.1   NOISE ESTIMATION

Estimation of noise from the degraded image is a challenging issue. Several techniques [73, 74] have been proposed to estimate noise. One such method uses wavelets. In a DWT applied image, HH subband mainly contains noise. An estimate of noise can be obtained by median measurement [67, 75]:

$$\hat{\sigma} = median\left(\frac{|W_{HH}|}{0.6745}\right) \tag{9.43}$$

where W_{HH} is the detail subband of wavelet transform. It is often denoted by $MAD(W)$, where MAD stands for "median absolute deviation." The method is tested on three test images (Square, Barbara, House). For experimentation, initially a noise-free "Square" image of 256×256 is considered. Several noisy versions of this image are synthetically generated by adding Gaussian noise. Equation 9.43 is then used to obtain noise parameter. It is observed that (Table 9.2) the estimated noise almost matches with the actual. Experiment is repeated on other images and results are tabulated. Figure 9.3 shows the graphical representation of results.

TABLE 9.2
Noise Estimation Using MAD

Actual	Estimated	Absolute (error)
Square (256 × 256)		
10	9.8738	0.1262
20	19.7197	0.2803
30	29.5589	0.4411
40	39.4031	0.5969
Barbara (512 × 512)		
10	11.0143	1.0143
20	20.6574	0.6574
30	30.3354	0.3354
40	40.0795	0.0795
House (256 × 256)		
10	10.0247	0.0247
20	19.7967	0.2033
30	29.6154	0.3846
40	39.4536	0.5464

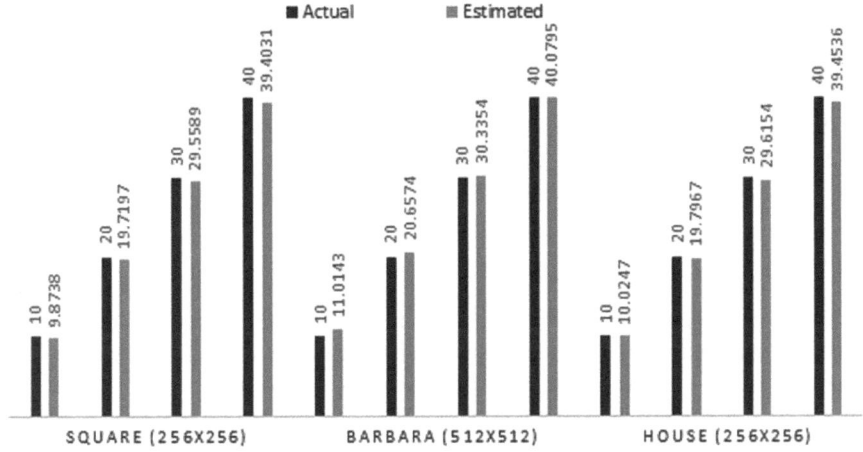

FIGURE 9.3 Graph of actual and estimated noise from three test images.

9.5.2 EVALUATION CRITERIA

Performance evaluation of denoising by suitable measure is very much necessary. For test cases, it is considered that the original clean image is available. Peak signal-to-noise ratio (PSNR) is used for quantitative analysis:

$$PSNR = 10 \times log_{10}\left(\frac{255^2}{MSE}\right) \tag{9.44}$$

where MSE is the mean squared error:

$$MSE = \frac{1}{N}\|x - \hat{x}\|^2 \tag{9.45}$$

where x is the original clean image and $\hat{x}$ is the restored image.

For objective analysis, mean structural similarity index (MSSIM) [76–78] is being used. It is considered to be very close to human visual perception [78]. If x and y are two images, SSIM between them is obtained using three parameters:

- Luminance similarity, $l(\cdot)$, uses local measures of mean of noise-free μ_x and noisy representations μ_y.
- Contrast similarity $c(\cdot)$, uses local measure of variance of noise-free σ_x^2 and noisy representations σ_y^2.
- Structural similarity, $s(\cdot)$, uses local measure of standard deviation of noise-free and noisy images and local measure of their covariance σ_{xy}. The expression for SSIM is given by:

$$SSIM(x,y) = \frac{(2\mu_x\mu_y + C_1)(2\sigma_{xy} + C_2)}{(\mu_x^2 + \mu_y^2 + C_1)(\sigma_x^2 + \sigma_y^2 + C_2)} \tag{9.46}$$

where C_1 and C_2 are constants and have default values $C_1 = (0.001L)^2$, $C_1 = (0.03L)^2$ [77] and L represents the dynamic range of pixel values $(L = 255 \; for \; gray \; level \; images)$. SSIM is calculated over block of size $M \times M$ (typically 8×8). Finally a mean SSIM (MSSIM) is computed by averaging all the SSIM values.

Denoising aims to suppress the noise and on the other hand to preserve details. Researchers have proposed several methods to achieve these goals.

9.5.3   SPATIAL DOMAIN DENOISING

In this section, the performance of various spatial domain denoising methods is compared. As quantitative measure, PSNR is used and MSSIM is used as benchmark for measuring structural similarity. For simulation, three test images from Figure 9.4 are considered, namely Lena (512 × 512), Child (256 × 256), and Boat (512×512). Noisy version of images is generated by adding Gaussian noise to these test images. Denoising is performed by using bilateral filter, NLM, TV minimization, K-SVD, and DAMRF methods. Figure 9.5 illustrates the denoising of the Lena image degraded by noise $\sigma = 20$. Figure 9.5(a) is the noisy representation of Lena image. Bilateral filter is applied to it and the result in Figure 9.5(b) shows that the noise is suppressed but edges are oversmoothed. Figure 9.5(c) is the resultant of nonlocal means filtering method. The reconstruction quality is better than bilateral filter. However, it can be observed that some of the fine details have got flattened due to smoothing effect. Figure 9.5(d) is the resultant of TV minimization. The default parameter settings of [79] are used for simulation. It suppresses the noise by oversmoothing the image. Result of K-SVD is given in Figure 9.5(e) and the result is drawn from [23]. It can be observed that the noise suppression is incomplete. The DAMRF method suppresses the noise to a good extent and provides better preservation of edges (Figure 9.5(f)).

Table 9.3 provides the quantitative analysis of test images (Lena, Child, and Boat), and Table 9.4 gives the mean structural similarity index values.

9.5.4   TRANSFORM DOMAIN DENOISING

Figure 9.6 shows the denoising of Lena image degraded with noise $\sigma = 20$ using various transform domain techniques. Both term-by-term and neighborhood methods

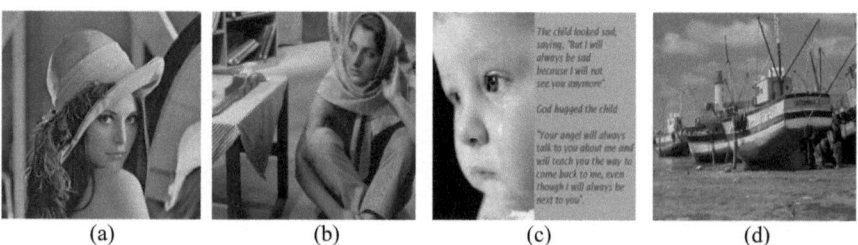

(a)   (b)   (c)   (d)

FIGURE 9.4   Synthetic images under consideration simulation. (a) Lena. (b) Barbara. (c) Child. (d) Boat.

FIGURE 9.5 Spatial domain denoising of Lena image with noise $A = 20$. (a) Noisy image. (b) Bilateral filter. (c) NLM. (d) TV minimization (e) K-SVD. (f) DA3-MRF.

are used for comparison. Soft thresholding is kill or shrink concept. It reduces the noise to certain extent. Since it is basically a smoothing operation, it leads to the reduction of edge quality (Figure 9.6(b)). At the regions of constant gray-level areas, it can also exhibit artifacts. Hard thresholding is a keep or kill process. Its ability to reduce the noise is competitively very less (Figure 9.6(c)). To reduce the artifacts due to soft and hard thresholding, TI denoising is used. It is achieved by using multiple cyclically shifted images and applying soft thresholding on each image. Shift is reversible and finally averaged to get the image (Figure 9.6(d)). Averaging reduces the artifacts and it produces better results than soft and hard thresholding.

TABLE 9.3

Quantitative Analysis PSNR (dB)

σ	Bilateral filter	NLM	TV minimization	K-SVD	DA3-MRF
Lena (512 × 512)					
10	31.0225	35.0225	32.8167	34.6731	33.9070
20	25.7036	31.8110	30.0191	30.271	30.5657
30	21.0037	29.6786	28.1614	29.6415	28.6599
Child (256 × 256)					
10	32.0162	33.3709	31.5488	34.1173	34.6350
20	25.9231	29.4667	27.5606	29.1971	29.5584
30	21.2698	26.8222	25.5928	28.0075	26.9475
Boat (512 × 512)					
10	30.3045	32.9235	31.2680	33.4342	32.3735
20	25.3543	29.6243	28.0730	29.2663	28.8240
30	20.8797	27.5636	26.1726	28.1307	26.9286

TABLE 9.4

Mean Structural Similarity Index (MSSIM)

σ	Bilateral filter	NLM	TV minimization	K-SVD	DA3-MRF
Lena (512 × 512)					
10	0.9311	0.9602	0.9413	0.9529	0.9566
20	0.8132	0.9207	0.8956	0.8744	0.9096
30	0.6257	0.8782	0.8502	0.8566	0.8573
Child (256 × 256)					
10	0.8930	0.9364	0.9264	0.9004	0.9380
20	0.5539	0.8557	0.8421	0.7294	0.8243
30	0.3374	0.7536	0.7631	0.7470	0.7302
Boat (512 × 512)					
10	0.9118	0.9508	0.9342	0.9562	0.9530
20	0.8315	0.8868	0.8576	0.8766	0.8912
30	0.6723	0.8309	0.7867	0.8438	0.8309

Results of noise suppression using bivariate shrinkage are shown in Figure 9.6(e). It suppresses noise but on the other hand artifacts will remain. The multi-spinning soft thresholding [60] applied on the images yielded in the good results (Figure 9.6(g)). The results are found to be better than that of soft and hard thresholding but similar to cycle spinning. In the methods discussed above (soft, hard, TI denoising, and multispinning soft) makes changes to each coefficient to achieve denoising without considering neighboring denoised better visualization, a zoomed

FIGURE 9.6 Denoising of Lena image degraded with noise Ã= 20. (a) Noisy image. (b) Soft thresholding. (c) Hard thresholding. (d) TI denoising. (e) Bivariate shrinkage. (f) Neighshrinksure. (g) Soft thresholding multispinning. (h) Neighshrinksure multispinning.

portion of each of the result is given in Figure 9.7. Simulations are continued on Barbara, Boat, and Child images corrupted by different noise levels and results are tabulated. Table 9.5 and Table 9.6 present the PSNR and MSSIM of denoising methods, respectively. It can be identified that the result of denoising of most of the methods is good at lower noise levels. With the increase in the amount of noise, few methods create oversmoothed representation and/or artifacts. The structural similarity is verified by using MSSIM.

Finally, some of the transform domain denoising methods are tested on real images of size 480×640 ("Traffic sign images presented in result are the courtesy of Dr. Hasan Fleyeh, Dalarna university, Sweden"). The amount of noise is determined by using MAD method and further the denoising is performed.

FIGURE 9.7   Zoomed version of Fig. 9.6(a)–(h), respectively.

9.6   DISCUSSION

This chapter addresses various denoising methods both in spatial and transform domains. Spatial domain methods use neighborhood smoothness criteria. Transform domain methods achieve denoising by separating the noise and signal components. Almost all methods yield acceptable results at lower noise levels. However, some of these methods fail in preserving edge information at higher noise levels. Literature also records methods that make use of a combination of spatial and transform domains, combination of two transforms to perform denoising. BM3D [80] utilizes nonlocal means applied in transform domain to achieve denoising. It makes use of grouping and collaborative filtering. BM3D depends on both local and nonlocal characteristics of natural images. A hybrid method that makes use of wavelet and bilateral filter has been developed in [81]. The method involves the application of soft thresholding in wavelet domain and the use of bilateral filter before and after wavelet transform. The results are found to be better than the individual methods involved in the denoising process. Kamilov et al. [82] introduced a new method based on wavelet for the implementation of total variation type denoising, which yields improved results compared to conventional wavelet cycle spinning and TV minimization methods.

An iterative method involving wavelet transform and TV minimization method is proposed in [83]. In every iteration, wavelet thresholding is applied and TV

TABLE 9.5
Quantitative Analysis PSNR (dB)

σ	Soft thresholding	Hard thresholding	Bivariate shrinkage	TI method	Multispinning soft thresholding	Neighshrinksure	Neighshrinksure multispinning
Lena (512 × 512)							
10	32.3150	29.6011	34.3201	33.2535	32.6530	34.2472	34.6043
20	28.7052	24.0665	31.1208	29.7462	29.1059	30.8608	31.1886
30	26.5465	20.7272	29.3444	27.5761	27.0880	28.9554	29.2425
40	24.8584	18.3769	28.0664	25.9683	25.5556	27.5880	27.8755
Barbara (512 × 512)							
10	29.9923	29.1045	32.1272	30.8175	30.4964	32.6344	32.8938
20	26.2656	23.6052	28.2390	27.0263	26.7327	28.6142	28.9220
30	24.2154	20.3476	26.1265	25.0307	24.7493	26.4020	26.7864
40	22.8747	18.0198	24.7781	23.7205	23.5018	25.0428	25.4011
Boat (512 × 512)							
10	30.7081	29.1364	32.3647	31.5439	30.9745	32.6133	32.8561
20	27.3385	23.8005	29.1245	28.1711	27.6139	29.0937	29.3193
30	25.3045	20.4874	27.2638	26.2126	25.7212	27.1258	27.3892
40	23.8757	18.1894	26.1318	24.8039	24.4262	25.8851	26.0895
Child (256 × 256)							
10	31.7450	29.6254	33.3376	32.7685	32.1558	33.9514	34.1507
20	27.9552	23.9770	29.1933	28.8937	28.3753	29.7568	30.1008
30	25.8517	20.6269	26.8330	26.7215	26.3669	27.4536	27.9423
40	24.3167	18.2755	25.4267	25.1156	24.9697	26.1831	26.5830

TABLE 9.6
Mean Structural Similarity Index (MSSIM)

σ	Soft thresholding	Hard thresholding	Bivariate shrinkage	TI method	Multispinning soft thresholding	Neighshrinksure	Neighshrinksure multispinning
				Lena (512 × 512)			
10	0.9274	0.9213	0.9555	0.9452	0.9469	0.9592	0.9615
20	0.8675	0.8042	0.9088	0.8999	0.8914	0.9171	0.9198
30	0.8129	0.6947	0.8686	0.8560	0.8384	0.8792	0.8797
40	0.7683	0.6049	0.8302	0.8181	0.7846	0.8412	0.8438
				Barbara (512 × 512)			
10	0.9173	0.9376	0.9598	0.9396	0.9446	0.9469	0.9659
20	0.8334	0.8368	0.9048	0.8688	0.8787	0.9179	0.9170
30	0.7672	0.7385	0.8557	0.8077	0.8182	0.8737	0.8680
40	0.7156	0.6594	0.8040	0.7589	0.7625	0.8339	0.8250
				Boat (512 × 512)			
10	0.9072	0.9250	0.9528	0.9273	0.9332	0.9556	0.9571
20	0.8240	0.8165	0.8926	0.8550	0.8618	0.8975	0.9009
30	0.7602	0.7086	0.8391	0.7974	0.7973	0.8462	0.8501
40	0.7039	0.6164	0.7959	0.7452	0.7430	0.7991	0.7986
				Child (256 × 256)			
10	0.8767	0.7813	0.9112	0.8767	0.8983	0.9147	0.9099
20	0.7788	0.5863	0.8448	0.7788	0.7824	0.8362	0.8082
30	0.7019	0.4615	0.7723	0.7019	0.6739	0.7703	0.7286
40	0.6486	0.2470	0.7115	0.6486	0.5772	0.7219	0.6681

minimization denoising in spatial domain is used. The procedure is repeated until convergence. Hybrid denoising that uses shrinkage in wavelet domain using stationary wavelets and nonlocal means in spatial domain in balanced manner to suppress the noise and to retain the edges is proposed in [84]. It is done in iterative manner with small amount of smoothing operation in each iteration in order to preserve the edges. A noniterative method of [84] is presented in [85]. An adaptive boosting technique proposed in [86] provides improved results in comparison with other spatial domain methods. It is nonadaptive and reduces noise level in each iteration. Further, machine learning methods that utilize filtering, genetic algorithm, and optimization algorithms are also proposed to achieve better denoising results [87].

FIGURE 9.8 Denoising of real image. (a) Image with unknown amount of noise. (b) Bivariate shrinkage. (c) Multispinsoft. (d) Neighshrinksure. (e) Neighshrinksure multispinning.

TABLE 9.7
Noise in Medical Images [88]

Sl. no.	Imaging methods	Type(s) of noise PDF
1	X-ray	Gaussian, Poisson
2	CT	Gaussian, Quantum
3	PET	Gaussian
4	SPECT	Gaussian
5	MRI	Gaussian, Rician, Rayleigh
6	Ultrasound	Speckle, Gaussian

Further, medical images are not exceptional to noise. Hence denoising is a part of preprocessing in medical images and will lead to better diagnosis. Medical images are obtained from X-ray, CT, ultrasound, positron emission tomography (PET), magnetic resonance imaging (MRI), and single photon emission computed tomography (SPECT). Major types of noise involve Gaussian noise, poison noise, speckle noise, and, in some cases, a combination of several noises [88]. Table 9.7 shows different types of noises encountered in various imaging methods.

Common assumption about noise is additive with zero mean or Poisson distribution. This assumption simplifies the filtering process [89]. Literature indicates several denoising methods from thresholding process to deep learning methods. Techniques involving deep learning methods [90–94] are expected to provide good results compared to other methods.

Satellite images are of widespread usage now a days. Applications like agricultural planning, city planning, disaster management, and military requirements are only few to mention. Satellite images are usually affected by noise due to system calibration, sensitivity of imaging sensor, thermal noise, and atmospheric noise. Few denoising methods use prior knowledge [95–97]. Having prior knowledge in all kinds of satellite images is not possible. Further, deep convolutional neural network (CNN) is proposed, which provides promising results. Denoising methods used both supervised and unsupervised learning methods based on images [98]. Several papers highlight multispectral and hyperspectral image denoising methods using multiresolution methods [98, 99].

9.7   CONCLUSION

Noise is inherent in images. It has a negative impact on other image processing techniques by causing deterioration in their final solution. Hence denoising is a prime preprocessing technique. This chapter gives a comparison of denoising methods both in spatial and transform domains. Here, most commonly known spatial domain methods and wavelet transform-based methods are discussed. Denoising is always a compromise between smoothing noise and preserving edges. Exploring machine learning methods for denoising specifically for biomedical images is expected to yield improved results. Denoising to obtain good images with better structural integrity is always a challenging task for researchers.

REFERENCES

1. G. E. Healey and R. Kondepudy (1994), Radiometric CCD camera calibration and noise estimation, in *IEEE Transactions on Pattern Analysis and Machine Intelligence*, 16(3), pp. 267–276, https://doi.org/10.1109/34.276126.
2. O. V. Michailovich and A. Tannenbaum (2006), Despeckling of medical ultrasound images, in *IEEE Transactions on Ultrasonics, Ferroelectrics, and Frequency Control*, 53(1), pp. 64–78, https://doi.org/10.1109/TUFFC.2006.1588392.
3. R. C. Gonzalez and R. E. Woods (2009), *Digital Image Processing*, 3rd edition, Pearson Prentice Hall.
4. J. M. Sanches, J. C. Nascimento and J. S. Marques (2008), Medical image noise reduction using the Sylvester–Lyapunov equation, in *IEEE Transactions on Image Processing*, 17(9), pp. 1522–1539, https://doi.org/10.1109/TIP.2008.2001398.
5. P. Gravel, G. Beaudoin and J. A. De Guise (2004), A method for modeling noise in medical images, in *IEEE Transactions on Medical Imaging*, 23(10), pp. 1221–1232, https://doi.org/10.1109/TMI.2004.832656.
6. J. S. Lee and K. Hoppel (1989), Noise modeling and estimation of remotely-sensed images, in 12th Canadian Symposium on Remote Sensing Geoscience and Remote Sensing Symposium, pp. 1005–1008, https://doi.org/10.1109/IGARSS.1989.57906.
7. G. R. K. S. Subrahmanyam, A. N. Rajagopalan and R. Aravind (2008), A recursive filter for despeckling SAR images, in *IEEE Transactions on Image Processing*, 17(10), pp. 1969–1974, https://doi.org/10.1109/TIP.2008.2002160.
8. R. Wallis (1976), An approach to the space variant restoration and enhancement of images, in Proceedings of the Symposium on Current Mathematical Problems in Image Science, Naval Postgraduate School, Monterey, CA, 10–12 November.
9. J. -S. Lee (1980), Digital image enhancement and noise filtering by use of local statistics, in *IEEE Transactions on Pattern Analysis and Machine Intelligence*, PAMI-2(2), pp. 165–168, https://doi.org/10.1109/TPAMI.1980.4766994
10. F. Jin, P. Fieguth, L. Winger and E. Jernigan, (2003), Adaptive Wiener filtering of noisy images and image sequences, in Proceedings 2003 International Conference on Image Processing (Cat. No.03CH37429), pp. III-349, https://doi.org/10.1109/ICIP.2003.1247253.
11. D. T. Kuan, A. A. Sawchuk, T. C. Strand and P. Chavel (1985), Adaptive noise smoothing filter for images with signal-dependent noise, in *IEEE Transactions on Pattern Analysis and Machine Intelligence*, PAMI-7(2), pp. 165–177, https://doi.org/10.1109/TPAMI.1985.4767641.
12. C. Tomasi and R. Manduchi (1998), Bilateral filtering for gray and color images, in Sixth International Conference on Computer Vision (IEEE Cat. No.98CH36271), pp. 839–846, https://doi.org/10.1109/ICCV.1998.710815.
13. M. Elad (2002), On the origin of the bilateral filter and ways to improve it, in *IEEE Transactions on Image Processing*, 11(10), pp. 1141–1151, https://doi.org/10.1109/TIP.2002.801126.
14. D. Barash (2002), A fundamental relationship between bilateral filtering, adaptive smoothing and the nonlinear diffusion equation, *IEEE Transaction on Pattern Analysis and Machine Vision*, 24(6), pp. 844–847.
15. W. C. K. Wong, A. C. S. Chung and S. C. H. Yu (2004), Trilateral filtering for biomedical images, in 2004 2nd IEEE International Symposium on Biomedical Imaging: Nano to Macro (IEEE Cat No. 04EX821), 1, pp. 820–823, https://doi.org/10.1109/ISBI.2004.1398664
16. L. I. Rudin and S. Osher (1994), Total variation based image restoration with free local constraints, in Proceedings of 1st International Conference on Image Processing, Vol. 1, pp. 31–35, https://doi.org/10.1109/ICIP.1994.413269.

17. T. F. Chan, S. Osher and J. Shen (2001), The digital TV filter and nonlinear denoising, in *IEEE Transactions on Image Processing*, 10(2), pp. 231–241, Feb. 2001, https://doi .org/10.1109/83.902288.

18. A. Buades, B. Coll and J. M. Morel (2004), *On Image Denoising Methods*, SIAM.

19. A. Buades, B. Coll and J. M. Morel (2005), Image denoising by non-local averaging, in Proceedings. (ICASSP '05). IEEE International Conference on Acoustics, Speech, and Signal Processing, Vol. 2, pp. ii/25–ii/28, https://doi.org/10.1109/ICASSP.2005 .1415332.

20. A. Buades, B. Coll and J. M. Morel (2008), Non-local image and movie denoising, *International Journal of Computer Vision*, Springer, 76, pp. 123–139, https://doi.org/10 .1007/s11263-007-0052-1.

21. A. Buades, B. Coll and J. M. Morel (2010), Image denoising methods, A new non-local principle, *SIAM Review*, 52(1), pp. 113–147, https://doi.org/10.1137/090773908.

22. M. Mahmoudi and G. Sapiro (2005), Fast image and video denoising via nonlocal means of similar neighborhoods, in *IEEE Signal Processing Letters*, 12(12), pp. 839– 842, https://doi.org/10.1109/LSP.2005.859509.

23. Marc Lebrun, Arthur Leclaire (n.d.), An implementation and detailed analysis of the K-SVD image denoising algorithm, *Image Processing Online*, pp. 96–133, https://doi .org/10.5201/ipol.2012.llm-ksvd

24. S. Geman and D. Geman (1984), Stochastic relaxation, Gibbs distributions, and the Bayesian restoration of images, in *IEEE Transactions on Pattern Analysis and Machine Intelligence*, PAMI-6(6), pp. 721–741, https://doi.org/10.1109/TPAMI.1984.4767596.

25. J. Besag (1976), Spatial interaction and the statistical analysis of lattice system, *Journal of the Royal Statistical Society Series B*, 36, pp. 192–236, https://www.jstor.org/stable /2984812

26. B. N. Aravind and K. V. Suresh (2014), MAP-MRF approach for image denoising, 2014 International Conference on Contemporary Computing and Informatics (IC3I), pp. 923–927, https://doi.org/10.1109/IC3I.2014.7019648.

27. B. N. Aravind and K. V. Suresh (2015), A discontinuity adaptive prior for image denoising, *International Journal of Computer Applications*, 110(2), pp. 14–19, https://doi.org /10.5120/19288-0710

28. S. Mallet (1998), *A Wavelet Tour of Signal Processing*, Academic Press.

29. M. Vetterli and J. Kovacevic (1995), *Wavelets and Subband Coding*, Prentice Hall.

30. M. N. Do (2000), *Directional Multiresolution Image Representation*, PhD Thesis, Swiss Federal Institute of Technology, Lausanne.

31. A. Pizurica (2002), Image denoising using wavelets and spatial context modeling, Master's thesis, Universiteit Gent.

32. L. C. Ludeman (1986), *Fundamentals of Digital Signal Processing*, Wiley India Edition.

33. N. Ahmed, T. Natarajan and K. P. Rao (1974), Discrete cosine transform, *IEEE Transaction on Computers*, 23, pp. 90–93.

34. E. J. Candes and D. L. Donoho (2000), Curvelets: A surprisingly effective nonadaptive representation for objects with edges, *Saint-Malo Proceedings*, pp. 1–10.

35. M. N. Do (2003), *Contourlets in Beyond Wavelets*, Academic Press.

36. M. N. Do and M. Vetterli (2005), The contourlet transform: An efficient directional multiresolution image representation, in *IEEE Transactions on Image Processing*, 14(12), pp. 2091–2106, https://doi.org/10.1109/TIP.2005.859376.

37. E. J. Candes (1998), Ridgelets: Theory and applications, Master's thesis, Stanford University.

38. F. Luisier, T. Blu and M. Unser (2007), A New SURE approach to image denoising: Interscale orthonormal wavelet thresholding, in *IEEE Transactions on Image Processing*, 16(3), pp. 593–606, https://doi.org/10.1109/TIP.2007.891064.

39. D. L. Donoho (1999), Wedgelets: Nearly minmax estimation of edges, *The Annals of Statistics*, 27, pp. 859–897.
40. G. Yu and G. Sapiro (2011), DCT image denoising: a simple and effective image denoising algorithm, *Image Processing* online, https://doi.org/10.5201/ipol.2011.ys-dct
41. Jean-Luc Starck, E. J. Candes and D. L. Donoho (2002), The curvelet transform for image denoising, in *IEEE Transactions on Image Processing*, 11(6), pp. 670–684, https://doi.org/10.1109/TIP.2002.1014998.
42. J. Ma and G. Plonka (2010), The curvelet transform, in *IEEE Signal Processing Magazine*, 27(2), pp. 118–133, https://doi.org/10.1109/MSP.2009.935453.
43. F. Luisier (2010), The sure-let approach to image denoising, Master's thesis, Ecole Polytechnique Federale De Lausanne (EPFL).
44. R. Eslami and H. Radha (2006), Translation-invariant contourlet transform and its application to image denoising, in *IEEE Transactions on Image Processing*, 15(11), pp. 3362–3374, https://doi.org/10.1109/TIP.2006.881992.
45. Y. Yan and L. Osadciw (2005), Contourlet based recovery and denoising through wireless fading channels, Conference on *Information Science and Systems*, pp. 16–18.
46. D. L. Donoho and I. M. Johnstone (1994), Ideal spatial adaptation by wavelet shrinkage, *Biometrika*, 81, pp. 425–455.
47. D. L. Donoho (1995), De-noising by soft-thresholding, in *IEEE Transactions on Information Theory*, 41(3), pp. 613–627, https://doi.org/10.1109/18.382009.
48. D. L. Donoho and I. M. Johnstone (1995), Adapting to unknown smoothness via wavelet shrinkage, *Journal of the American Statistical Association*, 90, pp. 1200–1224.
49. C. Taswell (2000), The what, how and why of wavelet shrinkage denoising, *Computing in Science and Engineering*, 2, pp. 12–19.
50. D. L. Donoho (1993), Wavelet shrinkage and w.v.d: A 10-minute tour, in *Progress in Wavelet Analysis and Applications*.
51. R. R. Coifman and D. L. Donoho (1994), Translation invarient denoising, in *Springer Lecture Notes in Statistics*, Springer-Verlag, pp. 125–150, https://doi.org/10.1007/978-1-4612-2544-7_9
52. A. K. Fletcher, K. Ramchandran and V. K. Goyal (2002), Wavelet denoising by recursive cycle spinning, Proceedings International Conference on Image Processing, https://doi.org/10.1109/ICIP.2002.1040090.
53. V. Strela, P. N. Heller, G. Strang, P. Topiwala and C. Heil (1999), The application of multiwavelet filterbanks to image processing, in *IEEE Transactions on Image Processing*, 8(4), pp. 548–563, https://doi.org/10.1109/83.753742.
54. T. D. Bui and G. Chen (1998), Translation-invariant denoising using multiwavelets, in *IEEE Transactions on Signal Processing*, 46(12), pp. 3414–3420, https://doi.org/10.1109/78.735315.
55. A. G. Bruce and H.-Y. Gao (1996), Understanding waveshrink: Variance and bias estimation, *Biometrika*, 83, pp. 727–745.
56. H.-Y. Gao and A. G. Bruce (1997), Waveshrink with firm shrink, *StatisticaSinca*, 7, pp. 855–874.
57. X.-P. Zhang and M. D. Desai (1998), Adaptive denoising based on SURE risk, in *IEEE Signal Processing Letters*, 5(10), pp. 265–267, https://doi.org/10.1109/97.720560.
58. Xiao-Ping Zhang (2001), Thresholding neural network for adaptive noise reduction, in *IEEE Transactions on Neural Networks*, 12(3), pp. 567–584, https://doi.org/10.1109/72.925559.
59. M. Nasri and H. Nezamabadi-Pour (2009), Image denoising in the wavelet domain using a new adaptive thresholding function, *Neurocomputing*, 72, pp. 1012–1025.
60. B. N. Aravind and K. V. Suresh (2012), Multispinning for image denoising, *International Journal of Intelligent Systems*, deGruyter, 21(3), pp. 271–291, https://doi.org/10.1515/jisys-2012-0012

61. W. Shengqian, Z. Yuanhua and Z. Daowen (2002), Adaptive shrinkage denoising using neighborhood characteristic, *Electronic Letters*, 38(11), pp. 502–503, https://doi.org/10.1049/el:20020352

62. T. T. Cai and B. W. Silverman (2001), Incorporating information on neighboring coefficients into wavelet estimation, *Sankhya*, Ser B63, pp. 127–148.

63. G. Y. Chen and T. D. Bui (2003), Multiwavelets denoising using neighboring coefficients, in *IEEE Signal Processing Letters*, 10(7), pp. 211–214, https://doi.org/10.1109/LSP.2003.811586.

64. G. Y. Chen, T. D. Bui and A. Krzyzak (2004), Image denoising using neighbouring wavelet coefficients, 2004 IEEE International Conference on Acoustics, Speech, and Signal Processing, pp. ii-9–17, https://doi.org/10.1109/ICASSP.2004.1326408.

65. G. Y. Chen, T. D. Bui and A. Kazyzak (2005), Image denoising using neighboring wavelet coefficients, *Integrated Computer Aided Engineering*, 12, pp. 99–107.

66. G. Y. Chen, T. D. Bui and A. Kazyzak (2005), Image denoising with neighbor dependency and customized wavelet and threshold, *Elsevier Pattern Recognition*, 38, pp. 115–124, https://doi.org/10.1016/j.patcog.2004.05.009

67. Z. Dengwen and C. Wengang (2008), Image denoising with an optimal threshold and neighboring window, *Elsevier Pattern Recognition Letters*, 29, pp. 1694–1697.

68. B. N. Aravind and K. V. Suresh (2011), Wavelet based image denoising using multispinning, 2011 Third National Conference on Computer Vision, Pattern Recognition, Image Processing and Graphics, pp. 118–121, https://doi.org/10.1109/NCVPRIPG.2011.32.

69. L. Sendur and I. W. Selesnick (2002), Bivariate shrinkage functions for wavelet-based denoising exploiting interscale dependency, in *IEEE Transactions on Signal Processing*, 50(11), pp. 2744–2756, https://doi.org/10.1109/TSP.2002.804091.

70. L. Sendur and I. W. Selesnick (2002), Subband adaptive image denoising via bivariate shrinkage, Proceedings. International Conference on Image Processing, Vol. 3, pp. 577–580, https://doi.org/10.1109/ICIP.2002.1039036.

71. L. Sendur and I. W. Selesnick (2002), Bivariate shrinkage with local variance estimation, in *IEEE Signal Processing Letters*, 9(12), pp. 438–441, https://doi.org/10.1109/LSP.2002.806054.

72. L. Sendur and I. W. Selesnick (2002), Bivariate shrinkage functions for wavelet-based denoising exploiting interscale dependency, in *IEEE Transactions on Signal Processing*, 50(11), pp. 2744–2756, https://doi.org/10.1109/TSP.2002.804091.

73. P. J. Rousseeuw and C. Croux (1993), Alternatives to the median absolute deviation, *Journal of the American Statistical Society*, 88, pp. 1273–1283, https://doi.org/10.1080/01621459.1993.10476408

74. B. R. Corner, R. M. Narayanan and S. E. Reichenbach (2003), Noise estimation in remote sensing imagery using data masking, *International Journal of Remote Sensing*, 4, pp. 689–702.

75. D. L. Donoho and I. M. Johnstone (1995), Wavelet shrinkage: Asymptopia? *Journal of Royal Statistical Society*, 57, Series B, pp 301–369.

76. Zhou Wang, A. C. Bovik, H. R. Sheikh and E. P. Simoncelli (2004), Image quality assessment: from error visibility to structural similarity, in *IEEE Transactions on Image Processing*, 13(4), pp. 600–612, https://doi.org/10.1109/TIP.2003.819861.

77. Z. Wang, E. P. Simoncelli and A. C. Bovik (2003), Multiscale structural similarity for image quality assessment, The Thrity-Seventh Asilomar Conference on Signals, Systems & Computers, Vol. 2, pp. 1398–1402, https://doi.org/10.1109/ACSSC.2003.1292216.

78. Z. Wang and A. C. Bovik (2009), Mean squared error: Love it or leave it? A new look at Signal Fidelity Measures, in *IEEE Signal Processing Magazine*, 26(1), pp. 98–117, https://doi.org/10.1109/MSP.2008.930649.

79. Pascal Getreuer (2012), Rudin-Osher-Fatemi total variation denoising using split Bregman, *Image Processing Online*, 2, pp. 74–95, https://doi.org/10.5201/ipol.2012.g-tvd

80. K. Dabov, A. Foi, V. Katkovnik and K. Egiazarian (2007), Image denoising by sparse 3-D transform-domain collaborative filtering, in *IEEE Transactions on Image Processing*, 16(8), pp. 2080–2095, https://doi.org/10.1109/TIP.2007.901238.

81. S. Roy, N. Sinha and A. K. Sen (2010), A new hybrid image denoising method, *International Journal of Information Technology and Knowledge Management*, 2, pp. 2364–2369.

82. U. Kamilov, E. Bostan and M. Unser (2012), Wavelet shrinkage with consistent cycle spinning generalizes total variation denoising, in *IEEE Signal Processing Letters*, 19(4), pp. 187–190, https://doi.org/10.1109/LSP.2012.2185929.

83. Y. Ding and I. W. Selesnick (2015), Artifact-free wavelet denoising: Non-convex sparse regularization, convex optimization, in *IEEE Signal Processing Letters*, 22(9), pp. 1364–1368, https://doi.org/10.1109/LSP.2015.2406314.

84. B. N. Aravind and K. V. Suresh (2017), Hybrid image denoising, 2017 International Conference on Electrical, Electronics, Communication, Computer, and Optimization Techniques (ICEECCOT), pp. 46–49, https://doi.org/10.1109/ICEECCOT.2017.8284524.

85. B. N. Aravind and K V Suresh (2019), Denoising: A dual domain method, *International Journal of Innovative Technology and Exploring Engineering (IJITEE)*, 9(1), pp. 1588–1592, https://doi.org/10.35940/ijitee.A4555.119119

86. Zhuang Fang, Xuming Yi and Liming Tang (2019), An adaptive boosting algorithm for image denoising, *Hindawi Mathematical Problems in Engineering*, Article ID 8365932, pp. 1–14, https://doi.org/10.1155/2019/8365932

87. Prabhpreet Kaur, Gurvinder Singh and Parminder Kaur (2018), A Review of Denoising Medical Images Using Machine Learning Approaches, *Current Medical Imaging Reviews*, 14(5), pp. 675–685, https://doi.org/10.2174/1573405613666170428154156

88. B. Goyal, A. Dogra, S. Agrawal and B. S. Sohi (2018), Noise Issues Prevailing in Various Types of Medical Images, *Biomedical & Pharmacology Journal*, 11(3), pp. 1227–1237.

89. Pierre Gravel, Gilles Beaudoin and Jacques A. De Guise (2004), A method for modeling noise in medical images, *IEEE Transactions on Medical Imaging*, 23(10), pp. 1221–1232.

90. Mufeng Geng, Xiangxi Meng, Jiangyuan Yu, Lei Zhu, Lujia Jin, Zhe Jiang, Bin Qiu, Hui Li, Hanjing Kong, Jianmin Yuan, Kun Yang, Hongming Shan, Hongbin Han, Zhi Yang, Qiushi Ren, Yanye Lu (2022), Content-noise complementary learning for medical image denoising, *IEEE Transactions on Medical Imaging*, 41(2), pp. 407–419.

91. C. Tian, L. Fei, W. Zheng, Y. Xu, W. Zuo, C. W. Lin (2020), Deep learning on image denoising: An overview, *Elsevier Neural Networks*, 131, pp. 251–275, https://doi.org/10.1016/j.neunet.2020.07.025

92. T. Zhao, M. McNitt-Gray, D. Ruan (2019), A convolutional neural network for ultra-low-dose CT denoising and emphysema screening, *Medical Physics*, 46(9), pp. 3941–3950, https://doi.org/10.1002/mp.13666

93. T. A. Song, F. Yang, Dutta (2021), Noise2Void: Unsupervised denoising of PET images. *Physics in Medicine & Biology*, 66(21). DOI: 10.1088/1361-6560/ac30a0

94. S. Cammarasana, P. Nicolardi, G. Patanè (2022), Real-time denoising of ultrasound images based on deep learning, Medical & Biological Engineering & Computing, 60, pp. 2229–2244.

95. W. He, H. Zhang, L. Zhang and H. Shen (2015), Hyperspectral image denoising via noise-adjusted iterative low-rank matrix approximation, *IEEE Journal of Selected Topics in Applied Earth Observations and Remote Sensing*, 8(6), pp. 3050–3061.

96. W. He, H. Zhang, L. Zhang and H. Shen (2016), Total-variation-regularized low-rank matrix factorization for hyperspectral image restoration, in *IEEE Transactions on Geoscience and Remote Sensing*, 54(1), pp. 178–188, https://doi.org/10.1109/TGRS.2015.2452812.

97. H. Fan, C. Li, Y. Guo, G. Kuang and J. Ma (2018), Spatial–spectral total variation regularized low-rank tensor decomposition for hyperspectral image denoising, *IEEE Transactions on Geoscience and Remote Sensing*, 56(10), pp. 6196–6213.

98. J. Song, J-H. Jeong, D-S. Park, H-H. Kim, D-C. Seo, J. C. Ye (2021), Unsupervised denoising for satellite imagery using wavelet SubbandCycleGAN, IEEE Transaction on Geoscience and Remote Sensing, 59(8), pp. 6823–6839.

99. N. A. Golilarz, H. Gao, S. Pirasteh, M. Yazdi, J. Zhou and Y. Fu (2021), Satellite multispectral and hyperspectral image de-noising with enhanced adaptive generalized Gaussian distribution threshold in the wavelet domain, *Remote Sensing*, MDPI, 13(1), https://doi.org/10.3390/rs13010101

10 Diabetic Retinopathy Diagnosis System Based on Artificial Intelligence

Nayana Hegde, Savitha Krishna, and Sunilkumar S. Manvi

CONTENTS

10.1 INTRODUCTION

Diabetic retinopathy (DR) is the most common type of diabetic complications, producing impaired eyesight and neurodegenerations in adult-age individuals. The number of elderly people with DR was forecast to be 103.12 million in 2020. In 2045, it is expected to rise to 160.50 million (Wejdan et al., 2020). No matter how severe their diabetes is, every diabetic patient can acquire DR. This disease is characterized by escalating retinal vascular abnormalities. It is estimated that 93 million individuals worldwide have DR, making it the leading cause of blind

DOI: 10.1201/9781003326830-10

sight among adults. The initial stages of DR are frequently asymptomatic, yet at this period neuronal retinal impairment and medically undetected microvascular abnormalities continue to progress (Wejdan et al., 2020; Mohamed Jebran et al., 2022). As a result, patients with diabetes should get routine eye exams because prompt identification and treatment of the problem are crucial. To lessen the burden of disease on the populace, detecting it in the early stage and treatment of DR are necessary (Mohamed Jebran et al., 2022; Mohsen Janghorbani et al., 2000).

Optometric physicians in basic hospitals and primary care doctors in community-based health centres have gradually received a substantial increase in the number of screening assignments for diabetic retina as a result of the growth of the medical consortium model. The quantity of duties is growing yearly, and the screening tasks are becoming more difficult. However, conventional idea of DR testing ways cannot adequately address all concerns of rising DR pervasiveness and blindness due to the rise in the quantitative total of persons having diabetes and the shortage of optometric physicians. With the development of the medical consortium model, numerous duties related to diseased retina examinations have gradually delegated to optometric physicians in general hospitals and medical practitioners in primary health service places. These duties will continue to grow and become more demanding over time. Fundus color image is a quick plus effective testing approach for fundus illnesses, and they are typically seen as the best and ideal for DR testing and further look into. Direct ophthalmoscope, fundus color imaging, fundus angiogram, etc., are examples of conventional DR diagnostic approaches.

Different types of lesions that present on a retinal scan can be used to identify DR. Soft and hard exudates (EX), hemorrhages (HA), and microaneurysms (MA) (Wejdan et al., 2020) are some of the important lesions that are briefed as follows:

- Soft exudates: Due to the swelling of nerve fiber these appear as round or oval-shaped white spots on the retina. They are frequently brought on by retinal alterations as a result of diabetes, hypertension, or obstructions in the blood arteries that supply the retina, including central retinal vein occlusion ((Wejdan et al., 2020; Mohamed Jebran Pendekal et al., 2022).
- Hard exudates: Because of the leakage of the plasma, bright yellow color spots are found on the outer layer of retina. They settle in the outer layers of the retina and are made up of lipid and proteinaceous substances. When they are deposited in the foveal region, these plaques frequently result in severe sight loss (Wejdan et al., 2020; Nandeeswar Sampigehalli et al., 2021).
- Hemorrhage: It appears with irregular margin on the retina with size greater than 125 micrometers. Superficial hemorrhages and deeper hemorrhages are the two common types observed. The symptoms of retinal hemorrhages can range from being undetectable to seriously impairing vision. Although vision issues are frequently transient, they can become permanent (Nandeeswar Sampigehalli et al., 2021).
- Microaneurysm: Because of the weakness of the vessel walls, it appears as small red rounds on the retina. It is considered as early stage of the MA. The

size of the dots is less than 125 micrometers. A retinal microaneurysm can be caused by any vascular illness or excessive blood pressure, but diabetes mellitus is the most common culprit (Wejdan et al., 2020; Nandeeswar Sampigehalli et al., 2021).

Depending on the presence of these lesions, we can classify the disease as follows. Absence of lesions is considered as no DR, presence of MA is mild DR, and definite bleeding in veins in DR and preretinal HA is considered as proliferative DR. Figure 10.1 illustrates the sample of DR stages. As shown in (a) disease severity is 0 and the image is considered as normal or no abnormalities. (b) Disease severity is 1 and it is known as mild DR or microaneurysms only. (c) Disease severity is 2 and it is known as moderate DR or greater compared to microaneurysms but lower to that of NPDR. (d) Disease severity is 3 and it is known as severe DR or it can be anything like: (i) in each of the four quadrants, there were more than 20 intraretinal hemorrhages; (ii) at least two unique venous beading quadrants; and (iii) strong IRMA in a single or more quadrants. (e) Disease severity is 4 and it is known as proliferative DR with either neovascularization or vitreous/preretinal hemorrhage. (f) Disease severity is 5 or considered chronic. It is known as macular edema. Permanent vision loss and macula damage are potential consequences.

Timely identification and treatments can avoid the majority of loss of vision caused by DR, according to scientific research (Decenciere et al., 2014). As a result, developed countries have implemented DR screening tests aiming at early detection, monitoring, and therapy. The majority of these methods rely on professionally qualified graders analyzing fundus pictures, which is commonly done via telemedicine. Furthermore, the clinical precision attained may not be ideal; additionally, extending

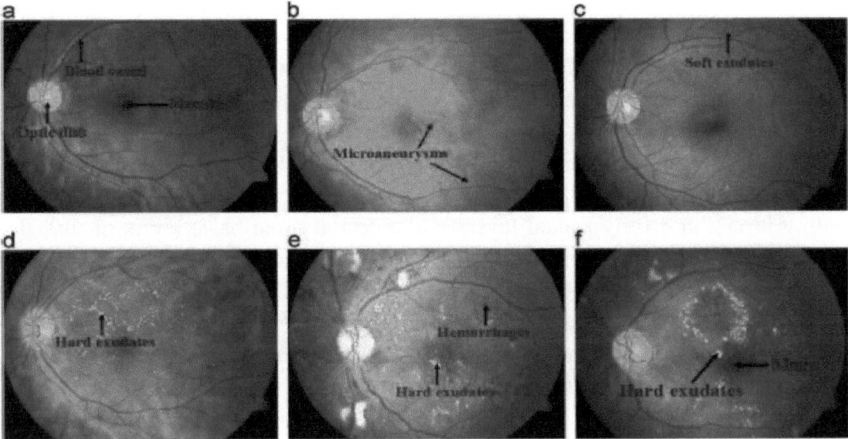

FIGURE 10.1   The DR stages: (a) normal retinal, (b) mild DR, (c) moderate DR, (d) severe DR, (e) proliferative DR, and (f) macular edema (Decenciere et al., 2014; Janapati et al., 2022a).

and maintaining such systems have proven difficult (Mohamed Jebran et al., 2022; Yau et al., 2012).

Artificial intelligence (AI) is a vast concept for the process of creating machines that mimic human brain activity. A subset of AI is called machine learning (ML). When a new information instance is presented, the computer knows how to respond. ML can aid in the preliminary diagnosis and therapy of a variety of dangerous diseases. Thus, diagnostic procedures could become much more inexpensive, accurate, and accessible (Siva Sundhara Raja and Vasuki, 2015; Jonathan et al., 2010). Deep learning (DL), the latest ML approach, has showed promise in image recognition, audio recognition, and natural language processing. It has produced solid results in several medical areas such as radiology and dermatology for ophthalmology in particular for medical imaging analysis. In lab settings, DL-based systems for detecting DR from retinal pictures have been tested. The use of advanced DL algorithms has resulted in increased diagnosis accuracy (Nandeeswar Sampigehalli et al., 2021) (Wenying Yang et al., 2010). In recent years, DR detection and classification have made extensive use of DL. Even with a large amount of input data, it can successfully learn its features by integrating sources that are different (Xue-Wen Chen and Xiaotiong Lin, 2014; Safi Hamid et al., 2018). Boltzmann machines, auto encoder, convolutional neural network (CNN) and sparse programing are some of the DL-based methods that are generally used in DR classification and detection. CNNs are predominantly made use in comparison to other approaches in medical image processing. They are very much advantageous for such processing (Yanming Guo et al., 2016; Srikanta Kumar Padhy et al., 2019; Mohamed Jebran et al., 2021; Scottish, 2014). CNNs were motivated by human perception, and their ideas are centered above basic mathematical process, called "convolution." In contrast to shallow neural networks, CNNs acquire 2D arrays as an input data. A CNN creates the characteristic map by sliding the majority of the main image and computing convolutions by making use of filters or kernels. The network's convolutional portion is sometimes referred to as "the feature extraction part," while the remaining portion is "the classification section." In an effort to categorize the images based on the created characteristics, the previous authors have condensed a single-dimensional matrix and as it went through, applied it for a deep neural network (DNN).

Figure 10.2 illustrates the various DL strategies. There are single data and single prediction with multiple data and multiple predictions possible. Convolution layers, pooling layers, and fully linked layers make up the three basic levels of the CNN architecture. In the CNN design, each layer has a distinct function. The first layer is convolutional layer that convolves an image and extracts features. The second layer is pooling layer, which decreases the dimension of the extracted attribute. Although there are several methods for pooling, the popular two methods are average pooling and max pooling. The fully connected layer describes the complete feature of the input image. ResNet (Nikos Tsiknakis et al., 2021), AlexNet (Amin Valizadeh et al., 2021), ImageNet (Amin Valizadeh et al., 2021), and Inception-V3 are some of the important pretrained CNN architectures. The most popular classification function used is SoftMax activation function (Nikos Tsiknakis et al., 2021; Amin Valizadeh et al., 2021).

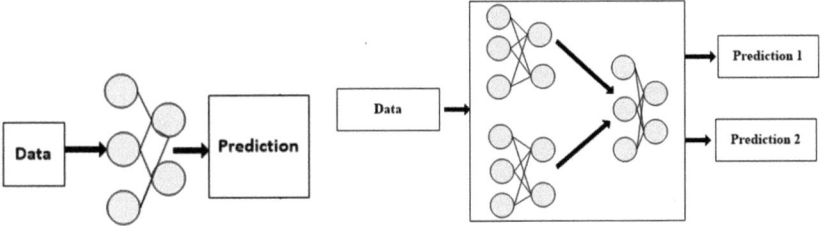

(a) One data source and with one prediction   b) One data source with multiple predictions

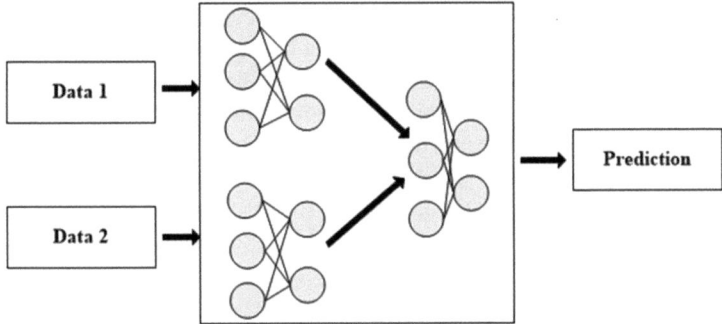

c) Multiple data source with single prediction

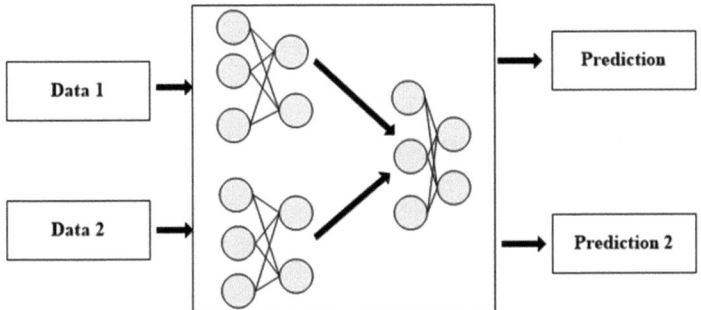

d) Multiple data source with multiple predictions.

FIGURE 10.2   Deep learning strategies with various data and outcomes: (a) one data source and with one prediction, (b) one data source with multiple predictions, (c) multiple data source with single prediction, and (d) multiple data source with multiple predictions.

The task of identifying and categorizing the DR images with DL algorithms begins by collection of datasets. Next step is application of required preprocessing methods for the improvement and enhancement of the images. Finally, these datasets are catered for DL algorithms to derive the attribute of images and categorize them as illustrated in Figure 10.3.

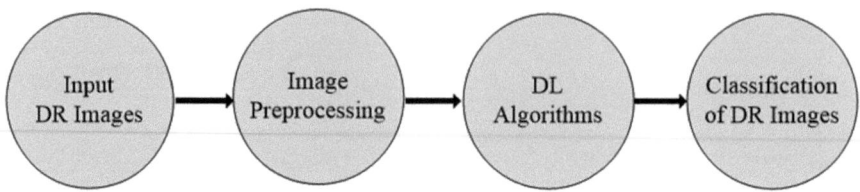

FIGURE 10.3   DR image classification process using DL algorithms.

For detecting retina DR and vessels there are several public datasets available. There are two basic sets of retinal images, namely, optical coherence tomography (OCT) and fundus color image. OCT images of retina are both two-dimensional or three-dimensional. They are captured with low-coherence light. Information provided by OCT images is shape and density of the retina. Fundus images are captured using reflected light and they are two-dimensional images (Abrham et al., 2010; Ravichander et al., 2022a). There are several fundus images datasets available, and they are listed in Table 10.1.

TABLE 10.1
DR Dataset Specifications

S.N	Image	Size of the image	Remarks
1	DIARETDB1 (Kauppi et al., 2007)	1,500 × 1,152 pixels	Includes 84 DR images
2	Kaggle (2013)	433 × 289 to 5,184 × 3,456 pixels	Poor-quality and incorrect labeled images
3	E-ophtha (Lam et al., 2018)	1,600 × 1,200 pixels	This includes E-ophtha EX and E-ophtha MA images
4	DDR (Decenciere et al., 2013)	1,500 × 1,500 pixels	Images are recorded from five stages of DR
5	DRIVE (Staal et al., 2004)	565 × 584 pixels	Contains seven different types of images
6	HRF (Budai et al., 2013)	3,504 × 2,336 pixels	Images provide blood vessel segmentation details
7	Messidor (Decenciere et al., 2014)	1,440 × 960 pixel	Images cover four stages of DR
8	STARE (Hoover et al., 2000)	700 × 605 pixels	Images provide blood vessel segmentation details
9	CHASE DB1(Owen et al., 2004)	1,280 × 960 pixels	Images provide blood vessel segmentation details
10	ROC [21]	768 × 576 to 1,389 × 1,383 pixels	Images interpret MA identification

10.1.1 Performance Measures

To analyze the accuracy of DL algorithms with regard to categorization, different assessment benchmark are utilized. Accuracy, area under the ROC curve (AUC), sensitivity, and specificity are frequently utilized metrics in DL (Zhang et al., 2019). The percentage of irregular image that is recognized as abnormal is known as sensitivity, and the percentage of normal images that is categorized as regular is known as specificity. AUC is a graph produced by contrasting specificity and sensitivity. The percentage of accurately classified images is investigated as accuracy (Parikh et al., 2008; Bolin and Lam, 2013; Jacob Shreffler and Martin R. Huecker, 2022). The equations for the calculation of these parameters are as follows:

Specificity: The percentage of real negatives between all participants who did not have sickness or state of disease is known as specificity. It is the test's or instrument's capacity to produce results that are within the standard extent or negative for a healthy individual. Equation 10.1 represents the specificity.

$$Specificity = \frac{True\ Negative}{True\ Negative + False\ Positive} \tag{10.1}$$

Sensitivity: It is the percentage of tests that come back as true positives for all patients with a disease. It is the capacity of an experiment or tool to return a positive outcome for a patient who is suffering from such illness. Equation 10.2 represents sensitivity.

$$Sensitivity = \frac{True\ Positive}{True\ Positive + False\ Negative} \tag{10.2}$$

The relationship between sensitivity and specificity is inverse: as sensitivity rises, specificity tends to fall, conversely.

Accuracy: The percentage of truly positive findings in the chosen population is represented by the accuracy value in numbers. Whether the test result is positive or negative, it is correct 99% of the time. Equation 10.3 represents accuracy.

$$Accuracy = \frac{True\ Negative + True\ Positive}{True\ Negative + False\ Positive + True\ Positive + False\ Negative} \tag{10.3}$$

True positive= Total image numbers that are grouped as disease
True negative = Total image numbers that are grouped as normal
False positive = Total normal image numbers that are groped as disease
False negative = Total diseased image numbers that are grouped as normal

Very specific tests will demonstrate that people without a finding do not have an illness, but distinctly oversensitive diagnosis would produce positive results for sick person. To give a complete picture of a diagnostic test, sensitivity and specificity should always be taken into account jointly.

10.1.1.1 Methods for Developing Models

In this section we mention some of the major DL methods for classification of DR. The methods list is as follows: (1) generic DL method; (2) training for models using patches; (3) using probabilistic output for segmentation; (4) attention map-based segmentation; (5) DL methods based on generative adversarial network (GAN); and (6) UNets (Parikh et al., 2008) (Bolin and Lam, 2013). Methods based on AI, and in particular DL, show promise for enhancing and advancing DR treatments. To ease the implementation of AI in clinical settings, however, a number of significant obstacles must be overcome.

10.1.2 IMAGE PROCESSING FOR ENHANCING DR IMAGES

Image processing is the approach applied for DR images to lower the noises and enhance the consistency of the image features. A number of important image processing methods adopted by the researchers is resizing images for a fixed resolution, image cropping to remove extra regions, normalization to change the intensity values of the pixels, and changing the color images to grayscale images. Methods used for removing noise are denoising with nonlocal means, median filter, and Gaussian filter. Technologies for image augmentation were utilized to enlarge the images. These methods include image translation, image flipping, and image contrast scaling, shearing, rescaling, and image rotation. The images can now be utilized as an input for the DL algorithms after being preprocessed (Dutta et al., 2018; Quellec et al., 2017; Orlando et al., 2018; Rakesh Sengupta, 2022b). We list some of the image preprocessing techniques in this section.

10.1.2.1 Contrast Enhancement

To start, any image processing or analysis pipeline uses contrast enhancement as a common preprocessing method to distinguish the foreground from the background. Histogram equalization, a straightforward technique for improving contrast in fundus images, boosts the image's overall contrast while ignoring the regional variances (Wei Zhang et al., 2019).

10.1.2.2 Denoise and Normalize

Denoising and normalizing are done to make the images noise less noticeable. Furthermore, it needs to be remembered that although a better denoising algorithm can remove more noise, the image's precise points may suffer as a result. Additionally, picture intensity normalization is used to normalize the data to a specific scale and prevent bias and lengthy training times from being introduced into the network (Qiqi Xiao et al., 2019; Ratul Ghosh et al., 2017).

10.1.2.3 Color-Space Conversion

Converting the colored images to another form of color images by using a single RGB channel increases the system's performance. As a consequence of ample information and strong incongruity in differentiation to the other two color channels, the green channels are frequently extracted by the fundus color images (Darshit Doshi et al., 2016).

10.1.2.4 Image Cropping and Resizing

Many of the datasets will include image that will differ in terms of aspect ratio plus resolution. Rescaling, cropping, image resizing, and removing dark regions will help to systematize the dimensions of the image. In some analysis, cropped images are used so that they are inscribed within retina in circles, and also analysis requires cropping the largest square image that is taken within retina (Carson Lam et al., 2018).

10.1.2.5 Augmentation

In order to improve image quality, color brightness augmentation, shifting, rescaling, rotating, shearing, flipping, and rescaling can be employed (Carson Lam et al., 2018; Gen-Min Lin et al., 2018).

10.1.3 METHODS FOR DETECTING DIABETIC RETINOPATHY

Numerous research works have tried to utilize DL to automatically detect as well as classify DR lesions. As per the classification technique employed, these methods can be distinguished into four groups of classifications. They are: (1) binary, 2) multi-level, (3) lesion-based, and (4) vessels-based classification.

10.1.3.1 Binary Classification

Binary classification means the DR dataset is classified into only two classes. In work (Xu et al., 2017), authors have classified the dataset images into normal images or DR images using the CNN algorithm. Authors have used Kaggle (2013) datasets for the classifications. Before feeding the data to the CNN algorithm, two preprocessing methods, namely image augmentation and image resizing, were applied. The authors have used CNN architecture, which consisted of eight convolutional layers, four max-pooling layers, and two fully connected layers. The authors have obtained 94% accuracy by using SoftMax [9] function for the classification of the last layer of CNN. In work (Dutta et al., 2018), authors have classified the images as the moderate stage of DR and mild stage of DR with the help of three types of CNNs training. Kaggle (2013), Diaret DB1 (Kauppi et al., 2007), and E-optha (Lam et al., 2018) were three different types of datasets considered. Image preprocessing methods used in their work are image resizing, cropping, and normalizing. Gaussian filter is applied for noise removal, and image augmentation procedures are adopted. The work was supported by the AlexNet function. In work (Esfahani et al., 2018; Ravichander et al., 2021a), the authors have used a CNN known by the name ResNet34 (Szegedy et al., 2016), which is available in pertained ImageNet database. The authors have applied some of the preprocessing techniques like image normalization, Gaussian filter, and weighted addition to improving the quality. The accuracy of the method observed by the authors is 85% and the sensitivity is 86%.

In work (PiresR et al., 2019) authors have used their own CNN architecture for the image classification. The new proposed CNN architecture consists of 16 layers, and it is similar to the one proposed in Simonyan and Zisserman (2015). The work involved twofold image cross-verification and resolution of multi-images. The datasets used for this work were Kaggle (2013) and Messidor (Decenciere et al., 2014). This work has balanced the class of data classification using image augmentation. The authors have achieved the ROC area under the curve with 98.2% accuracy by using the Messidor dataset.

10.1.3.2　Multilevel Classification

Multilevel classification will categorize the datasets into many classes of DR. The authors in work (Gulshan et al., 2016) have introduced a new model for detecting the DR using CNN. The classification was for diabetic macular edema (DME). The datasets used for the experiment setup were messidoor2 and eyepacs, which consisted of 1,748 images and 9,963 images in the test model. Before feeding them to CNN architecture, the following preprocessing was carried out. Image normalizing was done, and image resizing was done with 299 pixels. Trained with 10 CNNs, the authors have used linear average methods. Referable diabetic macular edema, moderate or worse DR, severe or worse DR, or totally gradable were the categories used to categorize the dataset images. Authors have achieved an accuracy of 95% and sensitivity of 97%. They did not, however, specifically identify the five DR stage images or non-DR images.

Work in Szegedy et al. (2015) integrated the CNN architecture with IDX-Device for detecting and classifying DR images. The authors have applied the image augmentation methods for the Messidoor2 dataset containing 1,748 images. To identify DR lesions as well as the healthy structure of the retina, their multiple CNNs were combined using a random forest classifier. The images in this work were categorized as having no DR, referable DR, or dangerous to the viewer's vision. The authors have achieved 98.6% efficiency. Sadly, they ignored the five DR stages and only took into account images of the moderate DR stage. In work (Pratt et al., 2016), to categorize images into five stages from the Kaggle (2013) dataset, the authors suggested an approach based on a CNN. The preprocessing steps performed involve color normalization and resizing the images with 512 × 512 pixels. Ten convolutional, eight maxpooling, and three fully connected layers made up their unique CNN architecture. The authors have made use of the SoftMax method with the help of 8,000 images. They have achieved results with accuracy of 75%, specificity of 95%, and sensitivity of 30%. Unfortunately, single dataset was utilized to test CNN's ability to recognize lesions within the image.

10.1.3.3　Categorization Based on Lesions

The research done to identify and categorize particular DR lesion types is outlined in this section. By combining DL techniques with area expertise to learn about features, the authors were able to identify only red lesions in DR pictures (Orlando et al., 2018). Following that, the photos were categorized using the random forest technique. The datasets used in the work were MESSIDOR, E-ophtha, and DIARETDB1, and the preprocessing procedures include expanding and removing the green band. The quality and resolution of the images were increased by using a Gaussian filter, an r-polynomial conversion, an inception operation, and numerous structural closure consequences. Then, red lesion patches were enhanced for CNN training and scaled to 32×32 pixels. The datasets, DIARETDB1, E-ophtha, and MESSIDOR, comprise 89 images, 381 images, and 1,200 photos, respectively. Four CONV layers, three pooling layers, and one FC layer make up their unique CNN. For the DIARETDB1 and E-ophtha datasets, they obtained competition metrics (CPM) of 0.4874 and 0.3683, respectively (Pratt et al., 2016).

A custom CNN is used for detecting the MA in DR images in work (Chudzik et al., 2018) by the authors. Three datasets were utilized for the experiment, namely DIARETDB1, E-ophtha, and ROC. The preprocessing techniques applied for enhancing the image quality are cropping and resizing with generating mask. Later for extracting the MA patches, random transformation methods were put in. The employed CNN has 18 convolutional layers, with a batch normalization layer, three max-pooling levels, three basic upsampling layers, and four skip connections between the two routes after each CONV layer. The work performed good with ROC of 0.35. The approach suggested by work (Adem, 2018) used a customized CNN with circular Hough transformation to detect exudates from DR pictures (CHT). The DiaretDB1 dataset has 130 photos, the DiaretDB1 dataset has 89 images, and the DrimDB dataset has 125 images. These three available datasets were used by the researchers. The datasets were all made into grayscale images. Then, adaptive histogram equalization functions and Canny edge detection were used. CHT then found the optical disc and eliminated it from the images. The photos, which have a resolution of 1152×1152 pixels, were fed into a custom CNN that has three CONV layers, three max-pooling layers, and an FC layer that employs SoftMax as a classifier. The accuracy of identifying excess fluid for DiaretDB00, DiaretDB1, and DrimDB was 99.17, 98.53, and 99.18, respectively.

10.1.3.4 Categorization Based on Vessels

Vascular fragmentation method is made use for diagnosis and to track the development of illnesses of the retina, e.g., glaucoma, DR, and high blood pressure. Numerous investigations into vessel fragmentation as a component of DR detection have been carried out. After the vessels have been removed, DR lesions are still visible in the image. Consequently, finding the remaining lesions in the photos helped identify and categorize DR images. The work in Yan et al. (2019) is able to detect red lesions once the vessel extraction is completed. The use of DL approaches in some vascular segmentation investigations is discussed in this section.

In work (Vengalil et al., 2016), the authors have modified the CNN architecture from RGB retina images; the retinal blood vessels can be extracted. From the dataset, they took 512×512 image patches, which they then fed to the CNN. The photos were then converted to binary values after they had applied a threshold. Eight CONV layers and three max-pooling layers make up the CNN. They have used HRF and DRIVE datasets to test the proposed method. They reported a 93.94% accuracy rate and a ROC area of 0.894. In work (Lu et al., 2018) the authors have used full CNN for the segments of the blood vessel in RGB image of the retina. The datasets used are STARE, HRF, DRIVE, and CHASE DB1. The preprocessing methods applied are flipping and cropping of images. After preprocessing, the images are fed to the CNN algorithm. After that, morphological operations were used to rebuild the vessel map.

They have 5 dilated CONV layers and 16 CONV layers in their CNN. In Oliveira et al. (2018), the authors utilized the stationary wavelet transform with full connected CNN for extracting blood vessel information. The datasets used were STARE, DRIVE, and CHASE_DB1. The preprocessing steps were image normalization and transformation. The result achieved an accuracy of 96.4% with the abovementioned datasets.

10.2   DL MODEL IMPLEMENTATION IN ACTUAL CLINICAL SCENARIOS

According to a review study given in Norah Asiri et al. (2019) and Usha Desai et al. (2022), there aren't many DL-based approaches, and to address this issue, enhanced DL techniques must be created. There have been several DL models created to this point, and some are even used in the circumstances of scientific decision-making in real-world settings. The fact that some have strongly navigated the authoritative procedure and received acceptance from pertinent international authoritative bodies, like the FDA, is even more significant. EyeArt, a commercial AI tool that has received FDA approval for use in DR screening, claims to have a sensitivity of 96% and a specificity of 88% for detecting more severe DR and a sensitivity of 92% and a specificity of 94% for detecting DR that poses a threat to vision (Eyenuk, 2021). EyeArt was also tested using 30,405 photos from the English Diabetic Eye Screening Programme's multicenter dataset, achieving good sensitivity for referable DR identification and a respectable specificity score (Heydon Peter et al., 2021). To identify referable DR, vision-threatening DR, and other eye-related disorders, Daniel Shu et al. (2017) suggested a DL-based approach. A total of 500 pictures from Singapore's National Integrated Diabetic Retinopathy Screening Program were used for the model's training and validation (SiDRP). In different clinical validation research, Jonathan Krause et al. (2018) and Janapati et al. (2021a) revealed lesser specificity values in comparison with medical professionals but greater sensitivity ratings. In order to take advantage of its high sensitivity while simultaneously making up for its lower specificity with the professional's second diagnosis, it is suggested that a human-supervised implementation of such a system will be ideal.

10.3   DISCUSSION

A total of 61 publications were reviewed for the current investigation. DL techniques were used in the research work discussed in related field of the DR screening system. Recent rise in the number of people with diabetes has made the request for trustworthy DR testing systems crucial. The issue of choosing trustworthy features for ML is solved by using DL in DR detection and classification; however, a lot of training data is needed. The majority of research made use of data augmentation to boost the number of images and avoid overfilling during the training stage. Additionally, a bias in the annotation of particular datasets may result from the grading of those datasets by a single expert. More advanced grading techniques are needed in order to eliminate this bias and create reliable actual information for the datasets. To improve the actual information and, consequently, the accuracy of the model, the obtained information should be assessed repeatedly by different researchers. Both the procedure and the effectiveness of the model might be affected by the information's poor-quality images. Subtle signs of retinopathy at a preliminary phase can be readily hidden due to low-resolution or hazy image and are effectively identified by AI models which are implemented based on training and evaluation datasets. The majority of the aforementioned datasets, however, are either insufficient or have an imbalance between their classifications. In light of this, there are a number of strategies to solve this problem, including the use of augmentation techniques, the generation of synthetic data using GANs, and the use of transfer learning to make use of the expertise of models that have been trained on huge datasets, like ImageNET.

10.4 INCORPORATING ARTIFICIAL INTELLIGENCE INTO RADIOLOGY CLINICAL PRACTICE

Future radiology practice could be greatly altered by AI, but there are still a number of problems that need to be overcome before this can happen. The role of various stakeholders in the advancement of AI for image analysis, the moral innovation and application of technology in health care, the proper affirmation of each established AI algorithm, the creation of efficient data sharing mechanisms, regulatory barriers to the approval of AI algorithms, and the creation of AI academic materials for both practicing radiologists and radiology students are some of these topics. The various types of bias that are frequently found in AI research should be known to radiologists, along with their potential impacts. It is necessary to establish strategies for efficient data sharing in order to develop, verify, and test AI algorithms. Understanding the fundamental ideas, potentials, and restrictions of AI is crucial for all radiologists. The advances of AI development are changing globally. The number of proper AI-related medical research works and experiments is the largest in the United States, which also tops the records of companies with the top VC funding to date. However, Asia, notably China, is experiencing the quickest growth, with Ping An's Good Doctor, the top online health management system, before only boasting more than 300 million members. Leading Chinese conglomerates and internet companies now have consumer-focused healthcare AI solutions there. Our knowledge of AI and its highest capability in health care, particularly in terms of how it will affect personalization, is still in its infancy. We would also incorporate imaging-based AI technologies, those that are currently in use in radiological, pathological, and ophthalmological fields of medicine, in this initial stage. As patients assume greater responsibility for their care in the second phase, we anticipate seeing an increase in AI remedies, which help patients move from healthcare center to home-based treatment, for example, by remote patient monitoring, virtual aides, or efficiency and enables using AI. With a growing emphasis on enhanced and scalable medicinal decision-support (CDS), few medical solutions are suggested. These solutions in a medical domain come from previous trials to facilitate the medicinal practices and have modified its mindset, cultures, and skill. Based on findings from third clinical studies, we expect to see more innovative approaches in clinical practice.

10.5 CONCLUSION

Automatic testing procedures acutely lessen the time required to diagnose, leading to time savings for ophthalmologists and reduced cost of treatment for the patient and also helping them to get treatment faster. For detecting DR at an early stage, automatic DR detection methods are essential. The types of damage that form on the retina dictate the phases of DR. Most common DL-based automated approaches for diagnosing and classifying DR have been studied in this chapter. We have given a brief overview of DL techniques and discussed the widely used, publicly available fundus DR datasets. The bulk of research uses CNN for the identification and categorization of DR images due to its efficacy. Additionally, the helpful methods that can be applied to identify and categorize DR utilizing DL have been covered in this chapter.

REFERENCES

A Hoover, V Kouznetsova, M Goldbaum (2000). Locating Blood Vessels in Retinal Images by Piece-wise Threshold Probing of a Matched Filter Response. *IEEE Transaction on Medical Images*, 19(3), 203–10, https://doi.org/10.1109/42.845178

A Oliveira, S Pereira, CA Silva (2018). Retinal Vessel Segmentation Based on Fully Convolutional Neural Networks. *Expert System Applications*, 112, 229–42, https://doi.org/10.1016/j.eswa.2018.06.034

Amin Valizadeh, Saeid Jafarzadeh Ghoushchi, Ramin Ranjbarzadeh, Yaghoub Pourasad (2021). Presentation of a Segmentation Method for a Diabetic Retinopathy Patient's Fundus Region Detection Using a Convolutional Neural Network. *Computational Intelligence and Neuroscience Volume*, Article ID 7714351, 14 pages, https://doi.org/10.1155/2021/7714351

C Lam, D Yi, M Guo, T Lindsey (2018). Automated Detection of Diabetic Retinopathy Using Deep Learning. *AMIA Joint Summits on Translational Science*, 147–55, https://www.ncbi.nlm.nih.gov/pmc/articles/PMC5961805/

C Szegedy, et al (2015). Going Deeper With Convolutions. In: IEEE Conference on Computer Vision and Pattern Recognition (CVPR), 1–9, https://doi.org/10.48550/arXiv.1409.4842

C Szegedy, S Ioffe, V Vanhoucke, A Alemi (2016). Inception-v4, Inception-ResNet and the Impact of Residual Connections on Learning. In: The Thirty-first AAAI Conference on Artificial Intelligence, 4278–84, https://doi.org/10.48550/arXiv.1602.07261

Carson Lam, Darvin Yi, Margaret Guo, Tony Lindsey (2018), Automated Detection of Diabetic Retinopathy Using Deep Learning, *AMIA Joint Summits Transl. Sci. Proc.*, 147–155. ISSN 2153–4063.

CG Owen, et al. (2009). Measuring Retinal Vessel Tortuosity in 10-Year-Old Children: Validation of the Computer-assisted Image Analysis of the Retina (CAIAR) Program. *Investigating Ophthalmol Vision Science*, 50(5), 2004–10, https://doi.org/10.1167/iovs.08-3018

D Siva Sundhara Raja, S Vasuki (2015). Automatic Detection of Blood Vessels in Retinal Images for Diabetic Retinopathy Diagnosis. *Computational and Mathematical Methods in Medicine*, https://doi.org/10.1155/2015/419279

Daniel Shu Wei Ting, Carol Yim-Lui Cheung, Gilbert Lim, Gavin Siew Wei Tan, et al. (2017). Development and Validation of a Deep Learning System for Diabetic Retinopathy and Related Eye Diseases Using Retinal Images from Multiethnic Populations with Diabetes, *J. Am. Med. Assoc.*, 318(22), 2211, https://doi.org/10.1001/jama.2017.18152.

Darshit Doshi, Aniket Shenoy, Deep Sidhpura, Prachi Gharpure (2016), Diabetic Retinopathy Detection Using Deep Convolutional Neural Networks. In: 2016 International Conference On Computing, Analytics and Security Trends (CAST), IEEE, ISBN 978-1-5090-1338-8, pp. 261–266, https://doi.org/10.1109/CAST.2016.7914977.

E Bolin, W Lam (2013). A Review of Sensitivity, Specificity, and Likelihood Ratios: Evaluating the Utility of the Electrocardiogram as a Screening Tool in Hypertrophic Cardiomyopathy. *Congenit Heart Dis.*, 8(5), 406–10. doi: 10.1111/chd.12083

E Decenciere, et al. (2013). TeleOphta: Machine Learning and Image Processing Methods for Teleophthalmology. *IRBM*, 34(2), 196–203, https://doi.org/10.1016/j.irbm.2013.01.010

E Decenciere, et al. (2014). Feedback on a Publicly Distributed Image Database: The Messidor Database. Image Analysis and Stereology , 33(3), 231–4, DOI: 10.5566/ias.1155

Eyenuk (2021). *Eyeart AI Eye Screening System.* https://www.eyenuk.com/. Accessed on 25/02/21.

G Quellec, K Charriere, Y Boudi, B Cochener, M Lamard (2017). Deep Image Mining for Diabetic Retinopathy Screening. *Med Image Anal*, 39, 178–93, https://doi.org/10.1016/j.media.2017.04.012

Gen-Min Lin, Mei-Juan Chen, Chia-Hung Yeh, Yu-Yang Lin, Heng-Yu Kuo, Min-Hui Lin, Ming-Chin Chen, Shinfeng D Lin, Ying Gao, Anran Ran (2018). Transforming Retinal Photographs to Entropy Images in Deep Learning to Improve Automated Detection for Diabetic Retinopathy, *J. Ophthalmol.*, 2018.

H Pratt, F Coenen, DM Broadbent, SP Harding, Y Zheng (2016). Convolutional Neural Networks for Diabetic Retinopathy. Procedia Computer Science, 90, 200–5, https://doi.org/10.1016/j.procs.2016.07.014

Heydon Peter, Catherine Egan, Louis Bolter, Ryan Chambers, John Anderson, Aldington Steve, M Stratton Irene, Peter Henry Scanlon, Laura Webster, Samantha Mann (2021). Prospective Evaluation of an Artificial Intelligence-enabled Algorithm for Automated Diabetic Retinopathy Screening of 30 000 Patients. *Br. J. Ophthalmol.*, 105(5), 723–728, https://doi.org/10.1136/bjophthalmol-2020-316594

J Lu, Y Xu, M Chen, Y Luo (2018). A Coarse-to-fine Fully Convolutional Neural Network for Fundus Vessel Segmentation. *Symmetry*, 10(11), 607, https://doi.org/10.3390/sym10110607

Jacob Shreffler, Martin R Huecker (2022). Diagnostic Testing Accuracy: Sensitivity, Specificity, Predictive Values and Likelihood Ratios. In: *StatPearls* [Internet]. Treasure Island (FL): StatPearls Publishing, Available from: https://www.ncbi.nlm.nih.gov/books/NBK557491/

JI Orlando, E Prokofyeva, M del Fresno, MB Blaschko (2018). An Ensemble Deep Learning Based Approach for Red Lesion Detection in Fundus Images. *Comput Methods Progr Biomed*, 153, 115–27, https://doi.org/10.1016/j.cmpb.2017.10.017

JJ Staal, MD Abramoff, M Niemeijer, MA Viergever, B van Ginneken (2004). Ridge-based Vessel Segmentation in Color Images of the Retina. *IEEE Transactions in Medical Images*, 23(4), 501–9, https://doi.org/10.1109/TMI.2004.825627

Jonathan E Shaw, Richard A Sicree, Paul Z Zimmet (2010). Global Estimates of the Prevalence of Diabetes for 2010 and 2030. *Diabetes Res. Clin.Pract.*, 87(1), 4–14.

JW Yau, SL Rogers, R Kawasaki, EL Lamoureux, JW Kowalski, T Bek, SJ Chen, JM Dekker, A Fletcher, J Grauslund (2012). Meta-analysis for Eye Disease [Meta-eye] Study Group. Global Prevalence and Major Risk Factors of Diabetic Retinopathy, *Diabetes Care*, 35(3), 556–564.

K Adem (2018). Exudate Detection for Diabetic Retinopathy With Circular Hough Transformation and Convolutional Neural Networks. Expert Systems with Applications, 114, 289–95, https://doi.org/10.1109/CCDC.2019.8833190

K Simonyan, A Zisserman (2015). Very Deep Convolutional Networks for Large-scale Image Recognition. In: International Conference on Learning Representations, https://doi.org/10.48550/arXiv.1409.1556

K Xu, D Feng, H Mi (2017). Deep Convolutional Neural Network-based Early Automated Detection of Diabetic Retinopathy Using Fundus Image. *Molecules*, 22(12), 2054, https://doi.org/10.3390/molecules22122054

Kaggle (2013). *Dataset* [Online]. Available, https://kaggle.com/c/diabetic-retinopathy-detection

MD Abrham, MK Garvin, M Sonka (2010). Retinal Imaging and Image Analysis. *IEEE Reviews in Biomedical Engineering*, 3, 169–208, https://doi.org/10.1109/RBME.2010.2084567

Mohamed Jebran Pendekal, Shweta Gupta (2022). An Ensemble Classifier Based on Individual Features for Detecting Microaneurysms in Diabetic Retinopathy. *Indonesian Journal of Electrical Engineering and Informatics (IJEEI)*, 10(1), 60–71, https://doi.org/10.52549/ijeei.v10i1.3522

Mohsen Janghorbani, Raymond B Jones, Simon P Allison (2000). Incidence of and Risk Factors for Proliferative Retinopathy and its Association With Blindness Among Diabetes Clinic Attenders. *Ophthalmic Epidemiol.*, 7(4), 225–241.

MT Esfahani, M Ghaderi, R Kafiyeh (2018). Classification of Diabetic and Normal Fundus Images Using New Deep Learning Method. *Leonardo Electron Journal Practical Technologies*, 17(32), 233–48, http://lejpt.academicdirect.org/A32/233_248.pdf

Nandeeswar Sampigehalli Basavaraju, Shanmugarathinam Ganesarathinam (2021). Early Detection of Diabetic Retinopathy Using K-means Clustering Algorithm and Ensemble Classification Approach, *Intelligent Engineering and Systems*, 488–496, https://doi.org/10.1109/ICCITechn.2015.7488129

Nikos Tsiknakis, Dimitris Theodoropoulos, Georgios Manikis, Emmanouil Ktistakis (2021). Deep Learning for Diabetic Retinopathy Detection and Classification Based on Fundus Images: A Review. *Computer Biology and Medicine*, 135, 104599, https://doi.org/10.1016/j.compbiomed.2021.104599

Norah Asiri, Muhammad Hussain, Fadwa Al Adel, Nazih Alzaidi (2019). Deep Learning Based Computer-aided Diagnosis Systems for Diabetic Retinopathy: A Survey, *Artif. Intell. Med.*, 99, 101701, https://doi.org/10.1016/j.artmed.2019.07.009

P Chudzik, S Majumdar, F Caliva, B Al-Diri, A Hunter (2018). Microaneurysm Detection Using Fully Convolutional Neural Networks. *Computer Methods Program Biomedical*, 158, 185–92, https://doi.org/10.1016/j.cmpb.2018.02.016

P Mohamed Jebran, Shweta Gupta (2021). Pre-Diabetic Retinopathy Identification Using Hybrid Genetic Algorithm-Neural Network Classifier. *Journal of Physics: Conference Series*, 1937 012033, IOP Publishing, https://doi.org/10.1088/1742-6596/1937/1/012033

Paisan Raumviboonsuk, Jonathan Krause, Peranut Chotcomwongse, Rory Sayres, Rajiv Raman, Kasumi Widner, JL Campana Bilson, Sonia Phene, Kornwipa Hemarat, Mongkol Tadarati (2018). Deep Learning vs. Human Graders for Classifying Severity Levels of Diabetic Retinopathy in a Real-World Nationwide Screening Program, arXiv preprint arXiv:1810.08290, https://doi.org/10.1038/s41746-019-0099-8

Qiqi Xiao, Jiaxu Zou, Muqiao Yang, Alex Gaudio, Kris Kitani, Asim Smailagic, Pedro Costa, Min Xu (2019). Improving Lesion Segmentation for Diabetic Retinopathy Using Adversarial Learning, in: International Conference on Image Analysis and Recognition, Springer, 333–344.

R Parikh, A Mathai, S Parikh, G Chandra Sekhar, R Thomas (2008). Understanding and Using Sensitivity, Specificity and Predictive Values. *Indian J Ophthalmol.*, Jan-Feb;56(1), 45–50. [PMC free article] [PubMed].

R Pires, S Avila, J Wainer, E Valle, MD Abramoff, A Rocha (2019). A Data-driven Approach to Referable Diabetic Retinopathy Detection. *Artificial Intelligence Med*, 96, 93–1, https://doi.org/10.1016/j.artmed.2019.03.009

Ratul Ghosh, Kuntal Ghosh, Sanjit Maitra (2017). Automatic Detection and Classification of Diabetic Retinopathy Stages Using CNN. In: 2017 4th International Conference on Signal Processing and Integrated Networks (SPIN), IEEE, ISBN 978-1-5090-2797-2, pp. 550–554, https://doi.org/10.1109/SPIN.2017.8050011.

Ravichander Janapati, et al (2022a). Towards a More Theory-Driven BCI Using Source Reconstructed Dynamics of EEG Time-Series. *Nano LIFE*, 12(02), 2250005.

Ravichander Janapati, et al. (2022c). Web Interface Applications Controllers Used by Autonomous EEG-BCI Technologies. *AIP Conference Proceedings*.Vol. 2418. No. 1. AIP Publishing LLC.

Ravichander Janapati, Vishwas Dalal, Rakesh Sengupta (2022b). Advances in Experimental Paradigms for EEG-BCI. Proceedings of the 2nd International Conference on Recent Trends in Machine Learning, IoT, Smart Cities and Applications. Springer, Singapore.

Ravichander Janapati, Vishwas Dalal, Rakesh Sengupta (2021a). Advances in Modern EEG-BCIsignal Processing: A Review, *Materials Today: Proceedings*.

Ravichander Janapati, et al. (2021c). Signal Processing Algorithms Based on Evolutionary Optimization Techniques in the BCI: A Review. *Computational Vision and Bio-Inspired Computing*, 165–174.

Ravichander Janapati, et al. (2020). Review on EEG-BCI Classification Techniques Advancements. *IOP Conference Series: Materials Science and Engineering*. Vol. 981. No. 3. IOP Publishing.

Ravichander Janapati, et al. (2020). Various Signals Used for Device Navigation in BCI Production. *IOP Conference Series: Materials Science and Engineering*. Vol. 981. No. 3. IOP Publishing.

RB Budai, A Maier, J Hornegger, G Michelson (2013). Robust Vessel Segmentation in Fundus Images. *International Journal on Biomedical Images*, 2013(1), https://doi.org/10.1155/2013/154860

ROC Dataset [Online].Available, http://roc.healthcare.uiowa.edu.

RP Kumar, SS Vandana, D Tejaswi, K Charan, R Janapati, U Desai (2022). Classification ofSS-VEP Signals using Neural Networks for BCI Applications, International Conference onIntelligent Controller and Computing for Smart Power (ICICCSP), pp. 1–6.

S Dutta, BC Manideep, SM Basha, RD Caytiles, NCSN Iyengar (2018). Classification of Diabetic Retinopathy Images by Using Deep Learning Models. International Journal of Grid and Distributed Computing, 11(1), 99–106.

Safi Hamid, Sare Safi, Ali Hafezi-Moghadam, Ahmadieh Hamid (2018). Early Detection of Diabetic Retinopathy. *Surv. Ophthalmol.*, 63(5), 601–608.

Scottish (2014). *Intercollegiate Guideline Network, Management of Diabetes: A National Clinical Guideline, Scottish Intercollegiate Guidelines Network, Edinburgh.*

SK Vengalil, N Sinha, SSS Kruthiventi, RV Babu (2016). Customizing CNNs for Blood Vessel Segmentation from Fundus Images. In: International Conference on Signal Processing and Communications, SPCOM 2016, p. 1–4, https://doi.org/10.1109/SPCOM.2016.7746702

Shu-I Pao, Hong-Zin Lin, Ke-Hung Chien, Ming-Cheng Tai, Jiann-Torng Chen, Gen-Min Lin (2020). Detection of Diabetic Retinopathy Using Bichannel Convolutional Neural Network. Journal of Ophthalmology, 2020.

Srikanta Kumar Padhy, Brijesh Takkar, Rohan Chawla, Atul Kumar (2019). Artificial Intelligence in Diabetic Retinopathy: A Natural Step to the Future. *Indian Journal of Ophthalmology*, 67(7), 1004–1009, https://doi.org/10.4103/ijo.IJO_1989_18

T Kauppi, et al. (2007). The DIARETDB1 Diabetic Retinopathy Database and Evaluation Protocol. In: Proceedings of the British Machine Vision Conference. pp. 1–10, https:// doi.org/10.5244/C.21.15

V Gulshan, et al (2016). Development and Validation of a Deep Learning Algorithm for Detection of Diabetic Retinopathy in Retinal Fundus Photographs. *American Medical Association*, 316, 2402–10, https://doi.org/10.1001/jama.2016.17216

W Zhang, et al. (2019). Automated Identification and Grading System of Diabetic Retinopathy Using Deep Neural Networks. *Knowledge Base System*, 175, 12–25, https://doi.org/10 .1016/j.knosys.2019.03.016

Wejdan L Alyoubi, Wafaa M Shalash, Maysoon F Abulkhair (2020). Diabetic Retinopathy Detection Through Deep Learning Techniques: A Review. *Informatics in Medicine Unlocked*, 20 100377, https://doi.org/10.1016/j.imu.2020.100377

Wenying Yang, Juming Lu, Jianping Weng, Weiping Jia, Linong Ji, Jianzhong Xiao, Zhongyan Shan, Jie Liu, Haoming Tian, Qiuhe Ji (2010). Prevalence of Diabetes Among Men and Women in China, *N. Engl. J. Med.*, 362(12), 1090–1101.

Wei Zhang, JieZhong, Shijun Yang, ZhentaoGao, Junjie Hu, Yuanyuan Chen, Yi Zhang (2019), Automated Identification and Grading System of Diabetic Retinopathy Using Deep Neural Networks, *Knowl. Base Syst.*, 175, 12–25, https://doi.org/10.1016/j.knosys .2019.03.016. ISSN 9507051.

Xue-Wen Chen, Xiaotong Lin (2014). Big Data Learning: Challenges and Perspective. *IEEE Access*, 2(1), https://doi.org/10.1109/ACCESS.2014.2325029

Y Yan, J Gong, Y Liu (2019). A Novel Deep Learning Method for Red Lesions Detection Using Hybrid Feature. In: Proceedings of the 31st Chinese Control and Decision Conference, CCDC 2019, p. 2287–92, https://doi.org/10.1109/CCDC.2019.8833190

Yanming Guo, Yu Liu, Ard Oerlemans, Sonyang Lao, Song Wu, Michael Lew (2016). Deep Learning for Visual Understanding: A Review. *Neuro Computing*, 187(1), 27–48, https://doi.org/10.1016/j.neucom.2015.09.116

11 Usability to User Experience with Interactive Systems in Biomedical Applications and Devices

Chitharanjan Billa and Murthy Chavali

CONTENTS

11.1 INTRODUCTION

Technology advancements in human interaction interfaces are taking a new trend in the past decade and humanity has greatly benefitted from these achievements. It should be noted that there is a lot to be achieved in both inventions and enhancements of existing devices both from a feature and usability standpoint. In this chapter, the importance of these applications and devices in pandemic situations, challenges of the current version of medical devices, and user-centered design that is essential from both usability and user experience standpoints are discussed [1–9]. As the current UX principles utilized by medical devices are primitive and obsolete, new principles are introduced to overcome these challenges. Privacy and cybersecurity are the primary challenges for any IT application, and it's a more important challenge for medical devices and the industry. Mitigation paths are discussed to circumvent

DOI: 10.1201/9781003326830-11

these challenges. Prominent features of next-generation medical application devices are discussed. A compelling use case in medical device innovation is Google Glass. This medical device allows the medical professional to review X-rays and CT scans directly in their glasses field. In the absence of this device, the physician needs to go to a different lab room or system room leaving the patient in the surgery room to get the details of the scan report or the image. Using this device, physicians could cut the surgery time and also complete the task most efficiently [10–16]. The success of this human-machine interaction could be directly attributed to the fact that the user-centered design was immensely used by Google starting from the device phase to the final delivery phase. In the human–computer interaction realm, it was observed that there were many design considerations that every HCI engineer should be well aware of and the software engineers should make every effort to apply them, but it was found these considerations are not observed at all.

The role of every engineer involved in the production of medical devices should focus on reviewing the universal design, usability of the device, accessibility of the device keeping in mind all the disabled users, and useful products as the important goal in the delivery of the medical devices. In this chapter, all the user-centred design principles are discussed in detail so that the engineers would acknowledge the importance of these principles and deliver products/medical devices for the betterment of human beings.

11.2 BIOMEDICAL APPLICATIONS AND MEDICAL DEVICES

A biomedical application or a medical device is expected for the purpose of diagnosis of disease or other conditions in the cure, mitigation, treatment, or prevention of disease. Medical devices are used for preventive purposes and monitoring purposes as the data gets transmitted at regular intervals to the central databases thus alerting medical agencies and patients and preventing casualties of the patients on a daily basis. Most biomedical applications or devices use sensors as primary inputs to measure certain parameters [17–19]. These parameters could be physical manifests such as temperature or chemical substance measurements such as glucose in the blood content. These sensor devices measure the required reading either physical or chemical and then relay back the signal to the device. The engine then deciphers, calculates, and displays the readings on the screen for the user/medical staff consumption.

There are four main principal characteristics of a medical device [20–24].

1. **Sensing capability**: The device should have sensing capability either optically or physically. Examples: sensing temperature via infrared rays, sensing glucose molecules in the blood, sensing oxygen levels etc.
2. **Calculating capability**: The device should be able to transform the sensed manifest (either physical or chemical) into a numerical data format. Example: Glucose molecules in the blood should be transferred into a readable number format so that the information could be used effectively.
3. **Relaying capability**: The device should be able to transmit/relay the data not only to the display screen (which is mandatory) but also to other

applications/devices as required. Example: relaying vital parameters to a cloud database so that the health staff could review and act upon the readings.
4. **Alerting capability**: Devices should relay an alarm, beep, or alert when the values are high or low than the desired values. Devices should allow the users to be calibrated for high and low values and adjustable..Example: A temperature measuring device would produce a beep sound when the patient's temperature is beyond a limit.

In all the above principles usability and user experience are pivotal. Hence, any medical device should keep usability and user experience in mind while designing the device starting from the manufacturer.

11.3 SIGNIFICANCE OF BIOMEDICAL APPLICATIONS AND DEVICES IN PANDEMICS

There is clear evidence that contactless medical devices played an important role in the recent COVID-19 pandemic situation, thus calling for new inventions to be proactive for any upcoming pandemics. In early 2020 when COVID-19 started unfolding and was declared a public health emergency, many governments across the world have taken steps for a timely and continued supply of high-quality medical devices to test and diagnose COVID-19. As most of the devices could not go through the standard process of testing and other standard processes, most of the governments issued emergency use authorizations (EUAs), which helped to gain instant access to various therapeutic and diagnostic medical devices across the world. For instance, in the United States, the Food and Drug Administration (FDA), the agency responsible for overseeing medical devices regularly, published the device list so that the public is aware of the device's availability.

11.4 CURRENT CHALLENGES IN MEDICAL DEVICES

Several studies across the research institutes, cybersecurity, and health auditors flagged errors that are primarily due to transmission errors and poor recording function in biomedical applications and medical devices, which are affecting prognosis outcomes in the patients. It was also observed that fewer errors or no errors were found when the data was directly read from RFID (radio) devices, barcoded, or when entered via a mobile application. The challenge to purging paper records and migrating or pushing the electronic data to computers or mobile applications is that there is anenormous shortage of these special medical devices and health enterprises must spend a lot of capital on these devices that are constrained by limited budgets. Another critical test is computers or mobile devices may have to undergo regular cybersecurity maintenance, sometimes daily maintenance, for antivirus and patch updates, which include Apple/Android/Windows/Linux upgrades. The US government mandates every company in the health industry to comply with HITECH Act (Health IT Act), which requires quality health records (such as electronic health records) in clinical settings at emergency rooms. This mandate helps insurance and government healthcare agencies such as Medicaid/Medicare to send payments to the beneficiaries quickly. The National Institute of Health (NIH) is encouraging health-related IT vendors with grants and incentives to modernize their equipment thus helping medical devices function properly.

11.5   USER-CENETRED DESIGN

In the earlier chapters, there is a clear illustration that the current version of biomedical applications and devices suffers from both design and user experience flaws. By adopting a user-cenetred design (UCD) methodology, the mentioned flaws could be rectified thus transforming the applications to a more usable state. Definition: UCD is a methodology used by design engineers and software engineers to guarantee that the team/company is creating products that meet customers' needs.

Design thinking is not a novel concept but a familiar concept that started in the late 1950s. A simple example would be the evolution of the electric switch, which has simple on/off functionality and used to be in push button mode. Taking several human needs and conveniences into consideration, the electric switch now transformed into a toggle mode electric device and further graduated to a mobile switch. In the same way, biomedical applications and devices consider human-centred or user-centred design as the main philosophy to deliver world-class devices. UCD of any application, website, or medical device website generally consists of four important elements, namely visibility, accessibility, legibility, and language.

As discussed earlier, UCD is derived from human–computer interaction, which is the concept or design methodology used in the manufacturing and production of human–computer interfaces. Earlier product design experience shows that by applying and leveraging UCD principles there is a very high chance of delivering good usable biomedical applications or devices. During the very early days when Amazon just launched its commercial website, the company emphasized tracking errors during online shopping events. The error numbers were gathered and analyzed and corrected in the subsequent releases of the product. Human–computer interaction mostly revolves around the usability of the product or device whereas UCD is directly related to the overall applicability and usefulness and ease of use of human–computer interfaces, which are primarily applications (including biomedical devices) and websites.

One important step in the UCD process is to incorporate end-user feedback and criticism during the design and development phase. Ignoring user criticism from a usability standpoint will lead to product failure, hence this step (incorporating user feedback) is essential. Having beta version releases with selected customer groups and taking their regular feedback would help in producing quality biomedical applications.

Principle of proximity: The principle of proximity is used in various scientific applications and devices to improve the usability and user experience of websites and applications. In the context of biomedical applications and devices, it's imperative to discuss Gestalt principles of proximity. The Gestalt principle of proximity states that in a visual object or a graphic, objects that are proximate to each other are regarded as part of the same object. This concept can be used to improve screen design and increase viewer comprehension. The best example would be menu items in a software application; if the items that are used frequently or the objects used often in a sequence in a pattern are placed in a group menu, it improves usability and enhances user experience. If the menu items are scattered, the user experience suffers and during the period the applications or devices (Figure 11.1) would be abandoned due to the time spent on searching for the correct functionality.

Figure 11.1 illustrates the proximity principle, where items are placed in groups on the right-hand side. The principle of proximity or grouping can have the greatest visual and usability impact just by organizing group items based on their functionality. This design principle is widely used in many software products such as Microsoft Office.

Principle of similarity: Research suggests that incorporating the law of similarity in visual designs by utilizing the similarity of shape, size, colour, and orientation in the elements of an interface to provide them with a perceived relationship has a better user experience. Ahead of this, users can compare and contrast the icons and other text items or screen elements within a group of similar elements. By adopting this principle in biomedical applications and devices, users can use the icons quickly without any confusion and delay. As a timely intervention in the patient situation is crucial in the outcome, the UI designs must be not complex and follow simplicity for quicker navigation across the device. Also in steps from a navigation standpoint, users and health professionals can quickly go to the next item or web element without any help or hesitation. An example would be, if similar icons are placed in a visual train kind of object, the medical professional can quickly use the similar buttons/icons and complete the required diagnostic tests or procedures.

Principle of familiarity: In human perception, the visual experience often boils down to a similar image, which was already registered in the human brain. The best example would be vehicle drivers driving the vehicle easily when navigating through a familiar route and the drivers would not check about the road closures, etc., before driving through the road. An important point in medical devices is an easy and quick

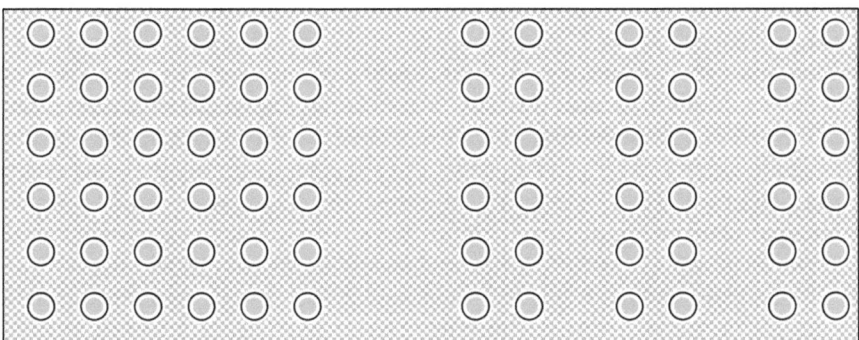

FIGURE 11.1 Principle of proximity.

navigation path as these are used in both critical and noncritical settings. When the user is very familiar with a certain navigation path or route, the usability becomes productive thus helping with the success of the biomedical application or medical device.

11.6 PROPOSED USER EXPERIENCE DESIGN PRINCIPLES FOR MEDICAL DEVICES

It is fundamentally important to keep the following UX design principles when building a new biological medical application or device. The first important principle would be Safety, and the next important principle would be Documentation. As the first point of reference, design engineers and UX engineers should consult standards such as IEC 62366-1 and IEC 62366-2 for guidance while building biological medical applications and devices from a usability engineering (UE) perspective.

UX is directly related to UE, which is focused on designing a user interface (UI) either on applications or devices that allows users to quickly interact with the device or application with minimal or no training and without obstacles or errors during the usage. The paradigm of intuitivism goes beyond smartphones and is also applicable to medical devices, hence design engineering principles should be adhered to.

As user safety takes prime importance, UE principles should be considered. Lack of adherence to these principles will add a risk that the products may fail to satisfy regulatory standards. As discussed, IEC 62366 is an international standard (ISO) that covers the application of UE to medical devices. Engineers and product owners should adhere to this standard, which helps streamline the process for specifications, analysis, development, and testing in the software lifecycle development during the product inception phase. IEC 62366 consists of two subsections:

1. IEC 62366-1: Application of UE (reference to medical devices)
2. IEC 62366-2: Regulation on the application of usability to engineering to biomedical devices

The success of most biomedical devices is closely linked to user interface quality. This is remarkably accurate in cases where there is significant market competition and the associated technology has been established, making user interface quality a leading factor in product diversity. The safety of most medical devices is also tightly correlated to user interface quality because design deficiencies may lead to and manifest errors either directly while using the device or indirectly during the transaction phase. This could lead to nonreversible consequences such as patient injuries and casualties.

Biomedical application and device manufacturers spend a tremendous amount of time designing and producing the equipment/devices, and most of the time these devices get very poor responses from the patients and operators from a usability and experience standpoint. At this juncture in the product, lifecycle manufacturers engage in user surveys and other opinion collection methods to gauge the experience and feedback from the end users. This step would be expensive to the manufacturer but to sustain the product these steps are necessary.

There are the proposed four UX design principles for improving and enhancing user experience and medical device usability:

1. **Grouped icons/menu items:** As discussed earlier, the principle of proximity plays a crucial role in the design aspect. Overcrowded icons and menu items in the display would not be received well in both the medical and patient communities. Members lose interest in using the devices as the screen would confuse the user.

2. **Textual simplicity:** Use consistent font and size across the biomedical application. This will give a sense of uniformity and simplicity while navigating and using the application. Simplify the words and limit excessive bold and underlines to avoid distraction during the usage. Any helpful information should be shown as cue items and should not be crammed into the main screen.

3. **Visual simplicity:** UX research suggests that using few colors in the application is favoured by most users as the devices are primarily used for critical diagnosis as well as for routine usage. UI engineers should be mindful of the colors so that these won't distract attention during usage. For example, using bold red colour for simple informational use would defeat the usability purpose. Limiting to a few standard colours and graying out non-essential information would mitigate the above flaw.

4. **Consistency:** End users prefer to have consistent language and options while selecting and cancelling (OK, Cancel) throughout the navigation. Any inconsistency either in text or colors will provide an impression that they have exited the application and using a different application (Figure 11.2).

Other principles and considerations: Biomedical applications and medical devices should not only be relying on UCD but also consider universal design principles. Universal design principles consider the user community including people with disabilities. The United States and most other countries mandated several laws to include users who have visual and hearing deficiencies. All US government applications (this includes ordinary websites and medical devices) should be following the accessibility laws so that all the user communities would be able to use them immediately. Most of the manufacturers during the design phase insist on a minimum viable product model, which helps to release the products in faster cycles.

The main missing feature during these types of software development cycles would be that the products would be usable by certain able people. The assumption that the users would be able to read/write and use it immediately is a wrong notion. The concept of universal design comes into the picture when all the user communities are accommodated. Human–computer interaction, which is the main design methodology for manufacturing medical devices, should account for the entire user base as this approach would avoid costly mistakes in treating the patients as well as averting accidents during the treatments. Enforcing accessibility standards in

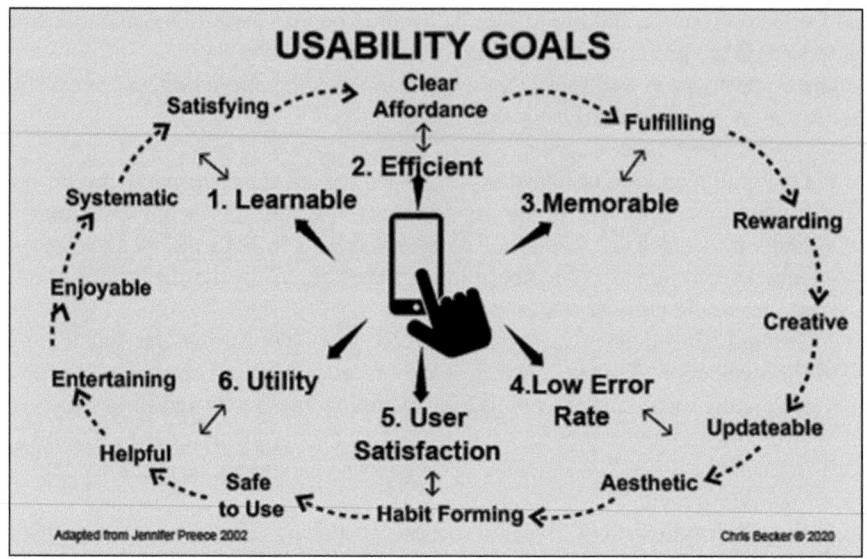

FIGURE 11.2   Usability goals summary—depicting various factors.

medical equipment is a requirement as per the US health code. All the equipment and facilities are to be fitted with appropriate devices and instructions so that people with disabilities would be able to use the equipment with ease. Hence considering these aspects during the design phase is very important while manufacturing biomedical applications and medical devices.

Design efficiency: Efficiency is one of the key components or contributors to evaluating the success of the medical device or software. Efficiency could be defined as the time taken either in seconds or minutes (in general) to achieve the task intended. It does not imply that the tasks that take longer time are less efficient but the calculations should be based on the instruments or devices that are similar. Concepts of Just-in-Time and Time-on-Task should be considered during the design process. This concept mainly deals with navigation during the usage of medical device interfaces. There should be thousands of calibration tests to be done before releasing the medical devices. For example, a medical device that is measuring blood glucose just displays a number, which might not be useful when compared to a device that shows green and red indicators, which alert the user if the levels are within the desired limits. As a principle, the small devices, handheld (smartphones), are intended to produce results quickly when compared to desktop computers or when there is big equipment such as X-rays/MRIs are involved.

Adaptability and know-how: One of the most important features of the medical device is user adaptability and if the device is intuitive, it means there is no need to spend more time adapting the medical device. These kinds of devices are also known as high-quality design devices. In the usability testing phase, these adaptability and learning times are measured when the users started using the device for the first time. Users' emotions such as likes and dislikes during usage should be carefully recorded as these provide important feedback on the device. Happy facial expressions provide positive feedback about adaptability.

The ability of the user to adapt the device quickly depends on the cues given by the manufacturer on device usage. There are many display design principles to be considered but it's worth noting 13 principles defined by Christoper Wickens. These principles of observation from a human visual standpoint should be used to create a compelling display design. These principles are intended to reduce errors, more adaptability with less learning curve (intuitive), and render user satisfaction and other usability benefits by adhering to these principles.

Observational principles:

1. The legibility of the display is of prime importance. If the device is audio-dependent, the voice must be relayed without any interference. If the icons, images, or text are distorted or not clear the device operator would find it difficult to operate or use in the clinical setting.
2. There should be a clear distinction between the adjacent buttons, images, or text, else it can create ambiguity during the selection. For example: if text 33333, 83338 is placed next to each other, users would find it difficult to distinguish when compared to a distinct and meaningful icon when placed next to each other.
3. Determination of a test should not be based on a particular color as the color perception would be different for each user and also the possibility of reporting a wrong color is possible. Hence single variable reporting should be avoided for medical devices. It is always better to decode the result to a perceivable number to avoid false positives or negatives.
4. Signal processing in the device that displays the final results should be following the user's expectations. If the device displays results that are not comprehendible by the user, adaptability fails.
5. Resource presentation (video and audio) could be done simultaneously or in a mixture. For example, while the results are displayed on the UI, if any audio signals are to be disseminated, it should be done simultaneously. A digital thermometer can produce a beep sound along the temperature display.

Thought principles

1. Visual response: There should be a distinctive display when showing high and low limits. Example: a red colour bar if the temperature is high and a green colour bar if the temperature is within the limits.
2. Movement related: If movement is to be measured, it should show in the form of a progress bar. If a dynamic parameter is to be measured (e.g., heart rate) a progress bar should be presented with progressing values.
3. Repeating signals: If the signal has to be presented multiple times based on the signal (e.g., heartbeat), it should be presented as a continuous positive signal and not as a repeating or redundant signal.

TABLE 11.1
Factors and Measures

Affecting factors/issues	Quantitative measures	Particular measures
Learnability: How fast the users can start using the device or the application starting from the first time and consequent uses. This is a very important factor to measure the usability of the medical device.	Measure time for the new users who started using the device and complete a particular task for which the device is manufactured.	Feedback from the users needs to be measured either in the form of a rating or percentage. This is during the learning phase of the system or device.
Effectiveness: This parameter measures how useful is the application or medical device in supporting the required tasks and also the reliability factor of the device.	Positive or successful performance of the planned tasks; this is a correct measure of productivity.	End users' rankings of the system'scapability to support their productivity and performance.
Efficiency: This is the measure of productivity by using the medical device after going through appropriate product manuals. There would be a bit learning curve initially and this time should not be included in calculating the efficiency.	Measure the time for the new users who started using the device to complete a particular task (after completing appropriate training and studying required documentation) for which the device is manufactured.	This is the users' ratings/ rankings and feedback compared to completing a particular with the new device when compared to the same task performed using current methods or devices.
Flexibility: Diversity of ways to complete the planned tasks.	This is the number of actual commands or clicks or swipes performed to achieve a particular planned task.	Users' rankings or rates of web applications or medical device's ability to provide different commands/paths to achieve a particular goal.
Memorability: How easily the users could use the application or device after the users start using the device after a period of absence.	This is a crucial measure of how well the user can relearn the system after a prolonged nonuse period.	User's rankings and ratings on the ability to remember system usage and how quickly the user can relearn the system in case of any gaps in using the system. This should be measured after a prolonged time of nonusage.
User satisfaction: This is the measurement that gives an idea about user satisfaction with using the instrument or device.	The best way to measure this parameter is how frequently the device or application is used and also how often each feature is used providing an overall idea about user satisfaction.	End users' feedback on their affinity toward the application/ device, their opinion about the device or application in general and its features.

TABLE 11.2
Example of Usability Questionnaire Scores after the Device Study (Sample Size = 6).

Items	M ± SD
1. Ease of using	1 ± 0
2. How simple is the system?	1 ± 0
3. Effectively complete tasks and scenarios	1 ± 0
4. Quickly complete tasks and scenarios	1 ± 0
5. Efficiently complete tasks and scenarios	1 ± 0
6. Comfortability in using the system	1 ± 0
7. Learnability	1 ± 0
8. Productivity confidence	1 ± 0
9. Error text messages	2.0 + 1.7
10. Recover from the messages	1 ± 0
11. System information	1 ± 0
12. Searching the information	1 ± 0
13. Understanding the information	1 ± 0
14. If the information helped to complete	1 ± 0
15. Information was organized	1 ± 0
16. Enjoyed using an interface	1 ± 0
Usability questionnaire overall	**1.05 + .10**

Focus principles

1. In UCD, user focus should be one of the primary considerations. Too many navigations across the application would divert the focus of the user.

This principle again relates to the principle of proximity that all the related icons should be nearby to cut down the speed of navigation while using the application.

2. Values or readings that are often corelated should be shown in one click/navigation so that there is no cost involved for multiple readings.

11.7   SECURITY IN MEDICAL APPLICATIONS AND DEVICES

Individual health information is a protected commodity by the law both in the United States and outside. Protected health information (PHI) is a constant subject of vulnerability, which is also known as "med jacking." News of med jacking surfaces almost every day as either old medical devices' vulnerability or a new potential threat. Biomedical applications and devices either in an enterprise setting or possessed by an individual are hyperconnected to other information systems via the internet. Medical professional heavily relies on the information disseminated from these systems to provide accurate and timely treatment to the patient community. The healthcare environment is so complicated that the connectivity spans from hospitals, manufacturers, devices, etc., which should work as intended. A cyberattack on anyone of the devices or facilities would have an irreversible and cascading effect on the entire health infrastructure and individuals.

Cyberattacks that occur on medical devices or settings generally consist of four types:

1. Large medical devices (X-ray machines, CT/MRI scanners, ultrasound devices etc.)
2. Health monitoring devices (smartwatches or electronic wrist/arm devices, etc.)
3. Personal wearables (glucose/insulin pumps that are externally attached to the patient and connected to the internet)
4. Embedded devices (pacemakers, etc., which are in the patient's body and transmit data to external servers)

Attackers initiate med jacking, which is intended to quickly penetrate medical devices, create control, and then use them as fundamental points to compromise and exfiltrate data from throughout the healthcare establishment. Once an attacker gets into the network and circumvents present security, they can cyber poison a medical device and establish an entrance subtly within the device for later access. The root cause of the vulnerability could be traced back to the manufacturer and mostly at the kernel level. In addition to the manufacturer's defect in the application or device, other causes include noncompliance to HIPAA standards as stipulated by the US Department of Health and Human Services (HHS).

The following steps are proposed to mitigate cyberattacks and security vulnerabilities in medical applications and devices:

1. **Immediate fixes:** Review security vulnerabilities as flagged by the report and informed by the security agencies. Fix current vulnerabilities on all devices, servers, and other access points and rescan the vulnerabilities after applying appropriate fixes.
2. **Review and implement HIPAA compliance:** Manufacturers and health enterprise settings should first review the current situation from a compliance perspective. There should be a particular focus on the data storage points such as databases, servers, network servers, email systems, data warehouses, EHR systems, etc.
3. **Current situation assessment:** A comprehensive security assessment should be done, which includes penetration tests, static analysis tests, and dynamic analysis tests. Password policies for every application and hardware component should be reviewed, and software security keys should be rotated regularly.
4. **Review and revise the security audit process:** Important factor in the security audit process would be if the biomedical applications or devices

are procured from reputed organizations. Unless needed, USB drives and other external media devices should be strictly restricted so that data and network would not be compromised. Any access to protected health information (PHI) should be role-based and multifactor authentication such as sending a token to a mobile device.

The Center for Devices and Radiological Health (CDRH), which is a branch within the US Food and Drug Administration (FDA), developed best practices for relaying cybersecurity vulnerabilities to enterprises and patients to provide timely and helpful information to consider when communicating with patients and caregivers about cybersecurity vulnerabilities.

Some of the prominent recommendations from the agency include:

1. Keep the content simple and easy to read, keep it timely, relevant, simple, and readable for all the audience including people with disabilities.
2. Indicate the potential risks introduced by the vulnerability and caution against the unknown effects of this vulnerability.
3. Provide appropriate links and resources that are both computer and mobile compliant so that users do not miss any key information due to truncation or font issues.

11.8 NEXT-GENERATION BIOMEDICAL APPLICATIONS AND DEVICES

With the advancement of new technologies and the advent of the 5G communication protocol, it's evident that the following features would be prominent in the next generation of biomedical applications and devices. The highly regulated environments and strict security protocols, although they are needed, slow down the production and release of new biomedical applications and devices. Despite the regulatory challenges the following features will be evident in the next generation of devices:

1. Hyperconnectivity: The devices that are isolated and running in standalone mode would be using cloud connectivity so that the medical data could be used across.
2. Mobile compliant: All the new biomedical applications and devices would be made mobile compliant so that they can be used with ease and carried easily.
3. Robotics: Medical robots, which are also classified as devices leveraging artificial intelligence, would be heavily used both for diagnosis and surgeries.
4. Identity proofing with biometrics: Privacy and security are always the primary concerns of medical devices. By leveraging biometrics these challenges could be mitigated.

Minimum viable product (MVP) is used mostly in any kind of product development including biomedical applications and medical devices. In this approach, product delivery is done in smaller chunks, which would be tied to the main product features. In the iterative model, the features are delivered on a regular basis and sent for testing and quality assurance. As most software developers prefer agile methodology, this works well within the team but there are several issues with this model. The main issue with this model is the definition of viability. If viable is defined by the manufacturer and not the end user, this MVP model will not help the end user.

The focus should be given to the user's needs and not the manufacturers' or owners' needs. Software developers who work in this model mainly focus on delivering the features defined in the MVP definition. This model will eventually lead to the production of very unmatured UX models. The focus of the developers should be on the value delivery to the user rather than focusing on the minimal approach defined by the organization. Although the MVP model allows faster delivery cycles and the products come out very soon to the market, this model could deliver very poor features to the end users. Hence, it is highly recommended that the MVP definition should be coming from the users and not the manufacturers.

11.9   CONSIDERATION OF IOT IN MEDICAL DEVICES FROM A USABILITY STANDPOINT

The Internet of Things (IoT) is the most talked about topic in the scientific community both in product manufacturing and research fields. Already IoT is making significant contributions to the medical community. It was also observed the medical industries are one of the key application areas in IoT and particularly related to medical devices. IoT is mainly used in the healthcare industry for quicker diagnosis, reliable results, and highly accurate results, especially in medical e-devices. One key feature of medical devices is the usage of RFID tags that are being effectively used to track patients. This is a major functionality that will enhance the applicability of medical devices developed on IoT technologies. The IoMT leveraged medical devices to provide a useful platform for both doctors and patients, connecting them and allowing more convenient communication.

Physicians can access patient details via screens and when needed and can even send electronic prescriptions to patients. Another example is patient appointment applications, which helped a lot during pandemic situations. As mobile health is becoming prominent, device usability is of utmost important feature either in application development or device production.

Prototyping medical devices: Prototyping is a way in which healthcare enterprises or organizations can demo the device, product, or feature in a controlled setting such as a development environment. Prototyping help industries not to spend a lot of money on a product to only know that the usability is not upto the users' needs. Users do not have to wait for the finished product but get a look and feel of the actual device. Based on the suggestions and feedback from the users, appropriate changes would be made to the product and can be released. The rapid disposition of the iterative design process in combination with prototyping can find designers developing low-reliability prototypes in the invention phase to help shape the direction of a product.

Low-fidelity prototyping conventionally involves noninteractive prototypes, often paper-based or wireframe-based diagrams or PowerPoint presentation, allowing the design engineers to create varying possible solutions to patient need, while not capitalizing huge amounts of time and effort to progress upon the application or device.

There are four prototyping process models that were widely used in the manufacture of biomedical applications: (1) incremental prototyping, (2) evolutionary prototyping, (3) throw-away Prototyping, and (4) extreme prototyping.

Whatever method of prototyping is deployed or used, it's imperative to be in a collaborative setting. In a collaborative prototype setting, there will be direct access to the user community and the feedback or survey results could be used for the better incorporation of the features from the feedback. Prototyping should be used in the design process of medical devices.

11.10 CONCLUSION

By leveraging appropriate UCD principles, the usability of the medical devices and biomedical applications would reach better adaptability and also the security principles that were discussed would provide safety to PHI and also the mitigation and security risk analysis and plan would help the manufacturers to be on the top of the issues. Medical devices often leverage interactive health technologies (IHTs) that are primarily designed to treat patients whether chronic or not. The success of these devices depends on how useful they are to the patients, and the easy-to-use factor plays an important role in determining the usability of the medical device. As discussed above using the principles of UCD during the application development process would result in world-class medical devices, which would be safe, accessible, effective, efficient, and easy to use. Manufacturers should keep the focus on tasks and users throughout the software development process as this will reduce the risk of the device being rejected by the user community, and it helps the patient community either for a self-user or in the clinical setting. When principles of UCD are employed, usability testing, techniques, icon labels, paths, and other web and display elements would be much more useful to the community. Iterative testing would help fix the known issues and incorporate feedback from the user community and provide an error-free final version to be deployed to the production environment. Prototypes should be evaluated very carefully as these provide valuable inputs for the next releases. This methodology allows incorporating end-user feedback to be used during design phases eventually leading to more usability and focus more on user-centered activities. This chapter also discusses various principles and methods such as analyzing the tasks, requirement analysis focusing on the end user needs, developing proof of concepts, evaluating other design patterns, and designing iterative test plans that would help produce excellent medical devices. By incorporating all the UCD principles discussed above, manufacturers can avoid time-consuming and costly cycles both in the manufacturing and software design processes. UCD is highly acclaimed as a design methodology to produce high-quality user-acceptable applications and devices, although this design methodology is widely accredited to health devices.

Frontline health community users such as physicians and physician's assistants should actively intervene in the feedback process thus helping in the quality of the products. It is imperative to note Gould and Lewis's fundamental principles of user-centred design, which are as follows:

1. Starting from the project inception, prime attention should be given to the user community and their feedback on the device.
2. Usability should be measured at constant intervals and reported to the design team.
3. Test the usability iteratively.

These principles were introduced over 25 years back and they are still used widely not only in commercial software applications but also in biomedical applications and devices. Manufacturers after releasing the biomedical application or device should conduct a randomized control as a pilot trial so that feedback from the users could be collected on the usability of the products. After a few more months or during the next release, a full-scale randomized controlled trial is warranted to test thoroughly for the efficacy of the product. These steps are currently missing for any biomedical devices as there is not much competition in this area and only a handful of manufacturers are producing the device. Due to few manufacturers in the field, user feedback and usability concerns go unanswered and this should be fixed immediately.

REFERENCES

1. Privacy and Cybersecurity Are Converging. Here's Why That Matters for People and Companies. Andrew Burt. *Harvard Business Review*. Retrieved 20 July 2021; https://hbr.org/2019/01/privacy-and-cybersecurity-are-converging-heres-why-that-matters-for-people-and-for-companies
2. Nikhil Bhalla, Pawan Jolly, Nello Formisano, and Pedro Estrel. (2016). Introduction to biosensors. Retrieved 20 July 2022. https://www.ncbi.nlm.nih.gov/pmc/articles/PMC4986445/. doi: 10.1042/EBC20150001
3. Emergency Use Authorization (2022). Food and Drug Administration, United States of America. Retrieved 17 August 2022. https://www.fda.gov/emergency-preparedness-and-response/mcm-legal-regulatory-and-policy-framework/emergency-use-authorization
4. User-Centred Design Basics (2022). United States Web Design System. Retrieved 01 August 2022. https://www.usability.gov/what-and-why/user-centered-design.html
5. Human-Computer Interaction (2022). Interaction Foundation Design. Retrieved on 02 August 2022. https://www.interaction-esign.org/literature/topics/human-computer-interaction
6. The Gestalt Principle of Proximity: In Action (2020). *Lukas Oppermann. UX Collective*. Retrieved on 02 August 2022. https://uxdesign.cc/how-to-enhance-your-design-with-the-gestalt-principles-of-proximity-a7828452058b
7. IEC 62366-1:2015. Medical Devices: Part 1: Application of Usability Engineering to Medical Devices. Retrieved 01 August 2022. https://www.iso.org/standard/63179.html
8. Edward Stull (2018). UX Fundamentals for Non-UX Professionals: User Experience Principles for Managers, Writers, Designers, and Developers, Chapters 5, 6; https://doi.org/10.1007/978-1-4842-3811-0

9. Manage an IT Accessibility/508 Program (2022). *General Services and Administration*, United States Government. Retrieved on 01 August 2022. https://www.section508.gov/manage

10. JIT Just-in-Time Manufacturing (2022). The University of Cambridge. Retrieved on 01 August 2022. https://www.ifm.eng.cam.ac.uk/research/dstools/jit-just-in-time-manufacturing/

11. Jan Gulliksen, Bengt Göransson, Inger Boivie, Stefan Blomkvist, Jenny Persson and Åsa Cajander (2010). *Key Principles for User-centred Systems Design*, 397–409. https://doi.org/10.1080/01449290310001624329

12. Preventing Medjacking (2021). *AJN, American Journal of Nursing*, October 2021 – 121(10), 46–50. doi: 10.1097/01.NAJ.0000794252.99183.5e

13. Best Practices for Communicating Cybersecurity Vulnerabilities to Patients (2021). Center for Devices and Radiological Health, United States Food and Drug Administration. Retrieved on 02 August 2022. https://www.fda.gov/media/152608/download

14. Minimum Viable Product (MVP) (2022). Gartner Publications (2022). Retrieved on 01 August 2022. https://www.gartner.com/en/marketing/glossary/minimum-viable-product-mvp-

15. Silpi Sarkar, Murthy Chavali (2020). Artificial Intelligence and Machine Learning approach towards COVID-19, *Nanomedicine & Nanotechnology Open Access*, 5(3), 190–200; https://doi.org/10.23880/nnoa-16000201

16. R. Janapati, et al. (2022). Towards a More Theory-Driven BCI Using Source Reconstructed Dynamics of EEG Time-Series. *Nano LIFE*, 12(02): 2250005.000

17. R. Janapati, et al. (2022). Web Interface Applications Controllers Used by Autonomous EEG-BCI Technologies. AIP Conference Proceedings. Vol. 2418. No. 1. AIP Publishing LLC.

18. R. Janapati, V. Dalal, and R. Sengupta. (2022). Advances in Experimental Paradigms for EEG-BCI. Proceedings of the 2nd International Conference on Recent Trends in Machine Learning, IoT, Smart Cities and Applications. Springer, Singapore.

19. R. Janapati, V. Dalal, and R. Sengupta. (2021). Advances in modern EEG-BCI signal processing: A review. *Materials Today: Proceedings* 80(6). DOI:<u>10.1016/-11j.matpr.2021.06.409.</u>

20. R. Janapati, et al. (2021). Progression of EEG-BCI Classification Techniques: A Study. *Inventive Systems and Control*, vol 1, 161–170. DOI:<u>10.1007/978-981-16-1395-1_13</u>

21. R. Janapati, et al. (2021). Signal Processing Algorithms Based on Evolutionary Optimization Techniques in the BCI: A Review. *Computational Vision and Bio-Inspired Computing*, 165–174 ICCVBIC 2020 Conference proceedings.

22. R. Janapati, et al. (2020). Review on EEG-BCI Classification Techniques Advancements. *IOP Conference Series: Materials Science and Engineering*. Vol. 981. No. 3. (1-7) IOP Conf. Ser.: Mater. Sci. Eng. 981 032019.

23. R. Janapati, et al. (2020). Various signals used for device navigation in BCI production. *IOP Conference Series: Materials Science and Engineering*. Vol. 981. No. 3. IOP Publishing.

24. R. P. Kumar, S. S. Vandana, D. Tejaswi, K. Charan, R. Janapati and U. Desai. (2022). Classification of SSVEP Signals using Neural Networks for BCI Applications. 2022 International Conference on Intelligent Controller and Computing for Smart Power (ICICCSP), pp. 1–6, doi: 10.1109/ICICCSP53532.2022.9862368.

12 Dynamic Energy Management of Hybrid Electric Vehicles (HEVs) Using Fuzzy Logic Controller

Sohan Das, Souvik Biswas, Sukrit Sarkar,
Tamal Maji, and Dr. Syamasree Biswas Raha

CONTENTS

DOI: 10.1201/9781003326830-12

12.1   INTRODUCTION

Pollution and global warming (Chan, 2007, 95(4)) are the major threat to the society, which are caused due to greenhouse gas emission from internal combustion engines (ICEs). Moreover, price of petrol and diesel is also skyrocketing, and the huge consumption of fossil fuels is bringing the scarcity of fuel in near future. As a promising solution, the demand for electric vehicles (EVs) in the market are increasing; however, the availability of charging stations, low-priced battery, and charge controller are limited, which imposes few restrictions to the implementation and utilizations of EVs (Rzaei et al., 2018, 67(11)). In that situation demand of hybrid electric vehicles (HEVs) is providing the optimum solutions of utilizing fuel and battery charging/discharging set in a same device; however, dynamic energy management (DEM) is very much essential for an efficient HEV. For this purpose, artificial intelligence-based approaches provide satisfactory results and several research works are also going on; however, optimum results are yet to be received.

In this connection, Chi et al. considered energy management of hybrid vehicles using artificial intelligence with different energy sources (Chi et al., 2013, pp.1765–1767). Here, two inputs such as fuel capacity and battery state of charge (SOC) are considered and several fuzzy rules are designed to control the electric motor. In this direction, Sarvestani et al. considered a novel optimal energy management strategy based on fuzzy logic for a HEV (Sarvestani and Safavi, 2009, pp.141–144). Here, the objective of the proposed work is to maximize the fuel economy by minimizing emissions, unlike conventional ICEs that are based on fuzzy logic for a parallel HEV. In this work, the fuzzy logic controller (FLC) is designed by considering battery SOC and desired torque for driving as two inputs. In this regards, Saib et al. presented that the HEV operates through fuel cell that is employed as a primary source and battery or supercapacitor for auxiliary power source using fuzzy logic for efficient energy management (Saib et al., 2017). Here fuel cell is provided for a steady-state power supply and a battery for transient power supply. In this connection, Zhang et.al proposed parallel hybrid electric vehicles (PHEVs) model that represents energy management through adaptive neuro fuzzy inference system (ANFIS) optimization algorithm (Zhang et al., 2017, pp.342–347). Here, the required torque of the clutch and the value of the SOC are considered as inputs in the FLC, which provide 51.2% of fuel economy. Further, regenerative braking using FLC by Mamdani-type fuzzy inference system is efficiently presented by Li etal. (Li et al., 2009, pp.1749–1754) for a HEV system. Here, two inputs such as brake pedal status and motor speed and as output regenerative braking torque are considered.

From the above research works, it can be observed that study on DEM for HEVs is going on by several researchers. To procure the above, artificial intelligence mainly in terms of FLC and ANFIS have been majorly utilized with several inputs such as required torque of the clutch, the value of the SOC, brake pedal status, motor speed, fuel capacity, etc.; in this connection, it is worthless to mention that battery plays a vital role in HEVs and therefore battery discharging in terms of SOC measurement up to a predefined threshold value is required to monitor carefully to avoid any failure before switching to fuel mode. Considerations of the issue are yet to receive in the literature in addition to the DEM.

In this work, a HEV model for DEM is developed in MATLAB 2016a platform incorporating FLC in addition to monitoring battery discharging. Here, two inputs of the FLC are chosen as the fuel capacity and battery SOC to generate output response in terms of either battery mode or fuel mode while prioritizing the battery mode utilizations. Further, battery discharging model is developed to switch the HEV from battery mode to fuel mode after attending a pre-threshold value. Here, FLC plays an effective role to control DEM.

12.2 DYNAMIC ENERGY MANAGEMENT OF HEV

12.2.1 INTRODUCTION

Utilization of HEVs in present scenario draws major attention having both the fuel and battery sources to drive the car. HEV also reduces environmental pollution. However DEM of HEV is a challenging task. In this work, DEM of HEV is considered utilizing FLC to utilize its available resources in an optimum and efficient way.

12.2.2 HYBRID ELECTRIC VEHICLE

The hybridization of vehicle is making a high demand because of increasing pollution. HEV is a special EV that operates through engine power and battery power. IC engine and electric motor that is to be powered by battery are amalgamated to make HEV run. It gives better fuel economy with respect to any conventional vehicle. There are three types of HEV. They are mentioned below:

- Series hybrid electric vehicle
- Parallel hybrid electric vehicle
- Series-parallel hybrid electric vehicle

12.2.3 DYNAMIC ENERGY MANAGEMENT

DEM is basically introduced in a system to manage the energy of that system in an efficient manner economically considering many constraints like temporary peak load demand, permanent energy savings, etc.

12.2.4 DYNAMIC ENERGY MANAGEMENT OF HEV

In recent days, due to excessive use of ICE engine-based vehicles, a significant environmental pollution is caused due to emissions of carbonmonoxide (CO), carbon dioxide (CO_2), nitrogen oxides, (NOx), unburned hydrocarbons (HCs), etc. The above issue raised the awareness to move toward the eco-friendly energy-efficient vehicles by the automobile industry such as battery electric vehicles (BEVs), HEVs/plug-in hybrid electric vehicles (PHEVs), and fuel cell vehicles (FEVs) (Chan, 2007,95(4)) (Rzaei et al., 2018, 67(11)). In this work, HEV is considered to study, which can be operated by both fuel and battery for its convenience. The core components of HEV are engine for fuel control, battery for electricity supply, and electric motor. Hence, efficient energy management by both sources is a major challenge for the HEVs. In this respect, the DEM can be explained as to control the flow of energy between the vehicle's battery system and the various elements such as the electric motor and the various power converters installed in the vehicle. Hence, the performance of HEV is dependent on the DEM strategy of the vehicle, which is controlled by different artificial intelligence-based controllers. In this work, FLC is incorporated to control DEM considering two major inputs such as fuel capacity and battery SOC to minimize pollutant emission and improve fuel economy by providing priority to battery utilizations. Further, a battery discharge system is also developed by maintaining a battery SOC threshold value, to switch to fuel mode to prevent sudden decay of battery SOC.

12.3 FUZZY LOGIC CONTROLLER

12.3.1 INTRODUCTION

A controller is a device that receives data from a measurement instrument, compares this data to setpoint, and, if necessary, signals a control element to take corrective actions. Controllers may perform complex mathematical functions to compare. Controllers always have the ability to receive input, to perform a mathematical function with input, and to produce an output signal. There are many controllers (Krishnaswami, 2012) such as proportional (P controller), proportional integral (P-I controller), proportional derivative (P-D controller), proportional integral derivative (P-I-D controller), and advanced controllers such as FLC, which are used in several industries.

12.3.2 FUZZY LOGIC CONTROLLER (FLC) FOR DEM

Several new controllers such as intelligent controllers have been applied for hybrid two-area control power system. Fuzzy logic is an important technology and a successful branch of automation and control theory, which provides good result in control of power system. Fuzzy logic is derived from mathematical algebra based on set theory.

From the previous section, it can be understood that DEM for a HEV needs an intelligent control mechanism. Here, FLC (Rajasekaran and Vijayalakhsmi Pai, 2011; Biswas, 2020) is utilized to handle DEM. Now, FLC in terms of Mamdani approach is applied here to design due to its widespread acceptance for intuitive and understandable process. The FLC employs a knowledge base expressed in terms of a fuzzy inference rules and a fuzzy inference engine to solve the problems. Fuzzy inference is also known as approximate reasoning, which refers to computational procedures that are used to evaluate linguistic description. There are four modules involved in the fuzzy system, i.e., fuzzy rule base, fuzzy inference engine, fuzzification module, and defuzzification module (Rajasekaran and Vijayalakhsmi Pai, 2011). These four modules are explained briefly below:

12.3.2.1 Fuzzy Rule Base

Fuzzy rule base consists of all the rules and the if-then conditions to control the decision-making system.

12.3.2.2 Fuzzification

This step takes place in fuzzifier. This step converts the inputs from crisp numbers to fuzzy sets so that the crisp inputs can be measured by the sensors and passed into the control system for further processing. Fuzzification splits the input signal into different steps like very low (VL), low (L), medium (M), high (H) and very High (VH).

12.3.2.3 Fuzzy Inference Engine

The fuzzy inference engine is an important part of the FLC. It helps to determine the degree of match between inputs and the rule base. In other words, it is necessary to apply the rules to fuzzy input for generating the fuzzy output. Depending on the inputs and corresponding outputs, rules are decided and inputs are fuzzified.

12.3.2.4 Defuzzification

After obtaining fuzzy output sets, it will go to defuzzifier where defuzzification takes place. The defuzzification process converts the fuzzy sets into crisp output value. In this regard, membership function is used to describe fuzziness graphically. The degree of truth in fuzzy logic is represented by membership function, which is not dependent on the elements of the fuzzy system.

Now, FLC for controlling DEM of HEV is explained in the block diagram as shown in Figure 12.1.

Further, the workflow of the FLC for solving DEM in HEV is explained in Figure 12.2.

In Figure 12.2, battery SOC and fuel capacity are considered as two inputs of FLC. If value of battery SOC is higher than predefined threshold value, vehicle will run in electric mode without charging battery and if not, it will switch to fuel mode. Now in battery mode, battery will start to discharge, which also needs to be monitored. The same process will continue until and unless the trip ends. Initially, if

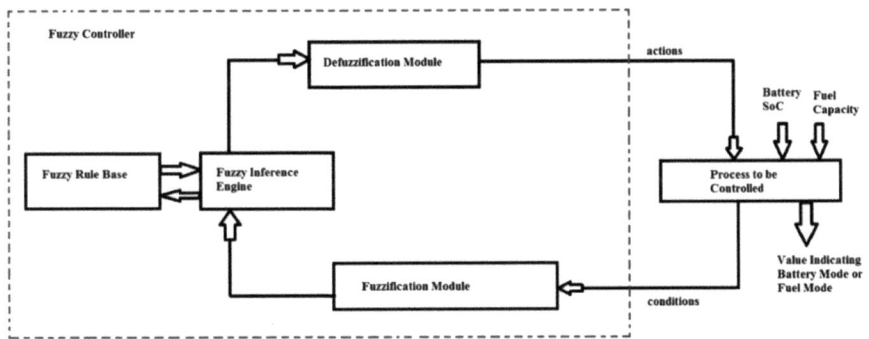

FIGURE 12.1   Block diagram of FLC used for HEV.

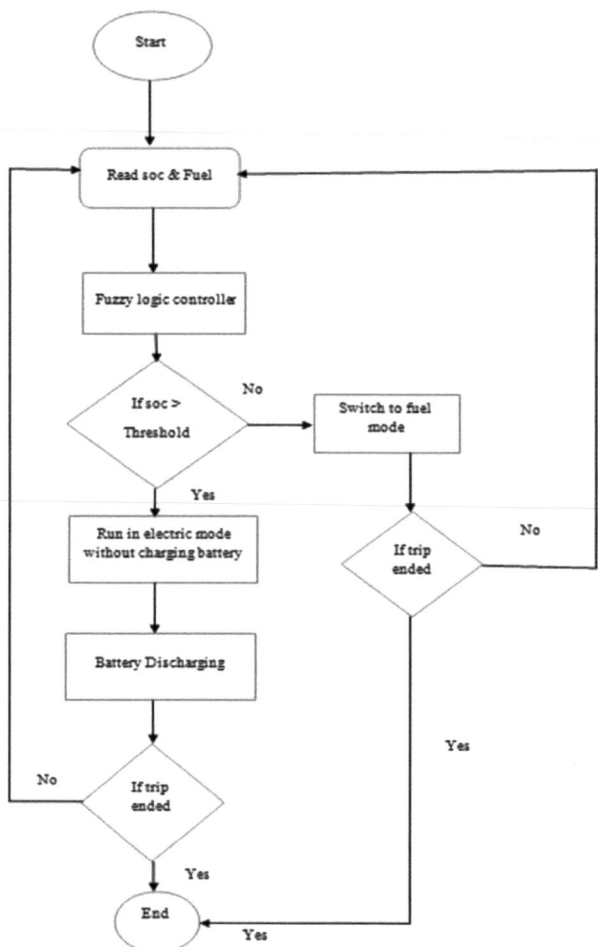

FIGURE 12.2   Flow chart of energy management.

sufficient battery charging is not available, the HEV will continue to run in fuel mode unless the trip ends.

Now the detailed parameters of the proposed FLC for DEM including battery discharging model are discussed in the next section.

12.4   TECHNICAL DETAILS OF THE PROPOSED WORK

12.4.1   FLC IMPLEMENTATION FOR DEM

As discussed, in this work FLC is considered to control DEM of HEV. Further, to make the HEV model more reliable battery discharging model is considered to switch the HEV from battery mode to fuel mode after attending a pre-threshold value. In this work, the FLC has considered two input parameters, i.e., battery SOC and fuel capacity and dynamic energy mode as output of the proposed HEV model. Here, triangular membership function is employed for both input and output due to its good performance and easy calculations compared to other membership functions as shown in Figure 12.3.

12.4.1.1 Inputs for FLC

To design the fuzzy inference engine, one input, i.e., battery SOC, is segregated into five stages: Very Low (VL), Low (L), Medium (M), High (H), and Very High (VH), as shown in Figure 12.4. As for example if x is assumed to be the parameter and the input functions $\mu(x)$ can be segregated into five stages, such as Very Low ($\mu_{VL}(x)$), Low ($\mu_L(x)$), Medium ($\mu_M(x)$), High ($\mu_H(x)$), and Very High ($\mu_H(x)$). The proposed parameters with range are presented by Equations 12.1–12.5:

$$\mu_{VL}(x) = \begin{cases} \dfrac{25-x}{25} \in 0 \le x \le 25 \end{cases} \tag{12.1}$$

$$\mu_L(x) = \begin{cases} \dfrac{x}{25} \in 0 \le x \le 25 \\[2mm] \dfrac{50-x}{25} \in 25 \le x \le 50 \end{cases} \tag{12.2}$$

$$\mu_M(x) = \begin{cases} \dfrac{x-25}{25} \in 25 \le x \le 50 \\[2mm] \dfrac{75-x}{25} \in 50 \le x \le 75 \end{cases} \tag{12.3}$$

$$\mu_H(x) = \begin{cases} \dfrac{x-50}{25} \in 50 \le x \le 75 \\[2mm] \dfrac{100-x}{25} \in 75 \le x \le 100 \end{cases} \tag{12.4}$$

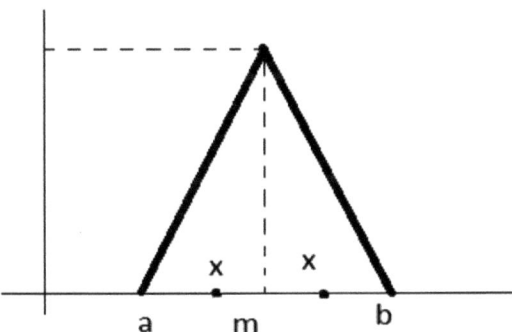

FIGURE 12.3 Triangular membership function.

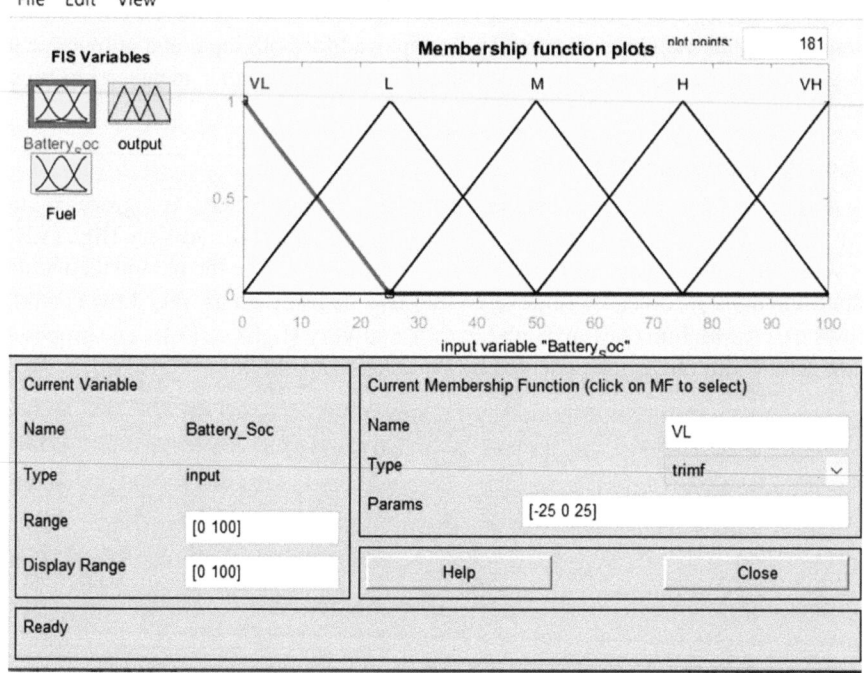

FIGURE 12.4 Battery SOC as an Input-1.

$$\mu_{VH}(x) = \left\{ \frac{x-75}{25} \in 75 \le x \le 100 \right. \tag{12.5}$$

Similarly, another input, i.e., fuel capacity, i.e., functions $\mu(y)$, can be segregated into five stages, such as Very Low ($\mu_{VL}(y)$), Low ($\mu_L(y)$), Medium ($\mu_M(y)$), High ($\mu_H(y)$), and Very High ($\mu_H(y)$). The proposed parameters with range are presented by Equations 12.6–12.10 and shown in Figure 12.5.

$$\mu_{VL}(y) = \left\{ \frac{25-y}{25} \in 0 \le y \le 25 \right. \tag{12.6}$$

$$\mu_L(y) = \begin{cases} \dfrac{y}{25} \in 0 \le y \le 25 \\ \dfrac{50-y}{25} \in 25 \le y \le 50 \end{cases} \tag{12.7}$$

$$\mu_M(y) = \begin{cases} \dfrac{y-25}{25} \in 25 \le y \le 50 \\ \dfrac{75-y}{25} \in 50 \le y \le 75 \end{cases} \tag{12.8}$$

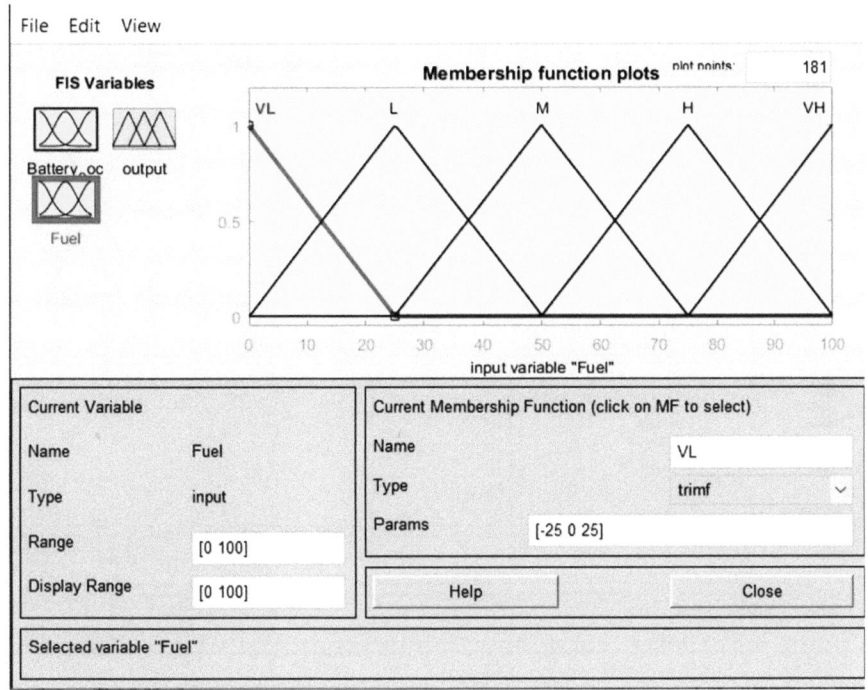

FIGURE 12.5 Fuel capacity as an Input-2.

$$\mu_H\left(y\right) = \begin{cases} \dfrac{y-50}{25} \in 50 \le y \le 75 \\ \dfrac{100-y}{25} \in 75 \le y \le 100 \end{cases} \tag{12.9}$$

$$\mu_{VH}\left(y\right) = \begin{cases} \dfrac{y-75}{25} \in 75 \le y \le 100 \end{cases} \tag{12.10}$$

12.4.1.2 Outputs for FLC

Finally, the output, i.e., DEM, in terms of battery SOC to fuel capacity of HEV has been framed considering six stages. These are Negative Low ($\mu_{NL}(z)$), Negative Medium ($\mu_{NM}(z)$), Negative High ($\mu_{NH}(z)$), Positive Low ($\mu_{PL}(z)$), Positive Medium ($\mu_{PM}(z)$), and Positive High ($\mu_{PH}(z)$), which are presented by Equations 12.11–12.16 and shown in Figure 12.6.

$$\mu_{NL}\left(z\right) = \begin{cases} \dfrac{16.67-z}{16.67} \in 0 \le z \le 16.67 \end{cases} \tag{12.11}$$

$$\mu_{NM}\left(z\right) = \begin{cases} \dfrac{z}{16.67} \in \quad 0 \le z \le 16.6 \\ \dfrac{33.34-z}{16.67} \in 16.67 \le z \le 33.3 \end{cases} \tag{12.12}$$

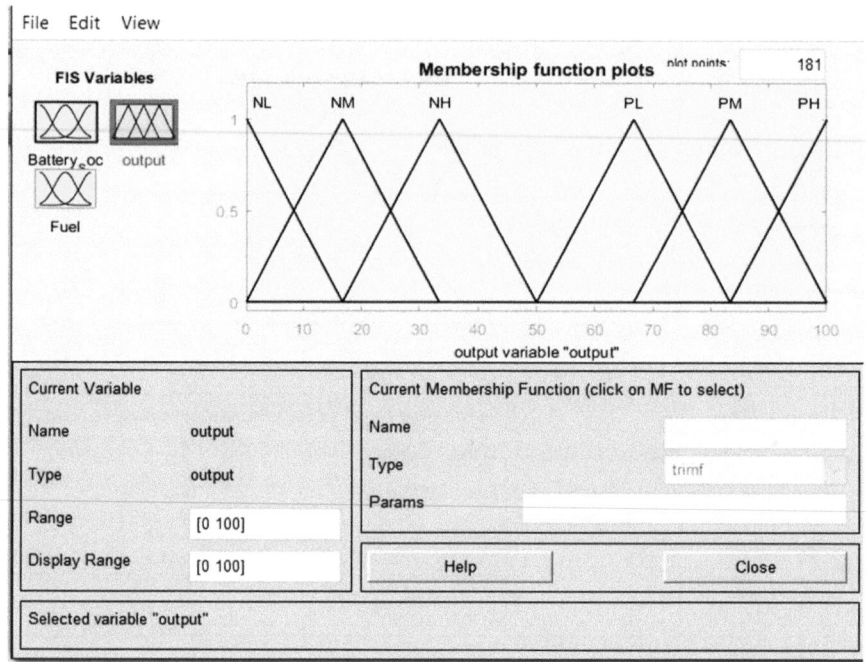

FIGURE 12.6   Output from FIS.

$$\mu_{NH}(z) = \begin{cases} \dfrac{z-16.67}{16.67} \in\ 16.67 \le z \le 33.34 \\[2mm] \dfrac{50.01-z}{16.67} \in\ 33.34 \le z \le 50.0 \end{cases} \tag{12.13}$$

$$\mu_{PL}(z) = \begin{cases} \dfrac{z-50.01}{16.67} \in\ 50.01 \le z \le 66.68 \\[2mm] \dfrac{83.35-z}{16.67} \in\ 66.68 \le z \le 83.35 \end{cases} \tag{12.14}$$

$$\mu_{PM}(z) = \begin{cases} \dfrac{z-66.68}{16.67} \in\ 66.68 \le z \le 83.35 \\[2mm] \dfrac{100.02-z}{16.67} \in\ 83.35 \le z \le 100.02 \end{cases} \tag{12.15}$$

$$\mu_{PH}(z) = \begin{cases} \dfrac{z-83.35}{16.67} \in 83.35 \le z \le 100.02 \end{cases} \tag{12.16}$$

12.4.1.3   Rule Base for FLC

Now, the vital part of the FLC is to be developed, i.e., fuzzy rule base. By using IF-THEN rule as shown in Table 12.1, total 25 rules are formulated in this work.

12.4.2 Battery Discharging Model

Further, to prevent sudden decay of battery SOC and make an efficient HEV model, a battery discharge system is developed as shown in Figure 12.7. In this model, a SOC threshold value is fixed where if the battery SOC goes below the threshold value, HEV will be automatically shifted to fuel mode. In Figure 12.7, battery discharge unit consists of 200V (nominal voltage), 100Ah lithium-ion battery (rated capacity), two MOSFETs (FET resistance 0.1Ω, internal diode resistance $0.01\ \Omega$), snubber resistance ($1e^5\ \Omega$), series RLC branch ($R = 0.05\ \Omega$, $L = 5.76e^{-2}$ H), ideal switch (internal resistance $= 0.001\ \Omega$, snubber resistance $1e^5\ \Omega$), current measurement, scope, bus selector, display. Here, the battery SOC is considered 37, and below this value HEV model will operate in fuel mode.

TABLE 12.1

If-Then Rule Base for DEM in HEV Proposed Model

	Battery state of charge				
Fuel	VL	L	M	H	VH
VL	NL	NH	PM	PM	PM
L	NL	NM	PM	PM	PM
M	NM	NM	PM	PM	PM
H	NH	NH	PL	PM	PH
VH	NH	NH	PL	PH	PH

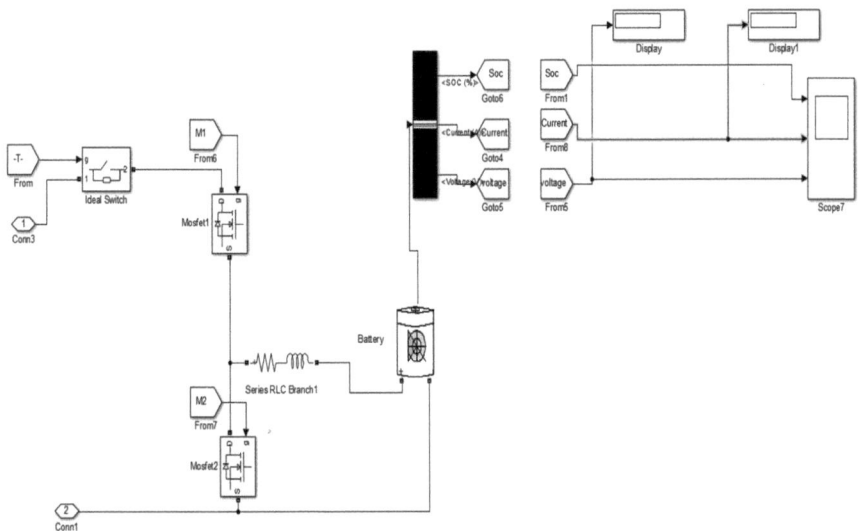

FIGURE 12.7 Battery discharge unit in MATLAB.

12.4.3 Proposed DEM Model with FLC and Battery Discharging Unit

Finally, the HEV model for DEM is developed and shown in Figure 12.8, which comprises of four blocks. The block III in the given Figure 12.8, fuel capacity and battery SOC as two inputs are entering to the FLC. Then the output goes to the display block. The output from FLC is taken through "Go to" block and it is given as input in "Switch" and "Switch1" through "From2" and "From3," both named Fuzzy, respectively. Now, the sections (II) and (IV) of the Figure 12.8 contain two switch blocks to check whether the output value is greater than 50 or not. In section II of Figure 12.8, reference current Ref_current via "Goto1 block" is implemented to set the amount of desired battery discharge current adding a constant block "Constant" and helps to measure difference current by using a Sum block. The difference goes through "PID Controller1" and "PWM Generator (DC-DC)" in a subsequent manner as shown in section (I) in Figure 12.8. Here the output signal from the "PWM generator" is bifurcated as Low state and High state using a "NOT gate" to trigger "MOSFET1" and "MOSFET2" because both can be turned on simultaneously.

12.4.4 Accuracy of the FLC

The accuracy of the FLC is determined by obtaining percentage error between the output from FLC (O_{FLC}) and output theoretically (O_{Theory}) and subtracted from 100 as given below in Equations (12.17–12.19)

$$Percentage\ error = \frac{\left|\left(O_{FLC} - O_{Theory}\right)\right|}{\left|O_{Theory}\right|} \times 100 \qquad (12.17)$$

$$Average\ percentage\ of\ error = \sum_{i=1}^{n} \frac{E_i}{n} \qquad (12.18)$$

Where i represents index of summation, n represents the upper limit of summation, and E is denoting the error of individual cases.

Accuracy (%) = 100 − (average percentage of error) (12.19)

Now, it is essential to consider result analysis for determining DEM of the proposed HEV model.

12.5 SIMULATION RESULT

After model formulation in MATLAB-2016a platform, different case studies considering 10 combinations of fuel level and battery SOC inputs are observed with FLC-based outputs as shown in Table 12.2. Further, to validate results theoretical calculations are also determined as shown in Table 12.2. Moreover, accuracy of the proposed FLC is also checked to justify the effectiveness of the proposed work.

From Table 12.2, it can be observed that one case such as battery SOC is 80 and fuel capacity is 40 which provides output from FLC 83.3, i.e., $\mu_{PM}(z)$ =PM directing to battery mode. This is due to the inputs such as the battery SOC value 80, lying

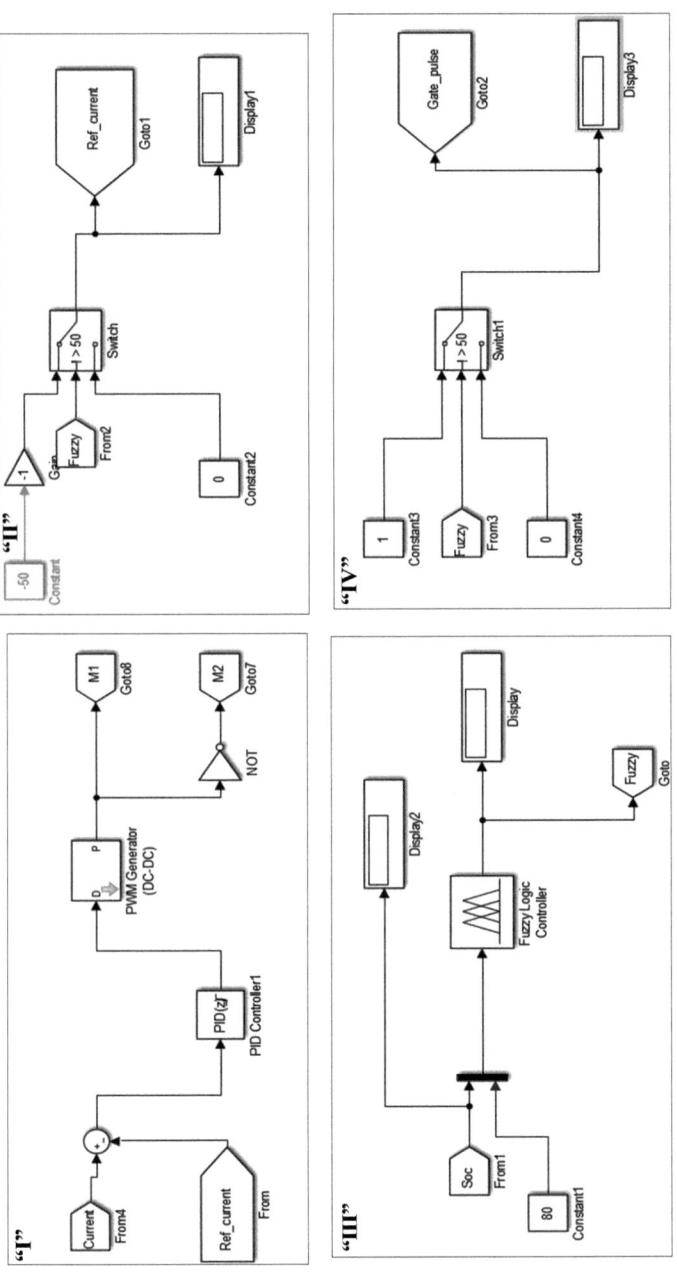

FIGURE 12.8 Proposed model with FLC and battery discharging unit.

between the range of 75 and 100, and overlaps between two membership functions of H (High) and VH (Very High). Now, considering Equations 12.4 and 12.5, results are obtained 4/5 for the membership function of High (H) and 1/5 for the membership function of Very High (VH). Similarly for the fuel capacity 40, which is lying between the ranges of 25 and 50, overlaps between two membership functions of L (Low) and M (Medium). By solving Equations12.7 and 12.8, the results are obtained 2/5 for the membership function of Low (L) and 3/5 for the membership function of Medium (M). As a result, four minimum values are fetched out comparing two possible combinations from the given two inputs. These values are 2/5, 3/5, 1/5, and 1/5, respectively. From those four values the maximum value 3/5 is chosen. Considering two input values, it is clear from the rule base (Table 12.1) that desired output should come in the range of Positive Medium (PM).

So the ultimate output (z) is obtained to put the maximum value after defuzzification as PM membership function ($\mu_{PM}(z)$), which is also the average of two values as shown in Equation 12.15. This way, total 10 numbers of cases are solved in detail, which is also validated by solving mathematical equations.

As for example, if the battery SOC is very high (VH) and fuel capacity is also very high (VH), the expected output will be in the range of positive high (PH), i.e., toward battery mode. To explain in detail, if battery SOC is 100 and fuel capacity is 85, the output is obtained 94.1, i.e., toward battery as output, and theoretically the output value is obtained 93.352, which is explained below. Moreover, if the battery SOC is 80(VH) and 40 (L) is the fuel capacity, the output is obtained as 83.3 toward the battery mode (PM).

TABLE 12.2
FLC-Based DEM Analysis for the Proposed HEV

Sl. no.	Battery SOC	Fuel capacity	Output from FLC (O_{FLC})	Output as theoretical value (O_{Theory})
1	20	70	29.3 (toward fuel)	33.34
2	35	65	48.5 (towards fuel)	43.342
3	55	30	83.3 (toward battery)	83.35
4	25	75	33.3 (toward fuel)	33.34
5	74	15	83.4 (toward battery)	83.35
6	90	10	83.3 (toward battery)	83.35
7	85	90	86.6 (toward battery)	93.352
8	80	40	83.3 (toward battery)	83.35
9	100	85	94.1 (toward battery)	93.352
10	55	85	71.8 (toward battery)	66.68
Accuracy of FLC (%)				95.991

- **Theoretical Value Calculation of the Output:**

A simple calculation is demonstrated here where assuming battery SOC is 80 and fuel capacity is 40. Taking these two values as inputs, two states are calculated as:

- **Battery SOC:**

$$\mu_H(X) = \frac{100-x}{25}\{75 \le x \le 100$$

$$= \frac{100-80}{25} = \frac{20}{25} = \frac{4}{5}$$

$$\mu_{VH}(x) = \frac{x-75}{25}\{75 \le x \le 100$$

$$= \frac{80-75}{25} = \frac{5}{25} = \frac{1}{5}$$

- **Fuel:**

$$\mu_L(y) = \frac{50-y}{25}\{25 \le y \le 50$$

$$= \frac{50-40}{25} = \frac{10}{25} = \frac{2}{5}$$

$$\mu_M(y) = \frac{y-25}{25}\{25 \le y \le 50$$

$$= \frac{40-25}{25} = \frac{15}{25} = \frac{3}{5}$$

And operator is considered here and min operator is used in rule base to evaluate the strength of each rule. The above four combinations cum strengths as denoted by S_1, S_2, S_3, and S_4 are determined below:

$$S_1 = Min\left(\frac{4}{5}, \frac{2}{5}\right) = \frac{2}{5}$$

$$S_2 = Min\left(\frac{4}{5}, \frac{3}{5}\right) = \frac{3}{5}$$

$$S_3 = Min\left(\frac{1}{5}, \frac{2}{5}\right) = \frac{1}{5}$$

$$S_4 = Min\left(\frac{1}{5}, \frac{3}{5}\right) = \frac{1}{5}$$

- **Defuzzification:**

For this, mean of max needs to be calculated. Hence:

$$\text{Max strength} = \text{Max}\ (\frac{2}{5}, \frac{3}{5}, \frac{1}{5}, \frac{1}{5}) = \frac{3}{5}$$

According to rule base, it is in PM range.

$$\mu_{PM}(z) = \frac{z - 66.68}{16.67}$$

Now,

$$\frac{3}{5} = \frac{z_1 - 66.68}{16.67} \qquad z_1 = 76.682$$

Or,

$$\frac{3}{5} = \frac{100.02 - z_2}{16.67}$$

$$z_2 = 90.018$$

$$z = \frac{z_1 + z_2}{2} = 83.35 \ \text{(theoretical value)}$$

Considering the overall simulation results, two surface views are noted and shown in Figures 12.9.

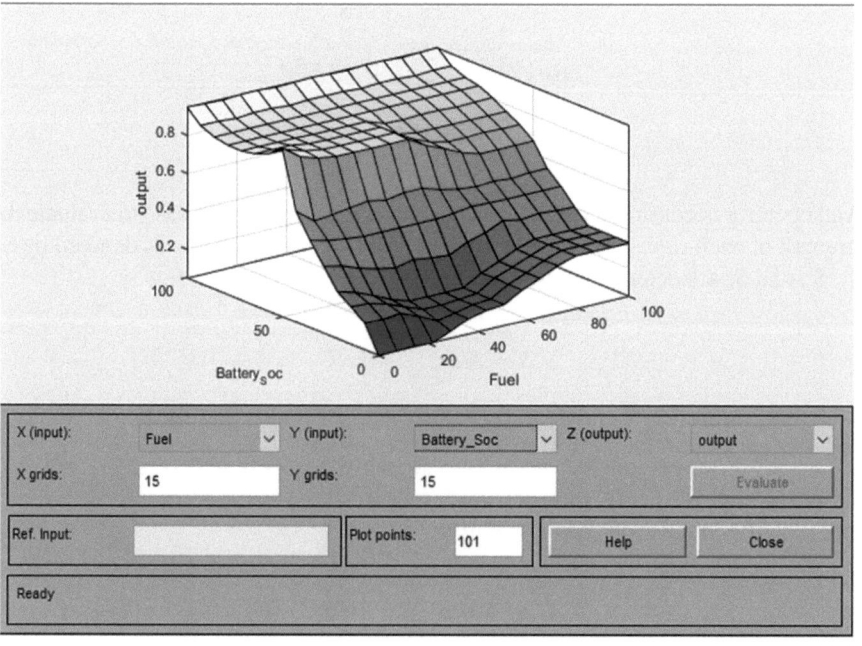

FIGURE 12.9 Surface view 1.

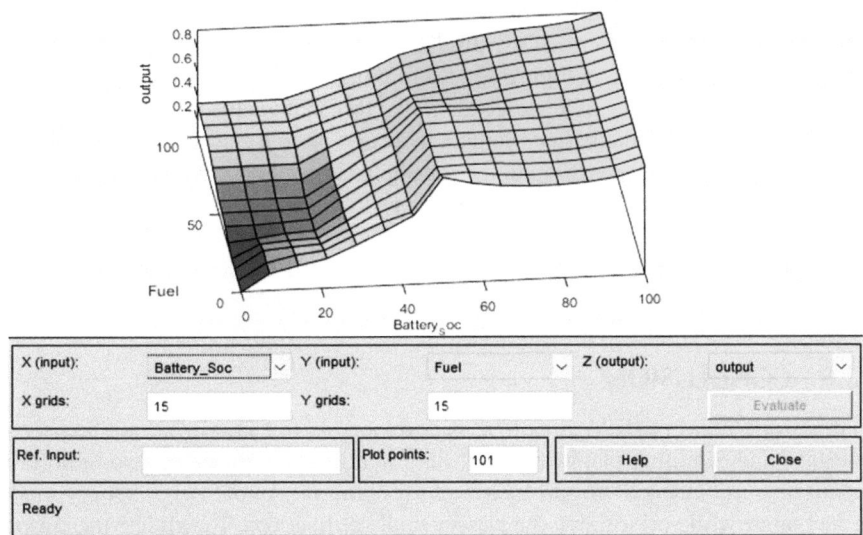

FIGURE 12.10 Surface view 2.

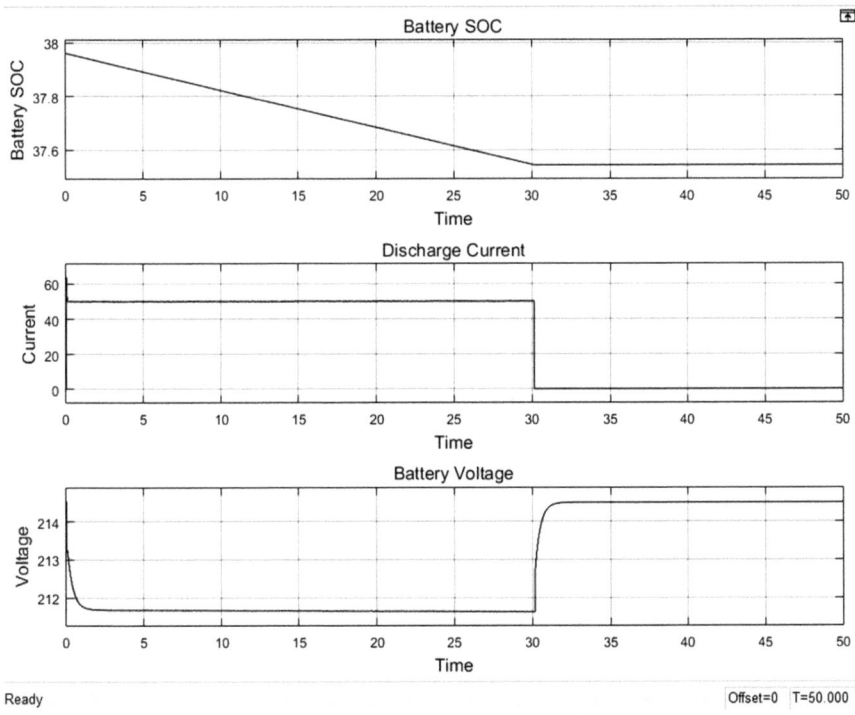

FIGURE 12.11 Graphical representation of battery discharging model.

Further, to improve the performance of HEV, a battery discharging model is proposed here as explained in Figures 12.7–12.8. By simulating the proposed model, three graphs such as battery SOC, discharge current, and battery voltage (up to down) are noted as shown in Figure 12.11. From Figure 12.11, it can be observed that when the value of battery SOC descends to the threshold value of 37, the discharge current becomes zero and battery voltage increases up to a certain value because of open-circuit path. After reaching to the threshold value of battery SOC, vehicle will run in fuel mode onward till the fuel level turns to minimum level.

Finally, the accuracy of FLC is determined as given in Table 12.2, which is obtained as 95.991%. This indicates that the design of FLC for DEM in HEV provides satisfactory output.

12.6   CONCLUSION

In this work, DEM of HEV using FLC is developed in MATLAB-2016a fuzzy logic interface and Simulink platform. Here, FLC is utilized considering two important inputs such as battery SOC and fuel level for obtaining output either battery mode or fuel mode while prioritizing the battery mode utilizations. To validate the performances of the FLC, its output is compared with the theoretical values. Further, % accuracy of the proposed FLC is determined, which shows satisfactory results. Since objective of the work is to lessen fuel consumption by utilizing the battery power for reducing pollution without compromising the performance of HEV, a battery discharge model is also included to switch it to fuel mode when battery SOC is reached to threshold value. The results in terms of graphical analysis showing the threshold value, below that value battery discharge current turned to zero and battery voltage increased due to open circuit. Hence, the proposed approach shows an effective path for DEM in HEV.

12.7   SOCIAL IMPORTANCE AND FUTURE PROSPECTS

India expects to cross 63 lakh unit marks per annum by 2027, according to a report by Indian Energy Storage Alliance (IESA) published on December 22, 2020, in *The Economic Times*. According to IESA report, the EV market is expected to grow at CAGR of 44% between 2020 and 2027 and hit 6.34 million unit annual sales by 2027. Similarly, the annual battery demand is expected to grow at 32% to hit 50GWh by 2027, of this 40 plus GWh will be on lithium-ion batteries.

ACKNOWLEDGMENT

A strong research environment with peaceful ambience and encouragement is the prime factors of a successful venture irrespective of the working field. The entire credit goes to the excellent and efficient personnel of our esteemed organization. The department of electrical engineering of Techno International New Town provides us such environment with all technical supports to conduct this research activity. So, the authors of the chapter would acknowledge it.

REFERENCES

Biswas (Raha), S.&Biswas, D. (2020). Fuzzy Controlled Demand Response Energy Management for Economic Microgrid Planning. IEEE International Conference on Power Electronics, Smart Grid and Renewable Energy, doi: 10.1109/ PESGRE 45664.2020.9070451.

Chan, C. C. (2007). The State of the Art of Electric, Hybrid, and Fuel Cell Vehicles. *Proceedings of the IEEE*, 95(4), 704–718.

Chi, H. R., Chui, K. T., Tsang, K. F., et al. (2013). Energy Management of Hybrid Vehicles using Artificial Intelligence. IEEE 2nd Global Conference on Consumer Electronics (GCCE), 1765–1767.

Krishnaswamy, K. (2012). *Process Control*, Second Edition. New Age International Publishers.

Li, X., Li, J., Xu, L., et al. (2009). *Power Management and Economic Estimation of Fuel Cell Hybrid Vehicle Using Fuzzy Logic*. IEEE, 1749–1754.

Rajasekaran, S.&Vijayalakshmi Pai, G. A. (2011). *Neural Networks, Fuzzy Logic, and Genetic Algorithms*, 15th ed., PHI Learning.

Rezaei, A., Burl, J. B., Zhou, B., et al. (2018). Catch Energy Saving Opportunity in Charge-Depletion Mode, a Real-Time Controller for Plug-In Hybrid Electric Vehicles. *IEEE Transactions on Vehicular Technology*, 67(11), 11234–11237.

Saib, S., Hamouda, Z.&Marouani, K. (2017). Energy Management in a Fuel Cell Hybrid Electric Vehicle using a Fuzzy Logic Approach. The 5th International Conference on Electrical Engineering – Boumerdes (ICEE-B).

Sarvestani, A. S.&Safavi, A. A. (2009). A Novel Optimal Energy Management Strategy Based on Fuzzy Logic for a Hybrid Electric Vehicle. ICVES, 141–144.

Zhang, X., Liu, Y., Zhang, J., et al. (2017). A Fuzzy Neural Network Energy Management Strategyfor Parallel Hybrid Electric Vehicle. The 9th International Conference on Modelling, Identification and Control, 342–347.

13 Predictive Maintenance and Anomaly Detection in Smart Manufacturing

Trisiladevi C. Nagavi, Sanchith Hegde,
V. Prajith, M. S. Sudeep and B. Rajesh

CONTENTS

13.1 INTRODUCTION

The rapid growth in the Internet of Things (IoT) is enabling the development of various kinds of applications. One such promising application is that of smart manufacturing. At its core, smart manufacturing is the utilization of IoT-enabled devices and automation in manufacturing processes to ensure that productivity remains optimal. A highly digitized shop floor could continuously collect and share data through

DOI: 10.1201/9781003326830-13

connected machines, devices, and production systems. Processing this data can provide insights to the manufacturing process and possibly help with the effective utilization of available capital.

Maintenance costs contribute significantly to the total operating costs of manufacturing or production facilities. Depending on the specific industry, ineffective maintenance or failure of even a single machine could cause immense losses to the manufacturing or production plant. Predictive maintenance allows manufacturers to significantly reduce maintenance costs, thus minimizing downtimes caused by sudden breakdowns and increasing the maintenance activities that can be scheduled in advance, based on actual machine health. It also allows manufacturers to replace machine components just before their failure, thus making the most out of their useful lifetime.

We suggest the use of anomaly detection to aid the problem of predictive maintenance in manufacturing. Different machine learning models are used, one to address the anomaly detection problem and the other to address the predictive maintenance problem. Specifically, autoencoders and isolation forests are utilized for solving the anomaly detection problem, and a model consisting of long short-term memory (LSTM) units in the primary layers of the neural network is used to solve the predictive maintenance problem. These models are then used to send out early warnings requesting maintenance via a mobile application whenever suspicious behavior is detected during the operation of the machines in a manufacturing plant. This helps greatly reduce or even eliminate unnecessary repairs, prevent catastrophic equipment failures, and reduce the negative impact of maintenance operations on the profitability of manufacturing plants.

The rest of the chapter is structured as follows. Section 13.2 presents the proposed framework. Further, Section 13.3 describes the dataset considered and elaborates on the implementation of the system. The different scenarios that were evaluated, along with the numerical results obtained, are discussed in Section 13.4. Section 13.5 summarizes the chapter and concludes it by discussing future works on this topic.

13.2   PROPOSED FRAMEWORK FOR PREDICTIVE MAINTENANCE AND ANOMALY DETECTION

13.2.1   Overview

The proposed system consists of the following components: (1) a machine learning model to detect when a sensor records anomalous values, which can be extrapolated to multiple sensors attached to a machine to decide upon its anomalous behavior; (2) an Android app to notify concerned users of the anomalies exhibited by a specific sensor, and machine where applicable, and to alert users about critically low predicted remaining useful life for a machine; (3) a machine learning model to predict the machine's remaining useful lifetime (or remaining time to maintenance), when a machine is confirmed to have displayed anomalous behavior; (4) a dashboard to display the current status of machines and all their sensors in near realtime; and (5) a HTTP API server to wrap around the aforementioned machine learning models and aid maintenance and management tasks such as registering new machines, sensors, and to verify proper functionality of the system.

The entire pipeline can be summarized using the framework depicted in Figure 13.1. Once the system is deployed, during the prediction phase, the latest sensor readings for a specific machine are read from a timeseries database, preprocessed and transformed into windows of fixed time intervals before being passed on to the anomaly detection model. If the anomaly detection model classifies the machine as anomalous, users are notified by means of a mobile application, and the preprocessed readings are passed on to the predictive maintenance model to predict the remaining useful lifetime of the machine. If the predicted remaining useful lifetime is critically low, users are again notified by means of the mobile application.

13.2.2 DATA COLLECTION

The data collection phase involves collecting metrics during the operation of the machines. The metrics collected (which in turn determine the sensors to be used) could include temperature, vibration, velocity, acceleration, voltage, and current, among others, although the exact kind of sensors used to collect these metrics would vary with the machine being monitored and its operational environment. The data gathered by the sensors results in each data point indexed by a timestamp, resulting in time series data. This time series data is stored in a time series database, from which the data is accessed during the training phase.

13.2.3 PREPROCESSING

The preprocessing of the time series data typically includes handling null or missing values and the like. Since the machine learning models employed in the system require continuous sequences of time to be provided as input, missing values cannot be just dropped. Instead, they must be filled with a certain value. The value used for filling missing data could just be a global static constant, could be the mean (or median) of the complete dataset, mean (or median) of a single series (readings for a single sensor in this case), or could be an "intermediate value" determined dynamically using values surrounding the data point with the missing value. Using neighboring values to dynamically estimate the missing value seems like a reasonable

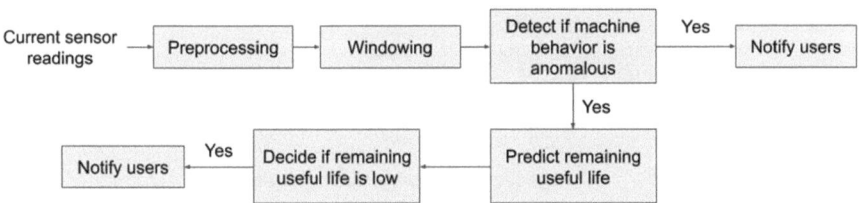

FIGURE 13.1   The framework of the proposed system which depicts the tasks involved in the system.

choice for this application. If the number of continuous missing values is greater than a certain user-configured threshold, then such sequences will have to be ignored.

13.2.4 WINDOWING

Once the data has been preprocessed, it has to be converted into windows of fixed time intervals (say 10 minutes, 2 hours, etc.) before being passed on to the model, either for training or for anomaly detection or prediction. The window size chosen must be the same for both training and anomaly detection or prediction, for ensuring proper functioning of the model.

13.2.5 ANOMALY DETECTION

The process employed for detecting anomalous machine behavior can be summarized using Figure 13.2. The data collection, preprocessing, and windowing steps are described in Sections 13.2.2, 13.2.3, and 13.2.4, respectively.

The training phase proceeds as follows. Since the data collected is time series data, a machine learning model that predicts a future data point by observing previous data points and their trends, such as an autoencoder, is employed for detecting anomalies, by training it on normal (nonanomalous) data for each metric. The training loss is calculated by using L1 loss, also known as the absolute error loss. Absolute error loss, as the name suggests, is the absolute difference between the predicted value and the actual value and is calculated for each value in a dataset. The aggregation of all the loss values, called the loss function, is the mean absolute error (MAE) in the case of absolute error loss. The MAE is calculated using Equation

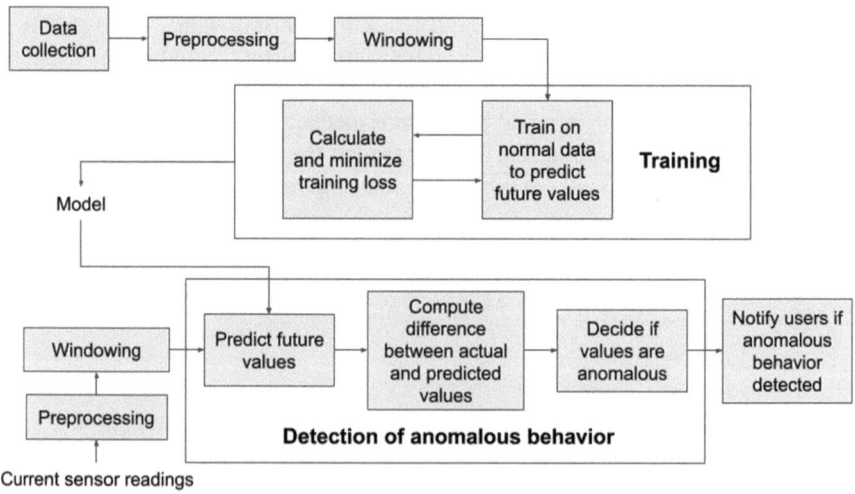

FIGURE 13.2 Flow for the detection of anomalous machine behavior.

13.1. The result of the training phase is a model that has learnt to predict future data points given a set of past data points.

$$MAE = \frac{1}{n} \sum_{i=1}^{n} \left| y_{actual} - y_{predicted} \right| \tag{13.1}$$

The detection phase proceeds as follows. Since the machine learning model is trained on nonanomalous data, it tries to predict future data points that are considered "normal," even if it encounters previous data points that are anomalous. If the difference between the predicted and actual values observed is greater than a certain threshold (which could either be a user configured one or determined dynamically based on a window) for a metric, the machine is said to display anomalous behavior with respect to the said metric. This can be extrapolated to multiple metrics to decide when a machine is displaying anomalous behavior on a whole and to obtain the machine's remaining useful lifetime.

13.2.6   MOBILE DEVICE NOTIFICATIONS

Once the machine learning model detects anomalous machine behavior, the concerned authorities are notified so that they can either attend to it or ignore the alert if the anomalous behavior was observed by the model because of an intentional change in the machine's environment. The notification can be sent via a variety of modes like emails, text messages, via an instant messaging service, or even via a mobile application. To deliver notifications via a mobile application, push notification providers can be utilized to send out notifications to users on various mobile platforms.

The process involved in sending push notifications to mobile devices with the aid of a push notification provider service after anomalous machine behavior is detected can be summarized using Figure 13.3. Once an anomaly is detected, the

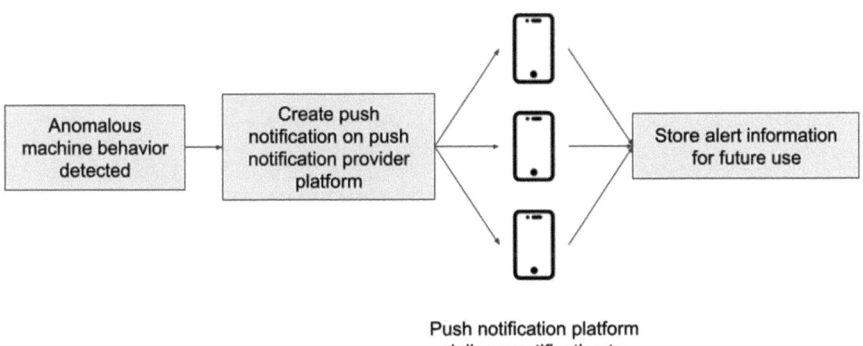

FIGURE 13.3   Flow for mobile device notifications once anomalous machine behavior is confirmed.

backend registers a push notification on the push notification provider platform, associating the notification with the machine that was found to have behaved anomalously. The platform in turn tries to send push notifications to all mobile devices that are subscribed to receive notifications about that machine. The push notification delivery to all subscribed mobile devices must be almost instant, provided they are connected to the Internet. The mobile devices may also opt to store information about notifications as and when they receive them, to keep track of all the previous notifications received. If required, analysis of any kind may also be performed using the stored information about the previous alerts received. The process involved in notifying mobile devices when the predicted remaining time to maintenance for a machine is critically low is quite similar to the process just described.

13.2.7   PREDICTION OF REMAINING USEFUL LIFETIME

The process employed for predicting remaining useful lifetime can be summarized using Figure 13.4. The data collection, preprocessing, and windowing steps are described in Sections 13.2.2, 13.2.3, and 13.2.4, respectively.

   During the training phase, a different machine learning model such as a LSTM model is trained on anomalous data from the dataset, to predict the time left before a machine supposedly fails and is to be scheduled for maintenance. An advantage of using an LSTM model is that it is able to capture seasonal patterns. This could include long-term seasonalities, such as yearly patterns, or short-term seasonalities, such as daily or weekly patterns. Another benefit that comes from employing an LSTM model is that it can triage the impact patterns caused by different categories

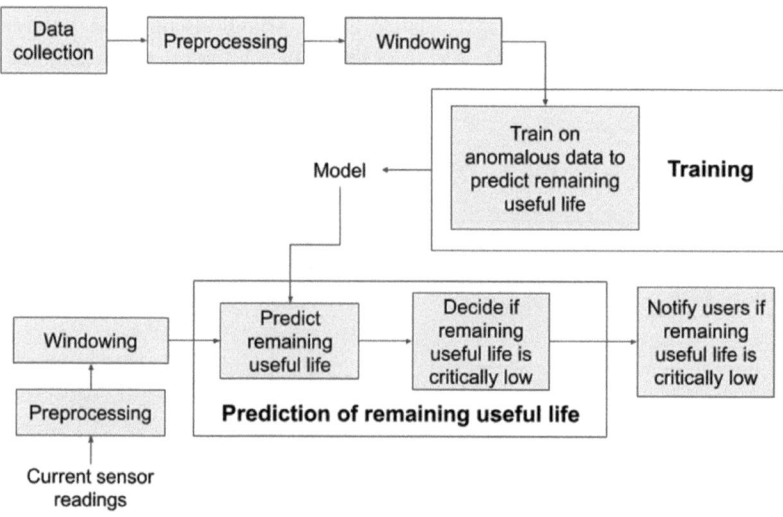

FIGURE 13.4   Flow for the prediction of the remaining useful lifetime of the machine.

of events. This is particularly true in case of events that impact demand for a particular commodity before, during, and after the occurrence of the event. The result of the training phase is a model that has learnt to predict the remaining useful lifetime of a machine given a set of past data points.

During the prediction phase, when it is confirmed that a machine is displaying anomalous behavior, the current sensor data is passed to the trained model to predict the remaining useful life of the machine. If the predicted remaining useful life is critically low, users are notified about it by means of the mobile application.

13.2.8 VISUALIZATION OF SENSOR READINGS AND MACHINE STATUS

To provide an overview of the machines and their sensor readings, the health or status of each machine is displayed based on their sensor readings on a dashboard. In addition to displaying the status of each machine, the readings and the variation of each metric over time are displayed in near real time, to provide detailed insights about each machine. In case a machine displays anomalous behavior, the remaining useful life of that machine is also displayed. A web-based dashboard is used to render all this information, utilizing one among the multiple observability or monitoring platforms available today.

13.3 PREDICTIVE MAINTENANCE AND ANOMALY DETECTION SYSTEM IMPLEMENTATION

Our system has been tested on a dataset collected from a real industrial machine. For the dataset considered in the study, an autoencoder and isolation forests are used for addressing the anomaly detection problem, and a LSTM model is used for addressing the predictive maintenance problem.

13.3.1 DATASET DESCRIPTION

The dataset used in the study contains sensor readings for a water pump situated in a town (Phantawee, 2019). The dataset contains readings from 52 sensors and an indication of the machine status, whether it was running normally, broken, or under repair (labeled as "recovering" in the dataset). Sensor readings are available for a duration of five months from the beginning of April 2018 to the end of August 2018, recorded at a frequency of one recording every minute, thus amounting to a total of around 220,000 readings. Unfortunately, the dataset doesn't contain any information about the kind of sensors that were used to record the readings, which prevents more specific analysis from being performed based on the sensor type. However, the proposed system was quite successful in detecting anomalies and predicting imminent machine failures, even when the same analysis procedure was carried out on each sensor, as can be seen in Section 13.4. A few of the readings present in the dataset are listed in Table 13.1 (numerical values rounded off to four digits after the decimal point for the sake of brevity).

TABLE 13.1

Sample of the Dataset Containing Timestamps, Data for 52 Sensors, and an Indication of Machine Status

timestamp	sensor_00	sensor_01	...	sensor_50	sensor_51	machine_status
2018-04-01 00:00:00	2.4654	47.0920	...	243.0556	201.3889	NORMAL
2018-04-01 00:01:00	2.4654	47.0920	...	243.0556	201.3889	NORMAL
2018-04-01 00:02:00	2.4447	47.3524	...	241.3194	203.7037	NORMAL
2018-04-01 00:03:00	2.4605	47.0920	...	240.4514	203.125	NORMAL
2018-04-01 00:04:00	2.4457	47.1354	...	242.1875	201.3889	NORMAL

13.3.2 PREPROCESSING

The preprocessing performed before both anomaly detection and prediction of remaining useful lifetime involves three major operations: normalization, transforming data into windows of fixed time intervals, and downsampling.

In a significant number of practical situations, machines in manufacturing facilities are subject to load balancing, to distribute the load among the running machines, whenever one of them has to be taken down for maintenance. Load balancing causes the range of values, which is otherwise considered "normal" for a metric, to vary significantly based on whether it is performed or not and the number of machines among which the load is distributed. To overcome this problem, normalization is performed on the dataset. Normalization is performed on each metric by dividing the current data point by the mean of previous values recorded over a fixed duration of time, for the metric in question. Normalization hasn't been performed on this dataset, since there were no visual indications of load balancing being performed for the time window captured in the dataset, and since the data with and without normalization provided similar results in both cases.

The next step involved in preprocessing is transforming the data into windows of fixed time intervals, to obtain it in a suitable format for the machine learning models to analyze. For the dataset used in the study, the previous 120 data points are considered, which amounts to the previous two hours' data, since data points were recorded at a frequency of one reading per minute.

The final preprocessing step is downsampling the data points in each window of data. Downsampling is the process of decreasing the time frequency of data, with the primary goal of shrinking the data while ensuring that the sample obtained is a fairly decent representative of the complete data. For downsampling, an algorithm called Largest Triangle Three Buckets (LTTB) algorithm proposed by Steinarsson (Steinarsson, 2013) is utilized. The algorithm tries to retain visual similarity between the downsampled data and the original dataset while also retaining the characteristics of anomalies in the downsampled data. In short, the LTTB algorithm divides all data points into buckets of approximately equal size and selects points from each bucket such that the triangle formed using points from three continuous buckets has the largest possible area. Downsampling is used when the volume of data points to consider is quite high but hasn't been used for the 2-hour time window chosen. Downsampling helps reduce the input size for the machine learning models, making it simpler, smaller, faster, and easier to understand and debug.

The preprocessing involved before the prediction of the remaining useful life is quite similar to the process mentioned before, except that instead of preprocessing all of the data, only the data before identified anomalies is considered during training and prediction.

13.3.3 Anomaly Detection

For the purpose of anomaly detection, the system utilizes autoencoders. Autoencoders try to replicate the input data in the output layer as much as possible, after compressing the data in the middle layers. A general structure of an autoencoder consists of encoding and decoding networks and an information bottleneck, which is a representation of the input learned through unsupervised training.

During training, the autoencoder used for anomaly detection is trained only on data for each metric, which is considered normal and doesn't contain any anomalies. Since the autoencoder is trained only on data without anomalies, it knows how to and tries to replicate only normal data, even when it is provided data with any anomalies. The configuration of the autoencode trained on each individual sensor is provided in Table 13.2.

Furthermore, another unsupervised machine learning model such as isolation forests is trained on the difference values output by the autoencoder during prediction, to identify anomalies without any human intervention.

During the anomaly detection step, for each metric, the autoencoder first predicts an expected value for the data by observing a window of the past data. The difference between the predicted value and the actual value that was recorded is then passed to the unsupervised model (isolation forests) to identify anomalies in the predicted values against the actual recorded values and possibly also provide a severity of the anomaly thus detected. The underlying principle is that the difference between the predicted value and the actual recorded value will be much higher for anomalous data points compared to normal data points. This logic is extrapolated to all the metrics associated with a machine, and the machine is labeled as displaying anomalous behavior if the fraction of metrics displaying anomalous behavior is greater than a set number.

13.3.4 Mobile Device Notifications

Notifying users of anomalous machine behavior and low predicted remaining useful life is facilitated with the aid of OneSignal customer messaging platform and an Android application to receive the notifications posted by the backend of the system. The OneSignal platform offers mobile and web push notifications, in-app messaging, SMS, and email options to engage with customers. The advantage that OneSignal provides is that clients can send messages to their customers using any device or operating environment, with a simple and easy-to-use user interface while being unaware of all the intricacies involved in the interaction with each targeted platform. On Android devices, OneSignal makes use of Firebase Cloud Messaging, which is natively available on devices with Google Play Services installed, thus minimizing the setup required to receive notifications on Android devices.

The architectural composition of the Android application developed to deliver notifications to users can be summarized by Figure 13.5. This mostly follows the architecture suggested in the official guide to Android app architecture page on the Android developer documentation (Google, n.d.).

The Android app consists of two architectural layers: (1) a user interface (UI) layer to display application data on the screen; and (2) a data layer that contains the business logic of the application and exposes application data. Whenever the application data present in the data layer changes, either due to user interaction or due to external input, the UI layer or the presentation layer is updated to reflect the changes. The UI layer consists of two components: (1) the user interface elements that render the data on the screen; and (2) state holders that hold data, expose it to the user interface, and handle logic. The business logic present in the data layer adds

TABLE 13.2

Configuration of the Autoencoder Model Used for Detecting Anomalies for Each Sensor

Layer (type)	Output shape	Param #
Linear-1	[-1, 1, 80]	9,680
ReLU-2	[-1, 1, 80]	0
Linear-3	[-1, 1, 50]	4,050
ReLU-4	[-1, 1, 50]	0
Linear-5	[-1, 1, 30]	1,530
ReLU-6	[-1, 1, 30]	0
Linear-7	[-1, 1, 15]	465
ReLU-8	[-1, 1, 15]	0
Linear-9	[-1, 1, 10]	160
Linear-10	[-1, 1, 15]	165
ReLU-11	[-1, 1, 15]	0
Linear-12	[-1, 1, 30]	480
ReLU-13	[-1, 1, 30]	0
Linear-14	[-1, 1, 50]	1,550
ReLU-15	[-1, 1, 50]	0
Linear-16	[-1, 1, 80]	4,080
ReLU-17	[-1, 1, 80]	0
Linear-18	[-1, 1, 120]	9,720
Total params		31,880
Trainable params		31,880
Nontrainable params		0

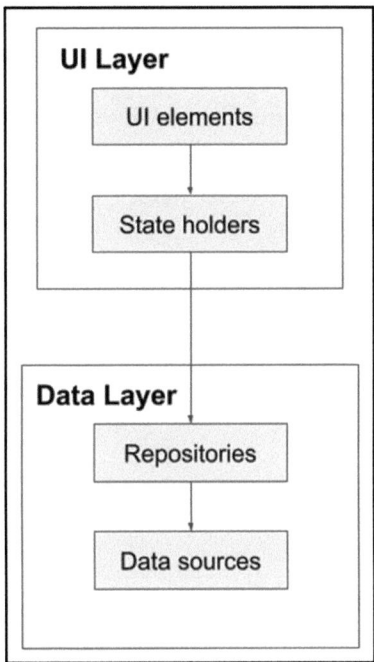

FIGURE 13.5   Architectural composition of the Android application.

value to the application, in that it contains rules that determine how the application creates, stores, and changes the data. The data layer consists of repositories, each of which contains zero or more data sources. Typically, each repository handles different types of data.

Within the application, two independent data layers are present: (1) one to handle the response received from the backend of the system; and (2) the other to retrieve the information about previous notifications received stored locally in a database on the Android device. On the UI side, the app consists of two independent UI layers: (1) one to display the previous notifications received on the device; and (2) the other to allow the user select his preferences on various things related to the system via a preferences screen. In addition to the data and UI layers, the app also consists of a service that periodically queries the Firebase server to learn of any new notifications targeting the application's users that might have been pushed. If there are any new notifications with matching criteria, the notification information is also stored locally in a database before alerting the user.

13.3.5   Prediction of Remaining Useful Lifetime

For the purpose of calculating the remaining useful life of a specific machine, the system utilizes a neural network with LSTM units forming the primary layers of the network. LSTMs are a special kind of recurrent neural networks (RNNs), which are

TABLE 13.3

**Configuration of the LSTM Model Used for
Predicting Remaining Useful Life of a Machine**

Layer (type)	Output shape	Param #
lstm (LSTM)	(None, 1, 128)	557,568
leaky_re_lu (LeakyReLU)	(None, 1, 128)	0
lstm_1 (LSTM)	(None, 1, 128)	131,584
leaky_re_lu_1 (LeakyReLU)	(None, 1, 128)	0
dropout (Dropout)	(None, 1, 128)	0
lstm_2 (LSTM)	(None, 64)	49,408
leaky_re_lu_2 (LeakyReLU)	(None, 64)	0
dropout_1 (Dropout)	(None, 64)	0
dense (Dense)	(None, 3)	195
Total params		738,755
Trainable params		738,755
Nontrainable params		0

capable of learning long-term dependencies; they are able to remember information for long periods of time. The remaining useful life or the remaining time to failure for a machine is simply computed as the duration between the predicted time of failure for the machine and the current time.

During training, the LSTM model is trained on a dataset containing fixed time windows of data from all sensors prior to each anomaly, to predict the absolute time of failure for the machine and in turn the remaining time to failure. In other words, the target variable for the LSTM model is the remaining time to failure. The configuration of the LSTM model trained is provided in Table 13.3.

Once the anomaly detection model identifies an anomaly in the machine behavior, a window of the data preceding the identified anomaly is passed on to the LSTM model to make a prediction about the remaining time to failure. If the predicted remaining time to failure is below a user-configured threshold (say 24 hours, 3 days, etc.), the remaining time to failure is labeled as being critically low, and the user is once again alerted by means of the Android application.

13.4   EXPERIMENTATION AND RESULTS

The trained anomaly detection models classified roughly 20% of the total data for each sensor in the dataset to be anomalous. Since the dataset doesn't contain any labels about which readings are considered to be anomalous, there is no sure shot way of calculating the accuracy of the anomaly detection model. However, the fact that the model classified data points a few hours prior to each machine breakdown recorded in the dataset as anomalous is reassuring. The trained predictive maintenance model provided a categorical accuracy of 88.07%, which indicates the fraction of data points in the test dataset for which the model correctly categorized the remaining time to failure.

Three separate instances from the pump sensor dataset were utilized to validate the developed anomaly detection and predictive maintenance models: (1) from 6:00 AM to 6:00 PM on August 24, 2018, when the machine was completely normal; (2) from 6:00 PM on July 24 to 6:00 AM on July 25, 2018, when a few of the sensors on the machine displayed anomalous readings; and (3) from 12 noon on April 29 to 12 midnight on April 30, 2018, when the machine displayed anomalous behavior.

In the first case, the anomaly detection model correctly predicted that the machine was completely normal and identified no anomalies. In the second case, the anomaly detection model did identify a few anomalies but labeled the machine as displaying normal behavior. In the third case, the anomaly detection model correctly detected that the machine indeed displayed anomalous behavior and the predictive maintenance model predicted that the machine is expected to fail within the next 12 hours with 99.9% confidence. Additionally, the system also sent a notification to mobile devices indicating the anomalous behavior and low remaining useful life.

Screenshots of the dashboard for each of the cases are included as Figure 13.6, Figure 13.7, andFigure 13.8, respectively. The blue line plots denote the sensor readings and indicate the variation of readings for each sensor over a period of time. Based on the time window selected by the user on the dashboard, the plots displayed on the dashboard are also updated accordingly. The red line plots denote the anomaly scores as calculated by the anomaly detection model, which in turn indicate how far off the predictions of the model were from the recorded sensor readings. It should be noted that the frequency of anomaly score calculations is much lesser compared to that of the sensor readings in the dataset, for the reason that the computation

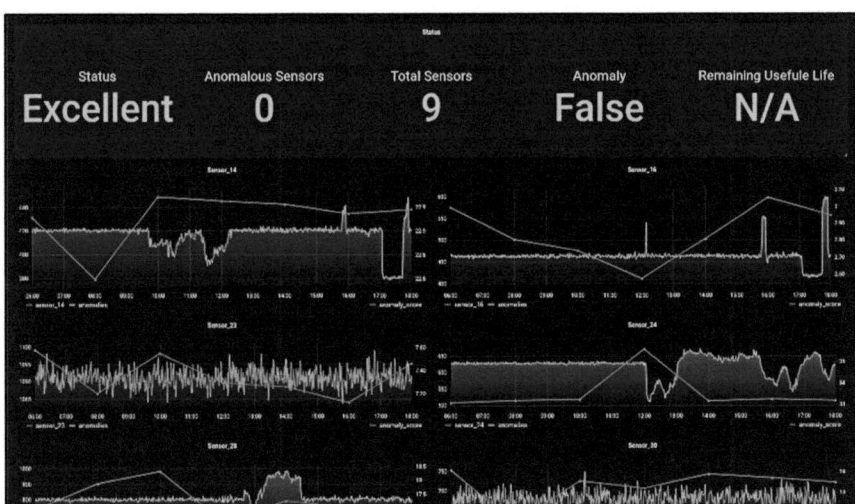

FIGURE 13.6 Screenshot of the dashboard when the machine displayed completely normal behavior.

involved becomes quite resource heavy when the number of sensors associated with a single machine increases, and that such a lower frequency (say once every 5 or 10 minutes compared to once per second) is feasible for most practical applications.

The data points displayed in orange color on the dashboard indicate the anomalous sensor readings, as identified by the anomaly detection model. Since the

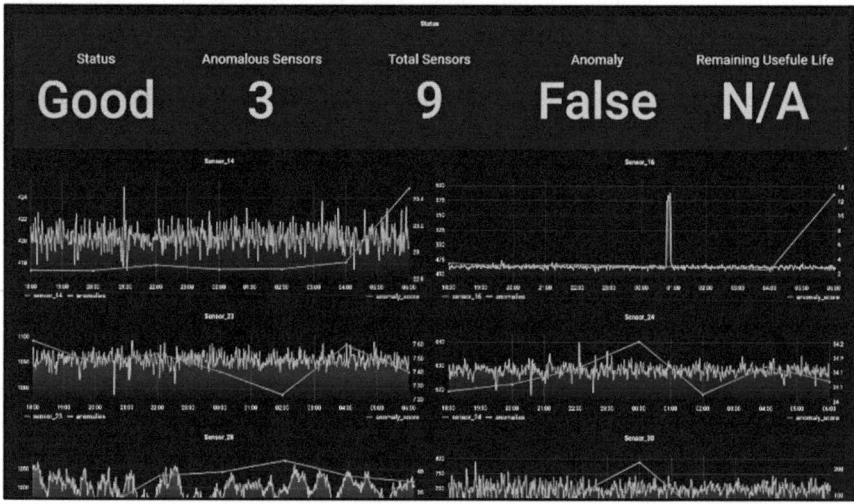

FIGURE 13.7   Screenshot of the dashboard when the machine displayed fairly normal behavior.

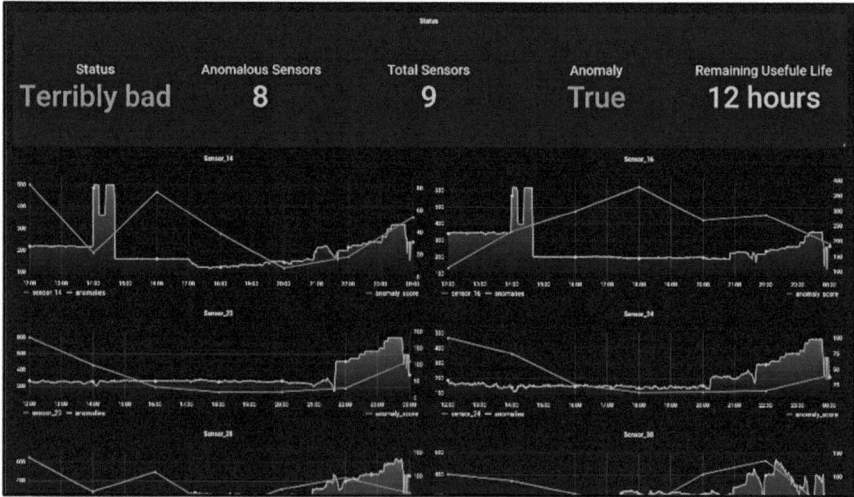

FIGURE 13.8   Screenshot of the dashboard when the machine displayed anomalous behavior.

information about which data points were classified as anomalies and the anomaly scores are stored in the database, the dashboard timeline can be zoomed out to understand the trends of anomalous behavior over any period of time. Additionally, the dashboard also contains a panel toward the top of the page displaying textual information about the machine including a machine status (mapped based on the fraction of anomalous sensors), the total number of anomalous sensors, a Boolean value indicating whether the machine is behaving anomalously or not as per the latest available sensor readings, and the remaining useful life, wherever applicable. For the purpose of this study, the machine will be labeled as behaving anomalously if more than half of its sensors are detected to be anomalous, in which case the remaining useful life is also calculated, as displayed in Figure 13.8. Further, the system was configured to notify users if the predicted remaining useful life is less than 24 hours, thus the system also sent notifications about low remaining useful life for the situation depicted in Figure 13.8.

Figures 13.9 and 13.10 include screenshots of the developed Android application. Figure 13.9 includes a screenshot of the settings screen of the application, listing out the machines that have been registered in the system. This list of machines is populated with the aid of an API endpoint implemented in the backend of the

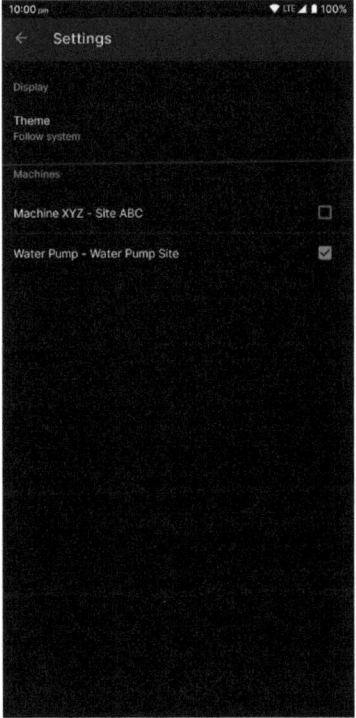

FIGURE 13.9 Settings screen of the mobile application, which provides the user an option to select the machines to receive alerts for.

FIGURE 13.10   Main screen of the mobile application, which displays all the previous alerts received by the user.

system. Checking the checkbox next to the entry for a machine subscribes the user to receive notifications related to that machine. The option to selectively receive push notifications for specific machines has been provided so that users are notified about only those machines that they are tasked with monitoring and maintaining and to prevent users being spammed with irrelevant notifications. Figure 13.10 includes a screenshot of the main screen of the application, which is displayed on launching the application. This screen lists all the previous notifications received by the user. This information is stored locally in a database on the Android device. The Android application can be extended to display analytics about the type and frequency of alerts received by the user over time, by making use of the stored information about the previously received notifications, if required.

The results presented in this section show that the developed predictive maintenance and anomaly detection system is able to adapt its behavior well to real-life scenarios.

13.5   CONCLUSION

With the ever increasing capabilities and complexity of machines used in the manufacturing plants, the maintenance effort spent on these machines also increases. To reduce the operational costs spent on the maintenance effort of these machines and to prevent equipment failures, a predictive maintenance and anomaly detection solution seems like a potential solution to tackle the problem. This idea was explored in this study by implementing a predictive maintenance and anomaly detection system with a mobile application for notifying users of suspicious behavior and a dashboard to display the current status of machines.

The developed system utilizes autoencoders and isolation forests to solve the problem of anomaly detection and a LSTM model to solve the problem of predictive maintenance. The anomaly detection and predictive maintenance models provide fairly accurate results for the test cases evaluated. The Android application aids in notifying the users of any inconsistencies in the machine behavior in almost real time. Finally, the dashboard helps users gain better insights about the performance of the machine, both at the present and for any time window in the past. Overall, we believe that the system will be quite useful in improving the maintenance situation in manufacturing facilities.

13.6   FUTURE ENHANCEMENTS

When a predictive maintenance and anomaly detection system for industrial machines is considered, the number of machines and in turn sensors that would have to be monitored will be huge in a medium- to large-scale manufacturing plant. This in turn greatly increases the amount of data flowing in toward the system, calling for more efficient and faster approaches for handling such amounts of data. In such a case, efficient software solutions need to be developed to process and analyze the incoming data in near real time, while keeping the resource usage fairly minimal, to ensure smooth functioning of the entire system.

ACKNOWLEDGMENT

The authors would like to extend sincere thanks to the management of Sri Jayachamarajendra College of Engineering, JSS Science and Technology University, Mysuru, Karnataka, India, for their whole-hearted support throughout the course of the project. Furthermore, the authors would like to thank the generous folks from Etag Software Solutions Private Limited for providing them access to data from one of their clients' locations to test the developed system against domain expertise on the subject matter and guidance at every step without which the study wouldn't have reached completion. Finally, the authors would like to thank their family and friends for their constant support throughout the duration of the study.

BIBLIOGRAPHY

Celik, M., Dadaser-Celik, F., &Dokuz, A. (2011). 'Anomaly Detection in Temperature Data Using DBSCAN Algorithm'. In Proceedings of the 2011 IEEE International Symposium on Innovations in Intelligent Systems and Applications (INISTA), Istanbul, pp. 91–95.
Chen, J., &Wang, D. (2016). 'Long Short-Term Memory for Speaker Generalization in Supervised Speech Separation'. In *The Journal of the Acoustical Society of America*, Vol. 141, No. 6, pp. 4705–4714.

Chen, T., &Guestrin, C. (2016). 'XGBoost: A Scalable Tree Boosting System'. In Proceedings of the 22nd ACM SIGKDD International Conference on Knowledge Discovery and Data Mining, San Francisco, pp. 785–794.

Decker, L., Leite, D., Giommi, L., &Bonacorsi, D. (2020). 'Real-Time Anomaly Detection in Data Centers for Log-Based Predictive Maintenance Using an Evolving Fuzzy-Rule-Based Approach'. In Proceedings of the 2020 IEEE International Conference on Fuzzy Systems (FUZZ-IEEE), pp. 1–8.

Ester, M., Kriegel, H.-P., Sander, J., &Xu, X. (1996). 'A Density-Based Algorithm for Discovering Clusters in Large Spatial Databases with Noise'. In Proceedings of the Second International Conference on Knowledge Discovery and Data Mining, Portland, pp. 226–231.

Fei, S.-w., &Zhang, X.-b. (2009). 'Fault Diagnosis of Power Transformer Based on Support Vector Machine With Genetic Algorithm'. In *Journal of Expert Systems with Applications*, Vol. 36, No. 8, pp. 11352–11357.

Google. (n.d.). 'Guide to App Architecture'. Retrieved from Android Developers Documentation: https://developer.android.com/topic/architecture

Hollenstein, L., Lichtensteiger, L., Stadelmann, T., Amirian, M., Budde, L., Meierhofer, J., ... &Friedli, T. (2019). 'Unsupervised Learning and Simulation for Complexity Management in Business Operations'. In *Applied Data Science: Lessons Learned for the Data-Driven Business*, pp. 313–331, Springer International Publishing.

Kabir, F., Foggo, B., &Yu, N. (2018). 'Data Driven Predictive Maintenance of Distribution Transformers'. In Proceedings of the 2018 China International Conference on Electricity Distribution (CICED), Tianjin, pp. 312–316.

Liu, F., Ting, K., &Zhou, Z.-H. (2008). 'Isolation Forest'. In Proceedings of the 2008 Eighth IEEE International Conference on Data Mining, Pisa, pp. 413–422.

Mobley, R. K. (2002). *An Introduction to Predictive Maintenance*, Butterworth-Heinemann.

Morselli, F., Bedogni, L., Mirani, U., Fantoni, M., &Galasso, S. (2021). 'Anomaly Detection and Classification in Predictive Maintenance Tasks with Zero Initial Training'. *Industrial IoT as IT and OT Convergence: Challenges and Opportunities*, Vol. 2, pp. 590–609.

Moschini, G., Houssou, R., Bovay, J., &Robert-Nicoud, S. (2021). 'Anomaly and Fraud Detection in Credit Card Transactions Using the ARIMA Model'. In Proceedings of the 7th International Conference on Time Series and Forecasting, Gran Canaria, 5.

Peres, R., Barata, J., Leitão, P., &García, G. (2019). 'Multistage Quality Control Using Machine Learning in the Automotive Industry'. In *IEEE Access*, Vol. 7, pp. 79908–79916.

Phantawee, N. (2019). 'Pump Sensor Dataset'. Retrieved from Kaggle: https://www.kaggle.com/datasets/nphantawee/pump-sensor-data

Plant Engineering. (2019, March). '2019 Maintenance Report'. Retrieved from Plant Engineering: https://www.plantengineering.com/wp-content/uploads/sites/4/2019/02/Plant-Engineering-2019-Maintenance-Report.pdf

Russo, S., Disch, A., Blumensaat, F., &Villez, K. (2020). 'Anomaly Detection using Deep Autoencoders for in-situ Wastewater Systems Monitoring Data'. *arXiv preprint arXiv:2002.03843.*,

Stack Overflow. (2022, June). 'Stack Overflow Developer Survey 2022'. Retrieved from Stack Overflow: https://survey.stackoverflow.co/2022/

Steinarsson, S. (2013). 'Downsampling Time Series for Visual Representation'. Master's thesis, School of Engineering and Natural Sciences University of Iceland, Reykjavik.

Theissler, A., Perez-Velazquez, J., Kettelgerdes, M., &Elger, G. (2021). 'Predictive Maintenance Enabled by Machine Learning: Use Cases and Challenges in the Automotive Industry'. *Journal of Reliability Engineering & System Safety*, Vol. 215, pp. 107864.

Wu, X. (2017, December 15). 'Metrics, Techniques and Tools of Anomaly Detection: A Survey'. Retrieved from https://www.cse.wustl.edu/~jain/cse567-17/ftp/mttad/index .html

Zhang, Y., Ding, X., Liu, Y., &Griffin, P. (1996). 'An Artificial Neural Network Approach to Transformer Fault Diagnosis'. *IEEE Transactions on Power Delivery*, Vol. 11, No. 4, pp. 1836–1841.

14 Impact of Digitalization and Artificial Intelligence (AI) on Diagnostic Pathology
Are We Looking at a Renaissance?

Asitava Deb Roy

CONTENTS

14.1 INTRODUCTION

This is an era where digitalization has creeped into every single task that we perform—be it transaction of money online, booking a cab on an app-based platform, asking a cloud-based voice service to play our favorite number, or consulting doctors online. And behind such digitalization, there lies an advanced technology platform that relies heavily on artificial intelligence (AI).[1] Although the onset and subsequent progress of digitalization in health care was slightly slow, the recent pandemic of COVID-19 that hit the world forced everyone in the healthcare industry to adopt digitalization to its maximum potential.[2] Healthcare digitalization has touched upon various aspects, viz. medical education, clinical consultations, radiological and pathological diagnosis, and also medical research.[3] Of these various fields, the impact

DOI: 10.1201/9781003326830-14

of digitalization and AI on diagnostic pathology has been notably significant. It has touched every aspect of a pathologist's life creating a human–machine interface—be it direct transfer of machine-generated reports to the software, highlighting the abnormal values for cross-check, ability to certify the reports online and make them available for the patients to view on their digital platforms, up to a level where whole slides of histopathology can be scanned and viewed through a software without even having the need for a microscope.[4–6]

However, despite its positive impact on the overall healthcare scenario, there are some challenges that need to be overcome. But before we succumb to the pessimism on AI, we should critically analyze the strengths, weaknesses, opportunities, and challenges of the same.

This chapter discusses the various domains of pathology where digitalization and AI have shown their impact with an additional analysis of the strengths, weaknesses, opportunities, and challenges of digitalization in pathology and finally concludes with a few recommendations to make implementation of digitalization in the field of pathology successful in our country.

14.2   A BRIEF HISTORY OF EVOLUTION OF PATHOLOGY

Pathology is a branch of medicine that deals with the study and diagnosis of diseases, and pathologists are those doctors who diagnose and describe the disease in patients by examining clinical samples in the form of patient's blood, body fluid, cells, or tissues. Pathology is one of those specialties in medicine where the pathologists do not examine or treat the patient directly but rather serve as consultants to other physicians in confirming their clinical diagnosis and planning a therapeutic regimen for the patients.[7]

The scientific method was originally used to the study of medicine in the Islamic Golden Age in the Middle East and the Italian Renaissance in Western Europe. The history of pathology can be used to explain this development.[8] The early doctors often considered the complete body when determining the root cause of a disease as well as gods and goddesses, stars, and celestial bodies in their orbits. Then the "four humors" and other hypotheses appeared, enslaving doctors for almost 2,000 years, only to give way in the last few hundred to the idea of organ-based disease and the development of anatomical pathology. In the latter part of the nineteenth century, the field of pathology underwent a revolutionary change with the development of microscopy and the breakthrough works of Morgagni, Bichat, and Virchow.[8,9] The invention of the microscope as a scientific tool was a blessing in the field of pathology. Histopathology, which had dominated pathology for only one and a half centuries, was born when concepts were refocused from organ to tissue to cell, even smaller. The invention of the microscope allowed histopathology to be practiced and led to numerous concomitant technical advancements required for contemporary practice, completely altering ideas of disease from focusing on whole organs to cells.[9,10] As a result, hand-cut, unstained slices of fresh tissue were initially evaluated. However, in the later decades of the century, frozen tissues, embedding methods, microtomes, a wide variety of stains, and significantly superior microscopes had replaced this primitive approach. As the twentieth century began, pathology research visibly picked up speed, the number of discoveries increased practically exponentially, and the number of discoverers also increased. Then, as the second millennium came to a close, potent newer technologies forced yet another reworking of our theories—from diseases based on cells to diseases based on genes to diseases based on specific molecules and their interactions. The pace of discovery and change has accelerated much more since the turn of the 20th century. Fixation, embedding, cutting, immunohistochemical staining, molecular methods, microscopy, and image processing have all advanced, leading to better diagnostic instruments and new, better, more accurate diagnoses.[11]

14.3 ARTIFICIAL INTELLIGENCE: A BRIEF HISTORY OF EVOLUTION

AI, a term initially coined by John McCarthy in 1956, has, of late, become quite familiar in the scientific community.[12] In general, the term means any work, comparable to that done by an intelligent human brain, being done by artificial means, viz. machines or equipments.[12]

The science of AI has experienced multiple cycles of anticipation, disillusionment, and funding loss since it was first established as a topic of study in 1956. However, it resulted in fresh ideas, achievements, and increased investment. Since its inception, AI research has tried and failed with a broad variety of approaches, including simulating human problem-solving, formal logic, vast knowledge libraries, and imitating animal behavior.[13] In the first 20 years of the 21st century, machine learning—which is largely grounded in mathematics and statistics—has dominated the field. Many challenging problems have been successfully solved using this method in both business and academia. The many subfields of AI research are centered on some specific goals and the application of some methods.[13,14] Some of the traditional goals of AI research include reasoning, knowledge representation, planning, learning, natural language processing, sensing, and the ability to move and manipulate objects.[14]

The first sentient robots and artificial entities first appeared in ancient Greek mythology.[13] A crucial turning point in humanity's search to comprehend its own intelligence was the development of the syllogism by Aristotle and its application of deductive reasoning. The history of AI as we know it now dates back less than a century, despite having deep and extensive roots.[13] The first mathematical model for creating a neural network was put forth in the 1943 publication "A Logical Calculus of Ideas Immanent in Nervous Activity" by Warren McCullough and Walter Pitts.[12,13] At the Dartmouth Summer Research Project on Artificial Intelligence in 1956, the term "artificial intelligence" was first used. The John McCarthy-led symposium is regarded as the origin of AI.[1,13] In 1959, while working at IBM, Arthur Samuel created the term "machine learning." In 2014, Google created the first self-driving automobile to pass a state driving test thanks to the steady and gradual growth of AI in the twenty-first century. Amazon's Alexa, a virtual assistant for smart homes, was also introduced that year.[14] Gato, an AI system developed by DeepMind and released this year in 2022, is capable of hundreds of activities, such as playing Atari, captioning photos, and stacking blocks with a robotic arm. Somewhere in the midst of all these developments, AI slowly creeped into the field of health care and diagnostics as well.[15]

14.4   AI AND DIGITALIZATION IN HEALTH CARE

In terms of medicine, 2021 World Health Organization research said that while implementing AI in the healthcare industry presents difficulties, the technology "holds considerable potential" as it may result in advantages such as better health policy and more accurate patient diagnosis. It is expected to bring about a paradigm shift in the field of health care.[16] Although, the apprehension among common people about AI replacing doctors or healthcare workers does not seem to be very realistic, yet it is certainly poised to bring about a foundational shift in the medical systems worldwide. It is quite a well-known fact that the healthcare sector has been unexpectedly sluggish to embrace the worldwide digital revolution, coming in at the bottom third of all industries in 2015.[16,17] The digital revolution in health care is still in its infancy compared to the advances we have witnessed in other sectors over the past ten years, such as banking, media, etc.[17] However, over time, new tools and technology have begun to spread throughout the healthcare system and show enormous potential for drastically altering how health care is delivered in the near future while also enhancing the effectiveness of patient care.[17] Every day, the medical sector produces a vast amount of data. Healthcare practitioners have long needed to employ technology to organize and stratify this data in order to make it easily accessible. The data that is available in the form of radiological images, electrocardiograms, laboratory tests, and electronic health records can be simply incorporated into algorithms to train AI models. Then AI can create its own algorithm, which can be incorporated into a system that transforms practical knowledge into useful tools. AI has aided the development of life-critical applications, rapid and precise diagnostics, and statistical analysis in health care. Like many other fields, the healthcare industry also had no other choice but to adopt digitalization to cope up with the pandemic situation, wherever possible. We here highlight some of the domains of health care that have seen the impact of digitalization/AI:

1. Medical education: The onset of digitalization in health care started with conducting seminars and conferences using online platforms and calling them webinars. Medical schools had to depend upon various online platforms to conduct classes for their students.[4, 18]
2. Clinical consultation: Clinicians started providing consultation to their patients with the help of telemedicine and online facilities as per guidelines issued from time to time during the pandemic.[19–21] Applications for smartphones also facilitate and ease the collection of data and patient monitoring.[22,23] Consultations in diabetes management and pain management have

proved to be quite beneficial.[24-26] There are nearly endless possibilities for other uses.[27,28]

3. Diagnosis: Radiologists took to reporting of films using various online software mostly for X-ray, CT scan, and MRI diagnosis.[29] Computer-aided detection has also demonstrated benefit in assisting radiologists to evaluate images more quickly and precisely for patterns connected to underlying diseases, such as for breast cancer during mammography screening.[30]

4. Medical research: By enabling the virtual screening of millions of compounds to perhaps boost the number of potential therapeutic leads, increased usage of technologies like machine learning may help reverse this trend. With the aim of reducing the time and costs needed to bring a medicine to market, digital solutions like clinical trial simulation, modelling and simulation, computer-assisted trial design, model-based drug development, and model-informed drug discovery and development may also start to replace some lab experiments.[31]

14.5   AI AND DIGITALIZATION IN PATHOLOGY

A very important part of healthcare services that could be partially digitalized was pathology. The idea of digital pathology is not new. Whole slide imaging (WSI) is a technique that allows for the scanning and conversion of whole glass slides or specific regions of slides into digital data files that can then be seen on a computer screen. The optical aspects of the scanning system, which are more or less comparable to those of glass slide microscopy, are what essentially determine the qualities of the presented image. By comparing the concordance of reporting between digital and glass slides, Wilbur et al. (2009) described a pilot study on the usefulness of digital pathology and found a respectable concordance of 91%.[32] Therefore, implementing digitalization in the domain of pathology for reporting slides was not a very difficult thing to achieve. The Royal College of Pathologists had established a guideline for remote reporting of digital pathology slides in response to the necessity for home reporting during the COVID-19 epidemic and in the future.[33] By contrasting glass with digital slides, they have suggested a self-directed validation procedure lasting one to two months. To enable reporting from home, the Department of Pathology at Universiti Kebangsaan Malaysia set up a computer with remote access.[3]

The study of "digital pathology" entails both the digitalization of histopathology slides using whole slide imaging (WSI) scanners and the computer analysis of those WSI.[34] The development of AI-based computational methods for digital pathology involves cross-disciplinary partnerships between pathologists, oncologists, computer engineers, and data scientists.[35]

In actuality, WSI is composed of two basic steps. The first step involves digitizing glass slides using specialized hardware (a scanner), which creates a sizable representative digital image (a "digital slide"). In the second method, these huge digital data are viewed and/or analyzed using specialist software (i.e., a virtual slide viewer).[35] In essence, a whole slide scanner is a robotically and electronically controlled microscope. This is connected to one or more highly specialized cameras that have cutting-edge optical sensors. Whatever its complexity, a WSI scanner must always have the following basic parts: a microscope with lens objectives, a light source (either bright field or fluorescent), robotics to load and move glass slides, one or more digital cameras to collect images, a computer, and software to handle, manage, and examine the digital slides.[36]

An image viewer is a term used to describe software used to traverse digital slides. It replicates the classic light microscopy experience in digital form by allowing users to examine and navigate (pan and zoom) virtual slides on a digital screen. Users of WSI viewers can examine virtual slides at different magnifications.[37] Additionally, viewers offer a wide range of extra features that conventional microscopes do not yet provide. These extra features range from simple ones like seeing slides at unusual magnifications, such as 1x, to more complex ones like image analysis, as well as annotation tools like measurements and text/object overlays.[38] Some systems additionally include teleconferencing features enabling several remote users to watch virtual slides at the same time. A few picture viewers enable synchronized navigation and simultaneous viewing of multiple slides.[39]

The author tried to perform a SWOC (strengths, weaknesses, opportunities, and challenges) analysis of the impact of digitalization in pathology and the results were as follows:[40]

14.5.1 Strengths

The main strengths of digitalization in pathology are:

1. Remote reporting: Several studies have shown that WSI has the same accuracy compared to routine light microscopy for primary diagnosis, although the use of virtual slides for primary diagnosis requires an extensive validation program. Whole slides can be scanned and sent anywhere to a pathologist with the help of software without any need for the pathologist to be present physically in the laboratory.
2. Review and second opinion: The use of virtual microscopy for diagnostic consultation (or e-consultation) is one of the most useful applications of virtual slide microscopy. It eases out logistics as no packing or mailing of slides is required; therefore, there is no fear of damage to slides. Slides can be easily shared by the pathologist with his/her peers or colleagues for any second opinion.
3. Storage: Scanned images of the slides can be stored without having to worry about slides getting broken or stains fading away. However, although storage is one of its major strengths, it also poses as one of its weaknesses (as discussed below) at times.
4. Ease of reporting intricate details: During reporting, pathologists can accurately measure size of the tumor, microscopic distance from surgical margins, areas of necrosis, mitotic count, and many such other important details.

14.5.2 Weaknesses

In India, although there have been efforts from many reputed institutions and diagnostic chains to establish the process of digitalizing pathology for reporting histopathology, but it is still in a very nascent stage and restricted to only those few institutions. It was a matter of concern as to why digitalization could not be fully implemented even after being described way back in 2009. The possible weaknesses/challenges of this technology that needed consideration are:

1. The cost factor: Currently there are more than a dozen vendors that are available in the market providing WSI platforms. High-end whole slide imaging equipment, which is quite expensive, is needed for the high throughput conversion of conventional glass slides to digital images and a good one costs anywhere between 50 lacs and 1 cr. For most of the institutions in India, cost becomes a deterring factor in implementation of WSI for reporting purposes.
2. The storage factor: Due to the great resolution of these photographs, each image file can be incredibly large—up to a few gigabytes. Because of this, maintaining the storage of these photographs could be costly.
3. The comfort factor: One of the reasons for failure in successful implementation also probably was the preference of pathologists to glass slide viewing. Many of our fellow colleagues are not tech savvy and therefore prefer to stick to the conventional glass slide viewing. Also, any requirement for recut/restain for a particular slide can be immediately ordered in the conventional system of reporting.
4. The technical factor: The implementation of digital pathology in cytopathology or histopathology units requires an electronic infrastructure along with a slide scanning instrument. The infrastructure requires efficient interaction between the information technology (IT) and the laboratory information system (LIS) personnel. However, there is not enough information and technology (IT) support available at every place to help the pathologists troubleshoot with the technical difficulties.
5. The manpower factor: Dedicated and trained technologists are required to scan the slides, upload the images, monitor the scan quality, perform initial screening, and do rescan if necessary. They also need to be trained in distribution of the image files to the pathologists in a manner that can be conveniently viewed by them. In a country like India, where we have to struggle to get technicians for routine histopathology, getting technologists trained in digital histopathology is a far cry.
6. The connectivity factor: In many places in India, it is always not possible to get a uniform and adequate speed of internet connectivity, which is essential for viewing of scanned digital slides. This becomes a real challenge as the slides take time to get downloaded and the turnaround time increases.
7. The standardization factor: Lack of standardization and availability of appropriate validation guidelines lead to lot of confusion in reporting digital images

in pathology. For WSI, "validation" means proving that digital slides and glass slides inspected under a traditional light microscope function diagnostically in a manner that is similar.[41] To evaluate intraobserver diagnostic concordance between digital and glass slides, examined at least two weeks apart, the College of American Pathologists (CAP) advises considering at least 60 routine cases per application. The cases chosen, each of which may contain multiple slides, ought to represent the variety and complexity of specimen types and diagnoses that are likely to be seen during ordinary operation.[42]

14.5.3 OPPORTUNITIES

Nonetheless, the current COVID-19 pandemic came as an opportunity to embrace the change. In this pandemic situation for the last two years we have realized that digitalization in pathology can be utilized as an opportunity to "work from home" for the pathologists.[1,4] The other opportunities or prospects in the field of AI in pathology could be:

1. Tissue folds in histopathology sections which may create difficulty in scanning the slides leading to artifacts may be overcome by way of color enhancement techniques.[43]
2. Software that constructs a fictitious entire mount section by stitching together tiny individual images has been developed by researchers. This, if successful, would not require expensive whole slide scanning systems.[44]
3. There are substantial implications for WSI from novel full slide image collection techniques using cutting-edge optics and mobile devices in the near future.[45]

14.5.4 CHALLENGES

However, things are not as simple and easy as they are thought to be. Despite this boom in technology, which has helped pathologists to report slides from a remote location, there are few areas where the presence of a pathologist cannot be ignored. These areas pose as additional challenges to digitalization of pathology, which need to be investigated:

1. Pathologists must be physically present for dissection of large tissue samples. Small biopsies may not need the presence of a pathologist during grossing, but pathologists or pathology trainees still need to gross huge specimens.[40]
2. Procedures like fine needle aspiration cytology (FNAC), bone marrow aspiration, biopsy, etc., usually require the presence of a qualified pathologist or a qualified medical practitioner for that matter.[40]
3. A frozen section biopsy that needs to be reported within 10–15 minutes may not be possible with the help of digitalization where proper IT support/internet connectivity is not in place.[46,47]

14.6 CONCLUSION

AI and digitalization have had significant contributions in the majority of industries,[48–56] health care being one of those. We have seen that digitalization has eased various aspects of health care starting from medical education to diagnosis and patient management.[57] However, although the positive impact of digitalization in health care, particularly in pathology, is definitely revolutionary, there are definite weaknesses and challenges that need to be addressed for it to be effective and implementable in developing countries like India. The following recommendations may be made, which should solve the majority of the problems related to implementation:

1. Adequate training of doctors and all other healthcare personnel including technologists in digital platforms
2. Easy access to IT support at places where digital services in health care have been implemented
3. Availability of cost-effective equipment for such services to be made available even in developing countries like India
4. Ensuring uninterrupted internet connectivity for easy scanning and downloading of slides for faster reporting
5. Proper validation protocols and guidelines for developing diagnostic algorithms and AI-based reporting in digital radiology and pathology

REFERENCES

1. Poola I. How Artificial Intelligence in Impacting Real Life Every day. *IJARnD* 2017;2(10):96–100.
2. Jazieh AR, Kozlakidis Z. Healthcare Transformation in the Post-Coronavirus Pandemic Era. *Front Med*. 2020 Jul 28;7:429.
3. Tan GC, Wong YP. Digital Pathology as a Solution for Working from Home. *Malays J Pathol* 2021;43(2):201.
4. Wong YP, Tan GC. Virtual Meeting in Pathology: Time to Adapt to the New-norm. *Malays J Pathol*. 2021 Apr;43(1):1.
5. Parwani AV, Feldman M, Balis U, et al. Telepathology. In: Pantanowitz L, Tuthill MJ, Balis U, editors. *Pathology Informatics: Theory and Practice*. Canada: ASCP; 2012:261–272.
6. Parwani AV, Feldman M, Balis U, et al. Digital Imaging. In: Pantanowitz L, Tuthill MJ, Balis U, editors. *Pathology Informatics: Theory and Practice*. Canada: ASCP; 2012:231–260.
7. Long E. *A History of Pathology*. New York: Dover Publications; 1965.
8. Malkin H. *Out of the Mist. The Foundation of Medicine and Modern Pathology During the Nineteenth Century*. Berkley: Vesalius Books; 1993.
9. Mauritz R. *Morbid Appearances. The Anatomy of Pathology in the Early Nineteenth Century*. Cambridge: Cambridge University Press; 1987.
10. Monti A. *The Fundamental Data of Modern Pathology. History, Criticisms, Comparisons, Applications*. London: The New Syndenham Society; 1900.
11. Race GJ, Tillery GW, Dysert PA. A History of Pathology and Laboratory Medicine at Baylor University Medical Center. *Proc (Bayl Univ Med Cent)*. 2004 Jan;17(1):42–55.
12. Russell SJ, Norvig, P. *Artificial Intelligence: A Modern Approach* (4th ed.). Hoboken: Pearson; 2021.

13. Nilsson N. *The Quest for Artificial Intelligence: A History of Ideas and Achievements.* New York: Cambridge University Press, 2009.

14. Wirtz, BW, Weyerer, JC, Geyer, C. Artificial Intelligence and the Public Sector: Applications and Challenges. *International Journal of Public Administration* 2018 24 July;42(7):596–615.

15. Mahajan P. *A Brief Introduction to AI in Healthcare in 'Artificial Intelligence in Healthcare'* (3rd ed.) New Mexico: MedMantra, 2022.

16. Bohr A, Memarzadeh K. The Rise of Artificial Intelligence in Healthcare Applications. In *Artificial Intelligence in Healthcare* Academic Press, 2020.

17. Park CW, Seo SW, Kang N, Ko B, Choi BW, Park CM, Chang DK, Kim H, Kim H, Lee H, Jang J. Artificial Intelligence in Health Care: Current Applications and Issues. *Journal of Korean Medical Science.* 2020 Nov 2;35(42), 40–42.

18. Dee FR, Meyerholz DK. Teaching Medical Pathology in the Twenty-first Century: Virtual Microscopy Applications. *J Vet Med Educ.* 2007;34:431–436.

19. Centers for Disease Control and Prevention. 2020. https://www.cdc.gov/coronavirus /2019-nCoV/index.html

20. World Health Organization Coronavirus Disease (COVID-19) Advice for the Public. 2020. https://www.who.int/emergencies/diseases/novel-coronavirus-2019/advice-for-public

21. Ministry of Health and Family Welfare, Government of India Telemedicine Practice Guidelines. https://www.mohfw.gov.in/pdf/Telemedicine.pdf

22. Galen LS van, Car J. Telephone Consultations. *BMJ.* 2018 Mar 29;360:k1047.

23. Greenhalgh T, Wherton J, Shaw S, Morrison C. Video Consultations for Covid-19. *BMJ.* 2020 Mar 12;m998.

24. Ghosh A, Gupta R, Misra A. Telemedicine for Diabetes Care in India during COVID19 Pandemic and National Lockdown Period: Guidelines for Physicians. *Diabetes Metab Syndr Clin Res Rev.* 2020 Apr 1;14, 12–14.

25. Ghosh A, Dutta K, Tyagi K, Gupta R, Misra A. Roadblock in Application of Telemedicine for Diabetes Management in India During COVID19 Pandemic. *Diabetes Metab Syndr.* 2020;14(4):577–578.

26. Bradwell H, Baines R, Edwards KJ, Stevens, S, Atkinson, K, Wilkinson, E, Chatterjee, A, Jones, RB. Exploring Patient and Staff Experiences With Video Consultations During COVID-19 in an English Outpatient Care Setting: Secondary Data Analysis of Routinely Collected Feedback Data. JMIR Form Res 2022; 6(3):1–13.

27. Iyengar K, Jain VK, Vaishya R. Pitfalls in Telemedicine Consultations in the Era of COVID 19 and How to Avoid Them. *Diabetes Metab Syndr.* 2020 September–October;14(5):797–799.

28. Shenvi EC, El-Kareh R. Clinical Criteria to Screen for Inpatient Diagnostic Errors: A Scoping Review. *Diagn Berl Ger.* 2015 Feb;2(1):3–19.

29. El-Kareh R, Hasan O, Schiff GD. Use of Health Information Technology to Reduce Diagnostic Errors. *BMJ Qual Saf.* 2013 Oct;22(Suppl 2):ii40–51.

30. Hadjiiski L, Sahiner B, Chan HP. Advances in Computer-aided Diagnosis for Breast Cancer. *Curr Opin Obstet Gynecol.* 2006 Feb;18(1):64–70.

31. Modeling and Simulation in Clinical Trials: Real Potential or Hype? [Internet]. *Applied Clinical Trials Online.* [cited 2022 Sep 12]. Available from: https://www.appliedclin icaltrialsonline.com/view/modeling-and-simulation-clinical-trials-real-potential-or -hype

32. Wilbur DC, Madi K, Colvin RB, Duncan LM, Faquin WC, Ferry JA et al. Whole-slide Imaging Digital Pathology as a Platform for Teleconsultation: A Pilot Study Using Paired Subspecialist Correlations. *Arch Pathol Lab Med.* 2009;133(12):1949.

33. Williams BJ, Brettle D, Aslam M, Barrett P, Bryson G, Cross S, Snead D, Verrill C, Clarke E, Wright A, Treanor D. Guidance for Remote Reporting of Digital Pathology Slides During Periods of Exceptional Service Pressure: An Emergency Response from the UK Royal College of Pathologists. *J Pathol Inform.* 2020;11:12.

34. Pantanowitz L, Valenstein PN, Evans AJ, et al. Review of the Current State of Whole Slide Imaging in Pathology. *J Pathol Inform.* 2011;2:36.

35. Ferreira R, Moon B, Humphries, Sussman A, Saltz J, Miller R, Demarzo A. The Virtual Microscope. *AMIA Ann Fall Symp.* 1996:449–453.

36. Gilbertson JR, Patel AA, Yagi Y. Clinical Slide Digitization: Whole Slide Imaging in Clinical Practice. In: Gu J, Olgilive RW, et al, editors. *Virtual Microscopy and Virtual Slide in Teaching, Diagnosis, and Research.* Boca Rotan, FL: Taylor and Francis; 2005.

37. Soenksen DG. A Fully Integrated Virtual Microscopy System for Ultra-fast Slide Scanning, Storage, Viewing, Annotation, and Processing. *Arch Pathol Lab Med.* 2003;127:805.

38. Amin W, Srintrapun SJ, Parwani AV. Automated Whole Slide Imaging. *Expert Opin Med Diagn.* 2008;2(10):1173–1181.

39. Farahani N, Parwani AV. Pantanowitz Whole Slide Imaging in Pathology: Advantages, Limitations, and Emerging Perspectives. *Pathology and Laboratory Medicine International* 2015;7:23–33.

40. Karn A, Priyadarshi A, Deb Roy A. A Review on Digitalization of Healthcare with SWOC Analysis of Digital Pathology in the Backdrop of COVID-19. *Global Journal of Research Analysis.* 2022;11(7):1–2.

41. Singh R, Chubb L, Pantanowitz L, et al. Standardization in Digital Pathology: Supplement 145 of the DICOM Standards. *J Pathol Inform.* 2011;2:23.

42. Pantanowitz L, Sinard JH, Henricks WH, et al. Validating Whole Slide Imaging for Diagnostic Purposes in Pathology: Guideline from the College of American Pathologists Pathology and Laboratory Quality Center. *Arch Pathol Lab Med.* 2013;137(12):1710–1722.

43. Bautista PA, Yagi Y. Improving the Visualization and Detection of Tissue Folds in Whole Slide Images Through Color Enhancement. *J Pathol Inform.* 2010;1:25.

44. Toth RJ, Shih N, Tomaszewski JE, Feldman MD, et al. Histostitcher™: An Informatics Software Platform for Reconstructing Whole-mount Prostate Histology Using the Extensible Imaging Platform Framework. *J Pathol Inform.* 2014;5:8.

45. Dander A, Baldauf M, Sperk M, Pabinger S, Hiltpolt B, Trajanoski Z. Personalized Oncology Suite: Integrating Next-generation Sequencing Data and Whole-slide Bioimages. *BMC Bioinformatics* 2014;15:306.

46. Evans AJ, Chetty R, Clarke BA, et al. Primary Frozen Section Diagnosis by Robotic Microscopy and Virtual Slide Telepathology: The University Health Network Experience. *Hum Pathol.* 2009;40(8):1070–1081.

47. Gifford AJ, Colebatch AJ, Litkouhi S, et al. Remote Frozen Section Examination of Breast Sentinel Lymph Nodes by Telepathology. *ANZ J Surg.* 2012;82:803–808.

48. Janapati R, et al. Towards a More Theory-Driven BCI Using Source Reconstructed Dynamics of EEG Time-Series. *Nano LIFE* 2022;12(02):2250005.

49. Janapati R, et al. Web Interface Applications Controllers Used by Autonomous EEG-BCI Technologies. AIP Conference Proceedings. Vol. 2418. No. 1. AIP Publishing LLC, 2022.

50. Janapati R, Dalal V, and Sengupta R. Advances in Experimental Paradigms for EEG-BCI. Proceedings of the 2nd International Conference on Recent Trends in Machine Learning, IoT, Smart Cities and Applications. Springer, Singapore, 2022.

51. Janapati R, Dalal V and Sengupta R. Advances in Modern EEG-BCI Signal Processing: A Review. *Materials Today: Proceedings* 2021;80(3):2563–2566.

52. Janapati R, et al. Progression of EEG-BCI Classification Techniques: A Study. *Inventive Systems and Control* 2021;204:161–170.

53. Janapati R, et al. Signal Processing Algorithms Based on Evolutionary Optimization Techniques in the BCI: A Review. *Computational Vision and Bio-Inspired Computing* 2021;1318:165–174.

54. Janapati R, et al. Review on EEG-BCI Classification Techniques Advancements. IOP Conference Series: Materials Science and Engineering. Vol. 981. No. 3. IOP Publishing, 2020.

55. Janapati R, et al. Various Signals Used for Device Navigation in BCI Production. IOP Conference Series: Materials Science and Engineering. Vol. 981. No. 3. IOP Publishing, 2020.

56. Kumar RP, Vandana SS, Tejaswi D, Charan K, Janapati R and Desai U. Classification of SSVEP Signals Using Neural Networks for BCI Applications. 2022 International Conference on Intelligent Controller and Computing for Smart Power (ICICCSP), 2022, pp. 1–6, doi: 10.1109/ICICCSP53532.2022.9862368.

57. Stathonikos N, Veta M, Huisman A, van Diest PJ. Going Fully Digital: Perspective of a Dutch Academic Pathology Lab. *J Pathol Inform*. 2013;4:15.

15 University Chatbot System Using Natural Language Processing

P. Pramod Kumar, Parupally Sindhu, Avula Jahnavi,
Bokka Prudvi Charan, and Sukka Prashanth

CONTENTS

15.1 INTRODUCTION

Technology plays a major role within the business. It is used in development of many wide range of functions and is applied in many various ways in different parts of the globe. Recently, many of the businesses are fantasized by computing. Artificial intelligence (AI) simulates the cognitive abilities of the humans. Chatbots are made to replace the human responses with the AI software, for being more related to humans. Chatbot is a software program that acts like a communication medium between the human and the machine. It is a virtual assistant that has become very widespread in past years essentially due to drastic advancements in the areas like AI, machine learning, and the other fundamental technologies such as natural language processing (NLP). Chatbots generally communicate with the humans using the interactive queries and answer them in a fun way. Recently, there has been a great increase in the chatter bot services, which has been made available for the development and improvement of the chatbot sector such as Clever bot and many others. Conversational agents have become more responsive, and the art of conversation between robots and humans over the past few years has improved tremendously.

DOI: 10.1201/9781003326830-15

In the chatbot system, we use artificial intelligence and NLP techniques to depict the user's queries. The main intention of the chatbot is to find out what the user intentions are and accordingly will provide a meaningful response by understanding patterns. This chatbot works as a user-friendly interface between the university students and the university management. This will be done with the help of behavioral analysis, which will allow chatbots to understand what user needs are and respond to the client queries accordingly.

This university chatbot system answers all the queries related to the university, which are asked by the users. This gives a great experience for the users while interacting with the software. This system makes university management and the university students to feel comfortable as they can communicate through the chatbot interface without any physical interaction.

15.2 LITERATURE REVIEW

Literature survey represents a body of text, which states the methodologies and technologies published on a particular research topic. It contains works on a particular topic carried out in a specific period of time. It aims to analyze all the important points of the methodologies and approaches. It states various methods and research works that are published.

P. Nikhila, K. Mounika, G. Jyothi, Dr. B. V. Ramana Murthy, and Mr. C. Kishor Kumar Reddy [1] have designed the chatbot using Artificial Intelligence Mark-up Language (AIML) to give response to the queries asked by users. AIML is designed to make or customize Alicebot that could support a chatbot application, i.e., ALICE free code [1].

Prof. Ram Manoj Sharma [2] proposed a college enquiry chatbot system, which was built using AI techniques. Chatbot understands the user's messages and respond to them. This system has modules such as Online chatbot, Online Noticeboards, etc. [3].

Pawar, Harsh, Pranav Prabhu, Ajay Yadav, Vincent Mendonca, Joyce Lemos [2], designed a chatbot using knowledge in the database. This proposed system has online enquiry and online chatbot system. The development of chatbot is done using different programming languages by creating a user-friendly graphical interface (GUI) to send and receive messages. Used Structured Query Language (SQL) for pattern matching, which is stored in program [2].

Nitesh Thakur, Sourabh Selote, Akshay Hiwrale, Prof. Namrata Mahakalkar, and Abhijeet Shinde [4] have proposed an AI chatbot using NLP techniques which is done in two ways: the first way is via written text and the second way is via verbal or voice communication. Written communication is much more easier than the verbal communication. This chapter introduces an interest in some emerging capabilities [4].

Prof. K. Bala, Mukesh Kumar, Sayali Hulawale, and Sahil Pandita [5] have proposed a chatbot for college management system using AI. This chatbot is built using AI techniques and algorithms [5].

Tarun Lalwani, Shashank Bhalotia, Ashish Pal, Shreya Bisen, and Vasundhara Rathod [6] have proposed "chatbot for educational institution." This chatbot is used by users like students, faculty, and parents to get the information about the institution. AIML and NLP are used for building these chatbots. AIML by using NLP can answer to user questions [6].

Sagar Pawar, Omkar Rane, Ojas Wankhade and Pradnya Mehta [7] proposed a "Web Based College Enquiry Chatbot with Results." This system is web application that provides answer to student's queries very effectively. Students should submit the query to the chatbot, which is used for chatting. If the answer is found invalid, then there is a system to declare the answer as invalid. These invalid answers can be deleted or modified by the administrator of the system. [7–9]

15.3 MOTIVATION

We got motivated to work on this project from the problem we faced in the real time in our university. University students face problems like knowing the university timings, university schedule, and other particular details about the university. So by personally experiencing this problem, we are proposing a chatbot to solve all the problems or queries of the university users. As we all know, chatbot is really a fashionable approach of communication between the user and the system text interface. We have a tendency to wish to use this system to implement university chatbot for an additional economical communication between the university students or users and administration. This chatbot was used for queries like university ranking, courses offered in different disciplines [10].

Our project to create a university chatbot would undoubtedly have a bright scope as this could bring a modification within the approach of interaction between university and the students. This project would inspire alternative universities still to develop internet chatbots to ease the communication between management and students [11].

15.4 PROPOSED METHODOLOGY

AI technology in chatbot will analyze user's queries about university activities and give responses to the user through web application. The trained data is obtained from the preprocessing of the information using NLP technique. The response is provided through a graphical user interface (GUI), which is the interface between the machine and user. Human can interact with this system using the GUI, which is created using tkinter GUI.

tkinter package is the standard Python interface in the Tcl/Tk GUI toolkit.

Our proposed system mainly consists of five significant phases, which include:

- Creating the text corpus (dataset).
- Pre-processing of the raw text.
- Build chatbot model (keyword matching—algorithms).
- Define the response function.
- Graphical UI (tkinter).

The NLP includes tokenization and lemmatization. After processing the text using NLP the training data is obtained. The model is trained using the data.

15.4.1   Dataset

Our dataset contains all the details related to the university through which chatbot responds to the user's queries.

15.4.1.1   Block Diagram

15.4.1.2   Natural Language Processing

The main aim of AI is to allow the machines to receive the information and interpret it in the same intelligent way as humans do. NLP is a popular subcategory of AI [15.1].

NLP is the study of how machines understand and analyze natural languages and produce meaningful information about the given text.

In other terms, NLP is a way to teach a machine how to read and understand human languages. Trained machine can extract the relationships between words and identify the entities in a sentence.

Data preprocessing is done using NLP techniques which includes:

- Tokenization
- Lemmatization

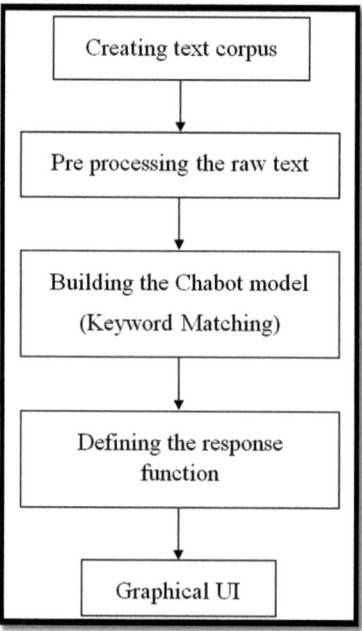

FIGURE 15.1   Block diagram.

15.4.1.2.1 Tokenization

Tokenization is a process of breaking the content into smaller units. Tokenization splits the raw data to words and sentences, which are known as tokens. The tokens that are generated after the tokenization give us an understanding of a specific situation and help us in building up the model for the NLP [15.2].

Consider the sentence "Creation of University chatbot System."

```
x = "Creation of University chatbot System"
After tokenization: ["Creation," "of," "University,"
"chatbot," "System"]
```

15.4.1.2.2 Lemmatization

Lemmatization is a technique that is used to reduce the words to a normalized form. It uses dictionary for mapping the various variants of any word to its root format or root word [15.3].

15.4.1.3 NLP Pre-processing Pipeline

Data preprocessing is done by following the below steps [15.4]:

1. First we perform tokenization on the data by which sentences are broken into words, which are called tokens.
2. Then we convert the words into lower.
3. The next step is lemmatization, which maps all the different variants of a word to a root word.
4. Removing special characters.

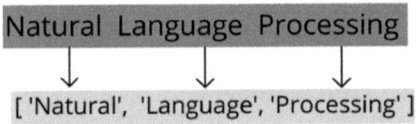

```
from nltk.tokenize import word_tokenize

word_tokenize("Hello there! My email is tanpengshi@gmail.com")
```

['Hello', 'there', '!', 'My', 'email', 'is', 'tanpengshi', '@', 'gmail.com']

```
raw=raw.lower() # converts to lowercase
sent_tokens = nltk.sent_tokenize(raw) # converts to list of sentences
word_tokens = nltk.word_tokenize(raw) # converts to list of words
```

FIGURE 15.2 Tokenization.

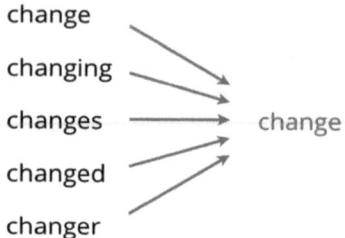

```
lemmer = nltk.stem.WordNetLemmatizer() # Lemmatization
def LemTokens(tokens):
    return [lemmer.lemmatize(token) for token in tokens]
remove_punct_dict = dict((ord(punct), None) for punct in string.punctuation)
def LemNormalize(text):
    return LemTokens(nltk.word_tokenize(text.lower().translate(remove_punct_dict)))
```

FIGURE 15.3   Lemmatization.

```
For example, Sentence = "What is your Name?"
Step 1: Tokenize.
["What," "is," "your," "Name," "?"]
Step 2: Convert to lower case.
["what," "is," "your," "name," "?"]
Step 3: Lemmatize.
["what," "is," "you," "name," "?"]
Step 4: Remove special characters.
["what," "is," "you," "name"]
```

15.4.1.4   Building the Chatbot Model

Building model is done using Term Frequency-Inverse Document Frequency and Cosine Similarity (TF-IDF) algorithms.

15.4.1.4.1   Term Frequency-Inverse Document Frequency Algorithm

TF-IDF algorithm distinguishes between very common words and rare words in the document. It measures the importance of a word in the corpus (dataset) by considering the frequency of the word that is occurring in the corpus and a document [15.5].

$$TF = \frac{Number\ of\ times\ a\ word\ "X"\ appears\ in\ a\ Document}{Number\ of\ words\ present\ in\ a\ Document}$$

$$IDF = \log\left(\frac{Number\ of\ Documents\ presents\ in\ a\ Corpus}{Number\ of\ Documents\ where\ word\ "X"\ has\ appeared}\right)$$

$$TF\ IDF = TF * IDF$$

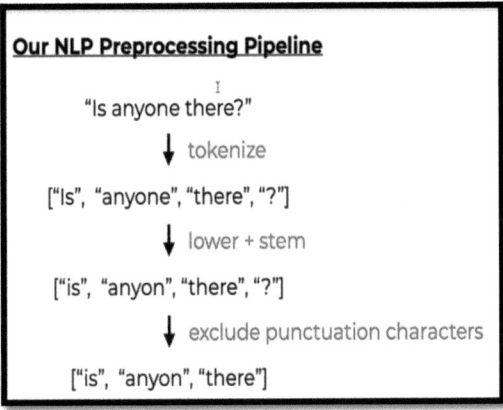

FIGURE 15.4 NLP preprocessing pipeline.

$$TF = \frac{Number\ of\ times\ a\ word\ "X"\ appears\ in\ a\ Document}{Number\ of\ words\ present\ in\ a\ Document}$$

$$IDF = log\left(\frac{Number\ of\ Documents\ present\ in\ a\ Corpus}{Number\ of\ Documents\ where\ word\ "X"\ has\ appeared}\right)$$

$$TF\ IDF\ =\ TF * IDF$$

$$w_{i,j} = tf_{i,j} \times log\left(\frac{N}{df_i}\right)$$

$tf_{i,j}$ = number of occurrences of i in j
df_i = number of documents containing i
N = total number of documents

FIGURE 15.5 TF-IDF representation.

15.4.1.4.2 Cosine Similarity

Cosine Similarity is used to measure the similarity between two documents.

Mathematically, it is used to measure the similarity between two vectors in a multidimensional plane [15.7]. Similarity between two vectors is represented as follows:

$$Similarity(p,q) = \cos\theta = \frac{p \cdot q}{\|p\|\|q\|} = \frac{\sum_{i=1}^{n} p_i q_i}{\sqrt{\sum_{i=1}^{n} p_i^2}\sqrt{\sum_{i=1}^{n} q_i^2}}$$

15.4.1.5 Results

Figure 15.8 shows the user interface that is developed using tkinter GUI toolkit. Using this interface, users can interact with the system. This GUI is very user-friendly.

Here in Figure 15.9, chatbot answers the user queries in test form, which are already present in the dataset having all the details about the university.

Chatbot answers the queries, which are asked by the users about the university. It provides all the details of the university by providing website links, etc. Here in Figure 15.10, chatbot is responding to the users queries by providing the required university links.

$$\text{Similarity}(p,q) = \cos\theta = \frac{p \cdot q}{\|p\|\|q\|} = \frac{\sum_{i=1}^{n} p_i q_i}{\sqrt{\sum_{i=1}^{n} p_i^2}\sqrt{\sum_{i=1}^{n} q_i^2}}$$

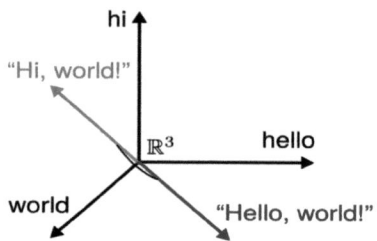

FIGURE 15.6 Cosine similarity vector representation.

```
sent_tokens.append(user_response)
TfidfVec = TfidfVectorizer(tokenizer=LemNormalize, stop_words='english')
tfidf = TfidfVec.fit_transform(sent_tokens)
vals = cosine_similarity(tfidf[-1], tfidf)
idx=vals.argsort()[0][-2]
flat = vals.flatten()
flat.sort()
req_tfidf = flat[-2]
if(req_tfidf==0):
    robo_response=robo_response+"I am sorry! what is it actually mean?"
    return robo_response
else:
    robo_response = robo_response+sent_tokens[idx]
    return robo_response
```

FIGURE 15.7 TF-IDF and cosine similarity implementation.

FIGURE 15.8 Output screen (GUI—graphical user interface).

FIGURE 15.9 Chatbot giving response as text about the university.

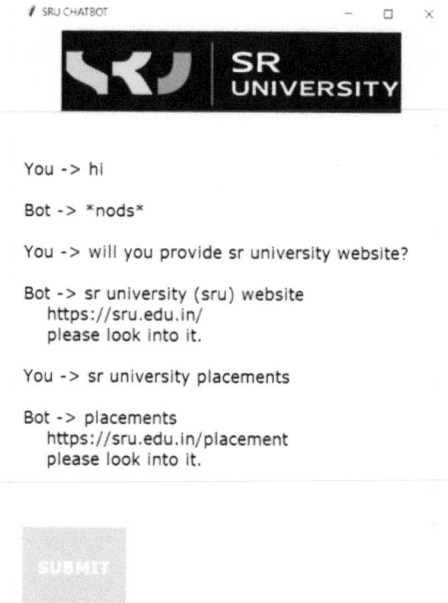

FIGURE 15.10   Chatbot giving response by displaying the university website and placements links.

15.5   CONCLUSION

University chatbot is useful in providing the information for the students or any users who wish to know the information about the college without a hassle. It instantly redirects them to the information they have asked for. In future, this project can be integrated with the university website to bring it to the use for all the university users. As this system will be user-friendly, all the students and other users can use this university chatbot very easily and comfortably.

REFERENCES

1. Nikhila, P., G. Jyothi, K. Mounika, "Chatbots Using Artificial Intelligence", Volume VIII, Issue I, January 2019, ISSN NO:2236-6124, pp 1–12.
2. Pawar, Harsh, Pranav Prabhu, Ajay Yadav, Vincent Mendonca, *"Engineering Technology (IJRASET)"*, ISSN: 2321–9653; IC Value: 45.98, SJ Impact Factor: 6.887, Volume 6, Issue IV, April 2018, pp 2494–2496.
3. Assistant Prof Ram Manoj Sharma, "College Enquiry Chat-Bot System", ISSN: 2455-3085 (Online), Volume-04, Issue 03, March-2019, pp 109–112.
4. Thakur, Nitesh, Akshay Hiwrale, Sourabh Selote, Abhijeet Shinde, "Application of artificial intelligence for university information system", *Universal Research Reports*, ISSN: 2348 – 5612, Volume 04, Issue 06, July–September 2017, pp 43–47.

5. Bala, K., Mukesh Kumar, Sayali Hulawale, Sahil Pandita, "Chatbot for College Management System Using A.I", *International Research Journal of Engineering and Technology (IRJET)* e-ISSN: 2395-0056, p-ISSN: 2395-0072, Volume 04, Issue 11, November 2017, pp 2030–2033.
6. Ranoliya, Bhavika R., Nidhi Raghuwanshi and Sanjay Singh, "Chatbot for University Related FAQs", 2017:1525–1530. 10.1109/ICACCI.2017.812605.
7. Pawar, Sagar, Omkar Rane, Ojas Wankhade, Pradnya Mehta, "A Web Based College Enquiry Chatbot with Results", *International Journal of Innovative Research in Science, Engineering and Technology*, ISSN(Online): 2319-8753, ISSN (Print): 2347-6710, Volume 7, Issue 4, April 2018, pp 3874–3880.
8. Tiwari, Amey, Rahul Talekar, S. M. Amey Patil, "College Information and General Science", ISSN 2091–2730, Volume 5, Issue 2, March–April, 2017, pp 131–137.
9. Janapati, Ravichander, et al., "Towards a More Theory-Driven BCI Using Source Reconstructed Dynamics of EEG Time-Series", *Nano LIFE*, Volume 12, Issue 02, 2022, p 2250005.
10. Janapati, Ravichander, et al., "Web Interface Applications Controllers Used by Autonomous EEG-BCI Technologies", AIP Conference Proceedings. Volume 2418, Issue 1. AIP Publishing LLC, 2022.
11. Janapati, Ravichander, Vishwas Dalal, Rakesh Sengupta, "Advances in Experimental Paradigms for EEG-BCI", Proceedings of the 2nd International Conference on Recent Trends in Machine Learning, IoT, Smart Cities and Applications. Springer, Singapore, 2022.

Index